Journey into the Promised Land

Daily Devotions for the New Millennium

Morris Cerullo

Unless otherwise noted, all Scripture references are from the King James version of the Bible.

Published by
Morris Cerullo World Evangelism

San Diego, California
Printed in the United States of America

Morris Cerullo World Evangelism
P.O. Box 85277
San Diego, CA 92186-5277

Morris Cerullo World Evangelism of Canada
P.O. Box 3600
Concord, Ontario L4K1B6

Morris Cerullo World Evangelism
P.O. Box 277, Hemel Hempstead,
HERTS HP2 7DH

Introduction

You are about to begin an exciting spiritual journey that will change your life forever.

Journey to the Promised Land is a classic devotional guide that features daily readings drawn from the prophetic, revelational, and inspirational teaching God has given me during more than half a century of ministry in the nations of the world.

The true illustrations, practical applications, and biblical meditations are the classic teachings that are the hallmark of this ministry – the New Anointing, Proof Producers, spiritual warfare, financial blessing – these are a few of the supernatural revelations that have impacted hundreds of thousands of people around the world and resulted in the raising up of the mighty spiritual force that has come to be known as God's Victorious Army.

This guide features a daily scriptural reading which accompanies each devotion and every reading includes a powerful concluding declaration of faith that will make you a doer of the Word and not just a hearer only. The monthly divider pages are illustrated with beautiful, historic photographs of the World Evangelism international ministry.

During the coming months, we will travel in the footsteps of the children of Israel as they crossed over Jordan – prepared for war – to take their promised land of Canaan. The monthly themes of this guide include:

JANUARY:	MARCHING INTO THE NEW MILLENNIUM
FEBRUARY:	OVERCOMING OBSTACLES
MARCH:	WINNING THE BATTLE FOR YOUR MIND
APRIL:	CONQUERING THE FLESH
MAY:	ACCESSING DIVINE PROVISION
JUNE:	PRODUCING THE PROOF
JULY:	OVERCOMING DEFEAT
AUGUST:	PENETRATING ENEMY LINES
SEPTEMBER:	DEALING WITH DECEPTION
OCTOBER:	PUTTING THE ENEMY UNDER YOUR FEET
NOVEMBER:	FULFILLING YOUR DESTINY
DECEMBER:	POSSESSING THE LAND

I encourage you to read through the entire book of Joshua before starting this devotional guide. Chapters 1-5 concern the preparation to enter in and crossing Jordan; chapters 6-12 describe the conquest of the land; chapters 13-22 detail the dividing of the land; and chapters 23-24 conclude with a tremendous message on faith by Joshua.

As we start our exciting spiritual adventure together, let me recap to where our journey begins. The children of Israel were slaves in Egypt for 430 years. Egyptian bondage is a picture of living life without God, in bondage to sin, and being dead spiritually. The book of Genesis closes with Joseph being buried in a coffin in Egypt – so it represents death.

At the end of these years of slavery, God raised up a man named Moses to supernaturally confront the wicked ruler, Pharaoh, and lead God's children out of bondage by the power of God and the blood of the lamb. This deliverance represents our salvation which was purchased by the blood of Jesus Christ.

After crossing a barren wilderness, God brought His people to the border of their promised land but they refused to enter in because of unbelief. They turned back into the wilderness and spent 40 years wandering aimlessly. This is a spiritual picture of the believer who refuses to enter into the promises of God and remains in the barren wilderness of unbelief and disobedience. These are believers who are miserable, complaining all the time, and wanting to go back to the "Egypt" of sin. Wilderness Christians.

After 40 years of wilderness wandering, the nation of Israel returned once again to the border of their promised land. This time, they acted in faith, entered in, conquered the enemy, and claimed all God had promised them. This is a spiritual picture of the believer who enters into the divine cycle of victory miracle living – claiming all God has for him.

As believers, we have inheritances and possessions in God. An inheritance is everything God has reserved for us through His Son, Jesus Christ. That includes every promise, victory, and blessing He has planned for us. A possession, however, is the portion of that inheritance we appropriate by faith.

A man named Joshua was the great spiritual leader who God raised up to lead His people into their promised land and appropriate what He had reserved for them. Joshua challenged the people with a question. This is a very penetrating question that I think each one of us must answer as we begin this devotional guide:

> *And Joshua said unto the children of Israel, How long are ye slack to go to possess the land, which the LORD God of your fathers hath given you?*
> *(Joshua 18:3)*

In Joshua chapter 1:1-9, Israel is preparing to enter the promised land. There are five principles we can draw from this passage that will prepare us to enter in to claim all God has for us during the coming year:

Now after the death of Moses the servant of the LORD it came to pass, that the LORD spake unto Joshua the son of Nun, Moses' minister, saying, Moses my servant is dead; now therefore arise, go over this Jordan, thou, and all this people, unto the land which I do give to them, even to the children of Israel. Every place that the sole of your foot shall tread upon, that have I given unto you, as I said unto Moses. From the wilderness and this Lebanon even unto the great river, the river Euphrates, all the land of the Hittites, and unto the great sea toward the going down of the sun, shall be your coast. There shall not any man be able to stand before thee all the days of thy life: as I was with Moses, so I will be with thee: I will not fail thee, nor forsake thee. Be strong and of a good courage: for unto this people shalt thou divide for an inheritance the land, which I sware unto their fathers to give them. Only be thou strong and very courageous, that thou mayest observe to do according to all the law, which Moses my servant commanded thee: turn not from it to the right hand or to the left, that thou mayest prosper whithersoever thou goest. This book of the law shall not depart out of thy mouth; but thou shalt meditate therein day and night, that thou mayest observe to do according to all that is written therein: for then thou shalt make thy way prosperous, and then thou shalt have good success. Have not I commanded thee? Be strong and of a good courage; be not afraid, neither be thou dismayed: for the LORD thy God is with thee whithersoever thou goest.

(Joshua 1:1-9)

Here are the principles:

Principle #1: All of life has prepared you for this moment. All the little twists and turns and the things that have happened that you haven't understood. They have all brought you to this moment in time – the time to enter in and claim your spiritual inheritance.

Principle #2: If we are to enter in, we must rely upon God and Him alone. All human props must be kicked out from under us. That is why God told Joshua, "*Moses, my servant is dead.*" Moses is dead, but God isn't!

I have preached it for years: The key to harvest is timing – knowing when and where God is going to move. It is time for you to enter into the promises of God relying solely upon God alone! The church's finest hour is just ahead. Your greatest personal spiritual hour is just ahead – so near that you may be even to reach out and touch it – it is that close!

You will not enter in through head knowledge. You will enter in through the manifestation of God's divine intention for His Church. Your breakthrough does not depend upon what you possess. It depends upon His divine capability. Your greatest hour is just ahead!

Principle #3: God is calling us to step out in faith in the face of impossibilities. Some of you may have just come through a very difficult year and are wondering how you can face a new one. But you can't turn back to the wilderness. You can't give up. You must cross Jordan even if it is flood season. You can't wait until a better season when it is easier or more convenient. You must cross it now – at the flood stage – if you want to witness God's supernatural power at work in your life.

Principle #4: We must move forth in faith. Faith is not passive, it is active. We can't just sit on the other side of the Jordan River and say "Well, God's given me the land so I'll just sit here and see what happens"–and do nothing about it. God's people had to cross the Jordan River. They had to fight the enemy, drive them out, and take possession of what was rightfully theirs.

Principle #5: The fifth principle is the principle of obedience. Day by day, as we study God's Word, meditate on these devotions, and make the accompanying faith declarations, we must do so with the predetermined commitment that whatever God says, we are going to obey.

Remember – these are not the typical "make you feel good" devotions. They are powerful, life-changing, hard-hitting strategies that will enable you to lay claim to all God has promised you and prepare you to face the challenges of the new millennium with total, 100 percent victory. This is one devotional guide that will truly change your life. You will never again be the same. So only one question remains...Are you ready to begin this great spiritual journey with me? Are you ready to begin your personal, spiritual *Journey to the Promised Land?*

If so, get on your spiritual marching boots. God's Victorious Army is moving out!

Faithfully God's servant,

Morris Cerullo

Table Of Contents

Section One

Theme: Marching Into The New Millennium

Month: January

Section Two

Theme: Overcoming Obstacles

Month: February

SECTION THREE

Theme: Winning The Battle For Your Mind

MONTH: March

Section Four

Theme: Conquering The Flesh

Month: April

Section Five

Theme: Accessing Divine Provision

Month: May

Section Six

Theme: Producing The Proof

Month: June

Section Seven

Theme: Overcoming Defeat

Month: July

Section Eight

Theme: Penetrating Enemy Lines

Month: August

SECTION NINE

Theme: Dealing With Deception

MONTH: September

Section Ten

Theme: Putting The Enemy Under Your Feet

Month: October

Section Twelve

Theme: Possessing The Land

Month: December

January

As a young man of fourteen years of age, God set before me an open door when He miraculously saved me and led me out of the Daughters of Miriam Orphanage. As you begin this year, God has shown me that He has set before you an open door that will lead you step-by-step to your miracle.

January

Marching into the New Millennium

This month launches the first thirty days of a brand new year, a brand new century, and a brand new millennium. Ahead of us lies the unknown. What trials will we face this year? What victories await us? What changes will occur throughout the world? How will our personal lives, our ministries, and our families be impacted in the coming months?

As we pause to reflect on the coming days of this year, we are much like God's people, Israel, who came to the border of their Promised Land. Ahead of them lay the unknown – a land they had not seen, an enemy they had not accessed. There were tremendous blessings ahead, but there were also some battles that awaited them on the other side of Jordan. The same is true for us.

God instructed Israel to follow closely behind the Ark of the Tabernacle, which was symbolic of His presence in the midst of His people. He told them to *"...come near unto it, that ye may know the way by which ye must go: for ye have not passed this way heretofore"* (Joshua 3:4).

As we think about the coming year, we too "have not been this way before." We do not know what joys and sorrows, victories and problems lie ahead. But one thing is for sure: We have the same words of assurance that God gave to Joshua at the border of Canaan:

> *Have not I commanded thee? Be strong and of a good courage; be not afraid, neither be thou dismayed: for the Lord thy God is with thee whithersoever thou goest.*
>
> *(Joshua 1:9)*

God has promised to be with us wherever we go. He has commanded us to be strong and of good courage and to not be dismayed. The presence of God is advancing before us into the coming year, just as the Ark led Israel into the unknown territory that lay before them. During the month of January, we will join Israel when they first came to the border of their Promised Land and turned back in defeat. Then we will return with them the second time, when they obeyed God's command to advance and entered in, conquered the enemy, and claimed their God-given possession.

As we follow in the footsteps of God's chosen people this month, we will draw many spiritual parallels that apply to our own lives. We will learn the dangers of murmuring and complaining, the assurance of God's promises, the importance of fostering a healthy fear of God, walking in one-hundred percent obedience, and developing spiritual vision.

There are some exciting times ahead for us as we look by faith to the future. God has tremendous victories planned for us in the coming days. There are some major challenges ahead also, but... *"...We are well able to conquer it...He will bring us into this land and give it to us..."* (Numbers 13:30; 14:8, TAB).

A Point Of Spiritual Destiny

Scripture reading: Exodus 3:1-10

The nation of Israel was at its greatest point of spiritual destiny. The decision they were to make that day would decide the fate of the entire nation. God had brought them to the land He had promised as an inheritance and commanded them to take possession of it.

When God led Israel into the Promised Land, He had already positioned them for victory (Numbers 13). He delivered them out of Egyptian bondage through miracles, signs, and wonders greater than any the world had ever known. He opened up the Red Sea and they marched across on dry land (Exodus 14:21-22). He miraculously led them by a pillar of cloud by day and a pillar of fire by night (Exodus 13:21-22).

He fed them manna from heaven (Exodus 16:4,15). He caused water to gush forth from the rock at Rephidim in Mount Horeb (Exodus 17:5-6). He gave them a great victory over their enemies, the Amalekites (Exodus 17:9-13). The Great I Am – God Almighty – had descended out of heaven and manifested Himself on Mount Sinai in flaming fire and billowing smoke.

They heard the great trumpet of God blasting across the plains, summoning them to meet with Him. They saw His glory. They heard Him speak in an audible voice. There on Mount Sinai, God joined Himself together with them in a blood-covenant relationship. They were bound together as one. God pledged Himself to be their God, to protect them, to prosper them, and to set them apart from the nations of the earth. Their enemies were His enemies. All that He had belonged to them.

He promised to bring them *"...to a land good and large, a land flowing with milk and honey – a land of plenty..."* (Exodus 3:8, TAB). He sent the Angel of the Covenant, the preincarnate Son of God, before them to lead, instruct, protect, and guide them and to drive out their enemies (Exodus 23:20-23). They were positioned for victory.

Pause today and think of all the things God has done for you in the past. Think of all the victories you have won and the prayers that have been answered. Today, you are at the border of a new land – a new year, a new vision, a new ministry, a new step of faith. God has positioned you for victory!

Make this declaration: "Lord, I am ready to enter into my spiritual destiny during the coming year. Thank you for your promise to go before me, leading, instructing, protecting, and guiding."

January 2

What Is Your Choice?

Scripture reading: Exodus 23:21-33

Approximately sixteen months after God delivered Israel out of Egyptian bondage, He brought them to the border of the Promised Land. From where they were camped, the Israelites could see rising before them above the hills which were part of the land God had promised them as an inheritance.

Can you imagine the excitement and joy that filled the hearts and minds of Moses and the people as they gazed toward the land of Canaan? Behind them were four hundred and thirty years of their sojourn in Egypt – the pain and sorrow, the cruel taskmasters, the bondage, mental anguish, discouragement, and defeat.

Before them, within a few days' journey, was the land of promise. Before them were all the blessings of the covenant – health, prosperity, daily provision, protection, everything they needed. Before them was the promised rest, peace, and safety from their enemies.

The Promised Land and all the covenant promises were their inheritance. Legally the land and promises belonged to them, but having the promises was not the same as taking possession of them. There was a condition and all the promises God made to Israel were based upon the fulfillment of that one condition.

God had told them: *"...if you will obey My voice in truth and keep My covenant, then you shall be My own peculiar possession and treasure from among and above all peoples; for all the earth is Mine. And you shall be to Me a kingdom of priests, a holy nation (consecrated, set apart to the God)"* (Exodus 19:5-6, TAB).

God sent the Angel of the Covenant before them and told them: *"...obey his voice, provoke him not; for he will not pardon your transgressions: for my name is in him. But if thou shalt indeed obey his voice and do all that I speak; then I will be an enemy unto thine enemies, and an adversary unto thine adversaries"* (Exodus 23:21-22).

God directed them to take possession of the land. There at Kadeshbarnea, Moses told them: *"...Ye are come unto the mountain of the Amorites, which the Lord our God doth give unto us. Behold, the Lord thy God hath set the land before thee: go up and possess it, as the Lord God of thy fathers hath said unto thee; fear not, neither be discouraged"* (Deuteronomy 1:20-21).

Their destiny was in their hands. They had a choice: To obey God and possess the land, or to disobey God, forfeit their inheritance. and reap the punishment of their disobedience.

You have a similar choice. What will you do?

Make this declaration: "I will obey God's voice and take possession of the land!

Your Destiny Is In Your Hands

Scripture reading: Exodus 13:1-20

God brought His people to the border of the Promised Land, but no one could enter unless their will was united to His will, unless they chose to obey and go in His Name and in His strength.

Today, God has brought His people to a new point in spiritual destiny. He has positioned us for victory. He has delivered us out of bondage, set us free, and given us power over sin. He has entered into an everlasting covenant with us that is sealed with the blood of Jesus. He has given us covenant promises that provide for all our needs. He has joined Himself to us as one and has given us His glory, all that He has and is. He has pledged Himself to be our God, to protect, prosper, and set us apart from the nations of the earth. He has sent His Spirit to live within us, giving us the power to obey Him. He has equipped us with the powerful weapons of spiritual warfare.

Behind us is our bondage to sin, the world and the flesh. Before us lies a new spiritual territory God has planned for us where we can live in a cycle of continual victory…

…Where we have taken possession of our spiritual inheritance and are partaking of the covenant promises.

…Where we are fully equipped and prepared to win the war and are exercising power and authority over Satan.

…Where we are pressing on into perfection, growing up unto the full stature of Jesus Christ.

Today, He has set this new spiritual territory before us and is saying, "Go up and possess it!" Our spiritual destiny is in our hands. We have a choice – to obey God and by faith take full possession of our spiritual inheritance and enter this new spiritual territory or to rebel and disobey God and, through our unbelief, forfeit our inheritance and live in defeat.

Those who enter this new spiritual territory, who are going to be greatly used of God during this great end-time harvest, are those who are walking in one-hundred percent obedience to Him. Do not be deceived. God will not pour out His power and glory upon those who are walking in rebellion and are disobedient to Him and His Word.

Satan's objective is to deceive you and lead you into rebellion against God and His Word so that you will not be able to take full possession of your spiritual inheritance and enter the powerful spiritual dimension He has planned for you. Your spiritual destiny is in your own hands!

Make this declaration: "I am leaving behind the bondage of sin, the world, and the flesh. I am looking forward to the new spiritual dimension God has planned for me where I can live in a cycle of continual victory."

January 4

WHICH VOICE WILL YOUR HEAR?

Scripture reading: Exodus 13:21-25

Moses sent twelve of the leaders of the tribes of Israel to explore the Promised Land. After exploring for forty days, they returned. The entire congregation was anxiously waiting to hear what these twelve leaders had to say. They told Moses, *"We came to the land to which you sent us; surely it flows with milk and honey; this is its fruit"* (Numbers 13:27, TAB).

Can you imagine the excitement building among the congregation as these twelve leaders showed them the fruit they had brought back from the land? Just as God promised, the land was rich and fertile. They brought back pomegranates and figs from the trees and a single cluster of grapes that was so large it took two men to carry it on a pole between them. At long last, their waiting was over. Soon they would be enjoying the fruit of the land and the blessings God had promised them!

The leaders continued, *"But the people who dwell there are strong, and the cities are fortified and very large; moreover there we saw the sons of Anak (of great stature and courage)"* (Numbers 13:28, TAB). At this report, fear began to spread among the Israelites and they began to express their fears and concerns.

Caleb encouraged the people and said, *"Let us go up at once and possess it; we are well able to conquer it"* (Numbers 13:30). The remaining ten leaders said, *"We are not able to go up against the people (of Canaan); for they are stronger than we are"* (Numbers 13:31). They began to spread a negative report that the land was full of giants and that it was a land that devours its inhabitants.

At this point, the people had heard three voices: The voice of God, the voice of faith and the voice of unbelief. God had spoken to them through Moses. He had promised to give them this land. He had promised to drive out their enemies before them. He had told them to go in and take possession of the land. No further discussion or consideration was necessary. All that was necessary was for them to hear the voice of God and in simple faith obey.

They heard the voice of faith. Caleb had said, "We are well able to conquer it." He and Joshua pleaded with the people not to rebel against God (Numbers 14:7-9).

They also heard the voice of unbelief. The remaining ten leaders had said, "We are not able." God's people had a choice to listen to the voice of God and the voice of faith and obey, or to listen to the voice of unbelief and rebel against God. Which voice will you listen to?

Make this declaration: "I will not listen to the voice of unbelief. I will listen only to the voice of God and act in faith upon what He says."

Don't Hesitate To Obey

Scripture reading: Hebrews 3:7-12

God had proven His faithfulness to Israel through mighty signs and wonders greater than anything the world had ever known. He had delivered them out of the land of their enemies. He had revealed Himself to them in lightning, fire and smoke on top of Mount Sinai. He had spoken to them in an audible voice. Yet, poised at the border of the Promised Land, at the crucial point of their spiritual destiny, they forgot God. They forgot the miracles. They forgot His promises.

Instead of listening to the voice of God, they listened to the voice of unbelief. They focused their eyes upon the giants and their own inabilities instead of looking to God and His power to deliver them. Their unbelief was the root cause of their disobedience. Had they rejected the voice of unbelief and refused to listen to the negative report, they would have obeyed God.

When God speaks to you, one of the first things Satan will do is create confusion, fear, and unbelief through other voices. There have been many times in my life when God has spoken to me giving me specific direction concerning a particular course He wanted me to follow in our ministry here at World Evangelism. Often, after I have shared the direction God gave me, Satan would begin to speak through other voices to get me off course:

- "The conditions aren't right. Maybe we should postpone this for a few months."
- "Are you sure about the timing on this?"
- "It's too dangerous to go into that country right now."
- "There just aren't enough funds to do that."
- "Do you think the people will respond to this need?"

There also have been voices that have tried to convince me to compromise what God has directed me to do. I have learned to stay on my face before God until I hear His voice. Then, once I know that God has spoken to me, I shut out all the other voices, regardless of whose they are, and obey God.

Satan's strategy is to cause you to rebel – to refuse or fail to act upon His Word, to cause you to disobey God when He speaks to you. God may speak to you and direct you to move out into some type of ministry. He may direct you to do something that you feel is beyond your ability. He may direct you to a particular Scripture and ask you to step out in faith and act on it. If you are facing financial problems, problems in your family or at work, He may speak to you with a *rhema* Word. If you hesitate to obey, you will give Satan an opportunity to speak through other voices and cause you to disobey God.

Don't hesitate – just obey.

Make this declaration: "I rebuke the voices of fear and unbelief that are causing confusion in my life. I will tune out all other voices except the voice of God."

January 6

ENTER INTO HIS REST

Scripture reading: Matthew 16:22-28

Just as Satan spoke through Peter to try to stop Jesus from obeying God's will (Matthew 16:23), he will speak through other people – even your closest friends and loved ones – to try and discourage you from following God's will. They'll say things like...

..."It can't be done."

..."Are you sure God told you to do that?"

..."Your situation looks hopeless, you might as well give up."

..."I don't see how you will ever get out of debt."

These voices will cause you to rebel and reject the voice of God like the children of Israel did.

After all God had done for Israel – the miracles He had performed in delivering them out of the hands of the Egyptians, leading them, feeding and providing for them during their journey – it is hard to understand why they listened to the negative report of the spies instead of believing that God would drive their enemies out as He had promised.

We wonder how they could possibly doubt God, yet there are many Christians today who react the same way. God has given us a written record of the great miracles He has done in behalf of His people. We have seen and experienced His love, provision, and healing power. But when we face circumstances in our lives that require us to act in obedience to God's Word, our hearts become filled with unbelief just like the children of Israel.

Paul warned the Hebrews:

> *See to it, brothers, that none of you has a sinful, unbelieving heart that turns away from the living God..."Today, if you hear his voice, do not harden your hearts as you did in the rebellion." Who were they who heard and rebelled? Were they not all those Moses led out of Egypt? And with whom was he angry for forty years? Was it not with those who sinned, whose bodies fell in the desert? And to whom did God swear that they would never enter his rest if not to those who disobeyed? So we see that they were not able to enter, because of their unbelief.*
>
> *(Hebrews 3:12,15-19, NIV)*

The word "rest" referred to by Paul here is not referring simply to a place of physical rest from all our labors. It is not a place where we are free from the problems and cares of this life. It is referring to that powerful spiritual dimension where those who enter have taken full possession of their inheritance. It is a place of peace and security – freedom from fear, worry, and stress – even in the midst of great adversity. It is a place of blessing and provision, a place of power and victory over our enemies, a place of joy and contentment, a place where we have ceased from our striving to walk in our own strength and are walking in God's strength. Enter into His rest today.

Make this declaration: "I will enter into God's rest today. I will experience peace and security. I will be free from fear, worry, and stress despite the circumstances around me."

Don't Forget What God Has Done

Scripture reading: Numbers 14:1-12

Israel, through their unbelief, disobeyed and failed to enter the rest God had planned for them. Paul warned the believers, *"Let us, therefore, make every effort to enter that rest, so that no one will fall by following their example of disobedience"* (Hebrews 4:11, NIV).

The only way to enter the powerful spiritual dimension God has planned for you is through faith and obedience. You must shut out every voice of unbelief that would tempt or deceive you into disobeying God. You must set your heart and mind to obey God, regardless of the cost.

The Israelites refused to listen to the voice of God, but chose instead to listen to negative reports. Their hearts were so filled with unbelief they began to grumble and murmur against Moses and Aaron. *"If only we had died in Egypt! Or in this desert! Why is the Lord bringing us to this land only to let us fall by the sword? Our wives and children will be taken as plunder. Wouldn't it be better for us to go back to Egypt?"* (Numbers 14:2-3, NIV).

Their grumbling and murmuring was directed at Moses and Aaron, but their murmuring was actually against God. It was an outward manifestation of the unbelief that was in their hearts. They had murmured and complained about a lack of food:

> *Would to God we had died by the hand of the Lord in the land of Egypt, when we sat by the flesh pots, and when we did eat bread to the full.*
> *(Exodus 16:3)*

They had murmured and complained that there was no water:

> *Wherefore is this that thou has brought us up out of Egypt, to kill us and our children and our cattle with thirst?*
> *(Exodus 17:3)*

They were not satisfied with the manna God gave them, and had murmured and complained that there was no flesh:

> *Who shall give us flesh to eat? for it was well with us in Egypt...*
> *(Numbers 11:18)*

Then, as they came to Kadeshbarnea and listened to the negative report concerning the giants in the land, they murmured and complained that they were going to die by the sword and their wives and children would be taken as plunder.

God heard their murmuring and was filled with wrath against them. Their constant murmuring was a means of accusing Him of not fulfilling His promises to them. They were turning their backs on Him and were forgetting what He had already done for them.

How about you? Are you murmuring and complaining because of your circumstances today? Are you turning your back on Him and forgetting what He has done for you in the past. If so...

Make this declaration: "I will not murmur and complain. I will not forget what God has done for me in the past."

Murmuring And Complaining

Scripture reading: 1 Corinthians 10:1-13

Satan will use murmuring and complaining against God to defeat you. Murmuring is an outward expression of unbelief. It reveals a lack of trust in God's goodness, His love, and His provision.

There are many Christians today who have fallen into this snare of the enemy. When they face problems in their finances, in their families, or on the job, instead of trusting God – believing that He will meet their need, and praising Him for what He has already done and is going to do for them – they begin to murmur. They may not speak it audibly, but in their hearts they are thinking, "Why has God allowed this to happen to me? I'm tired of never having enough money to meet our needs. Why do our children always have to do without things they want? Why doesn't God answer and meet our needs? Has He forgotten us?"

On the job they murmur about the amount of work they have. They murmur and grumble about their salaries, the working conditions, and that they aren't being treated fairly.

The Israelites murmured about the carnal desires and appetites of their flesh. They were not satisfied with the manna God provided for them. They were ungrateful for what God gave them and longed for the fish, melons, leeks, onions, and garlic they had enjoyed in Egypt.

Today, there are Christians who murmur about the type of food they have to eat. They murmur about their lack of clothes, the type of car they drive, their furniture, or the house they live in. We must recognize murmuring as a trick of the enemy to keep us living in defeat. When we murmur, we are doubting God's provision and care for us. Moses told the Israelites, *"The Lord has heard your grumblings which you murmur against Him, what are we? Your murmurings are not against us, but against the Lord"* (Exodus 16:8, TAB). When we murmur and grumble, we are not murmuring against people or circumstances. We are murmuring against God and He hears us. Not only does He hear us, but our unbelief hinders us from acting in faith upon His promises and we cannot receive what we need.

Paul used the example of the Israelites to warn the Corinthians not to grumble. He said, *"And do not grumble, as some of them did and were killed by the destroying angel. These things happened to them as examples and were written down as warnings for us, on whom the fulfillment of the ages has come"* (1 Corinthians 10:10-11, NIV). Just as murmuring kept the Israelites out of the Promised Land, murmuring and complaining will keep you from entering the spiritual dimension of God's rest where there is blessing, provision, power, and victory.

Make this declaration: "I will learn from the example of Israel. Words of murmuring and complaining are no longer part of my vocabulary."

WHAT WILL YOU DO?

Scripture reading: Numbers 14:13-23

Murmuring is an outward sign of unbelief and it will keep you living in defeat. Just as murmuring led the Israelites further into rebellion against God, it will lead you into rebellion and disobedience. When Satan tempts you to murmur about your circumstances, recognize it as one of his strategies to defeat you and, instead of murmuring, begin to praise and thank God for His faithfulness toward you. Release your faith and trust God to meet your need.

Because of their unbelief, the Israelites rebelled and disobeyed God when they first came to the border of their Promised Land .

- They refused to hear the voice of God.
- They rejected His Word.
- They refused to act in faith on what He had promised.
- They refused to obey.

God had told them to go and take possession of the land, but instead of going forward they rebelled and turned back. Before them was the land God promised. It belonged to them. It was a rich, fertile land with green valleys. It was a land where God promised to bless and multiply them above the nations of the earth. It was a land where God had promised to give them health and prosperity, yet, because of unbelief, they turned back. They said, *"...Wouldn't it be better for us to go back to Egypt? And they said to each other, We should choose a leader and go back to Egypt"* (Numbers 14:3-4, NIV).

Not only did they refuse to hear the voice of God, they rejected Him as their leader. God sent the preincarnate Son of God to go before them to lead, instruct, and guide them. As long as they were obedient to Him, God promised to give them victory over their enemies. They refused to follow God and obey Him, however, and chose instead to follow their self-will. They chose to go back into Egyptian bondage rather than obey God. They chose to satisfy the carnal desires of their flesh instead of pressing forward in faith and taking possession of the land.

To take possession of their inheritance, it would be necessary for Israel to face the giants, to fight against their enemies, and drive them out. Because of their unbelief, they feared their enemies more than they feared God. Instead of going into battle and taking dominion over their enemies, they chose to go back to their bondage and slavery in Egypt. There on the verge of victory, Israel retreated. They forfeited their inheritance and were defeated.

Poised at the threshold of a new year – a new millennium – what will you do?

Make this declaration: "I will hear God's voice. I will receive His Word. I will act in faith upon what God has promised and obey."

Choosing To Obey

Scripture reading: Numbers 14:24-35

God has positioned you for victory. He has set before you new spiritual territory and is telling you, "Go up and possess it!"

- It is time for you to walk in the power God has planned for you.
- It is time for you to press on into the perfection that is possible in Christ.
- It is time for you to walk in one-hundred percent victory over Satan, the flesh, and the world.
- It is time for you to enter into a place of total trust and faith in God that will enable you to look at every circumstance absolutely free from worry, knowing that God will give you victory.

Your spiritual inheritance, including all the covenant promises of God, is yours. Legally, they belong to you. However, having the promises is not the same as taking possession of them. To enter into this new spiritual territory, you must be willing to walk in obedience to God and His Word. You must be willing to go into battle and take dominion over the enemy in your life.

Satan will do everything possible to get you to turn back and disobey God. He will try to get your eyes on the giants of your weaknesses and inabilities. He will make you think you are too weak spiritually, that you have too many problems, and that your faith is weak.

Israel's victory over their enemies was not based upon their human strength and abilities. It was not dependent upon the strength or weaknesses of their enemies. Their victory was dependent upon God's ability. They were to go into battle against their enemies, but it was God Who was going to drive out the enemy and give them victory.

Israel was defeated because they had their eyes focused upon their own limited strength instead of upon the power of Almighty God and His faithfulness to keep His promises and give them victory.

God has set before you a new spiritual territory. He has told you to take possession of it. His one requirement is obedience. Your ability to walk in one-hundred percent obedience to God is not dependent upon your own limited strength. It is dependent upon God's ability. If you keep your eyes on your own strength, you will be tempted to turn back.

By faith, Caleb looked beyond the giants, beyond his own limited strength and saw the power of God. He was convinced that God would deliver them out of the hands of their enemies, just as He had promised. He said, *"We are well able to conquer it … if the Lord delights in us, then He will bring us into this land and give it to us …"* (Numbers 13:30; 14:8, TAB). You must now, by faith, begin to look beyond your own limited strength and keep your eyes upon God's ability. You must choose to obey. You must yield yourself to God and walk in obedience in His strength!

Make this declaration: "I will walk in the power God has planned for me. I will press on into the perfection that is possible in Christ. I will walk in one-hundred percent victory over Satan, the flesh, and the world."

WHAT GOD REQUIRES

Scripture reading: Deuteronomy 4:1-10

God has brought you to a new point in your spiritual destiny. All that God requires of you is to hear His voice and obey. God has positioned you for victory, but you must move forward by faith and obedience in the fear of God.

One major cause for Israel's continual rebellion and disobedience against God was that they did not fear Him. On Mount Sinai, God revealed Himself to the Israelites in flaming fire and billowing smoke and spoke to them in an audible voice out of the fire so they would fear Him.

As they were preparing to go in to take possession of the Promised Land, Moses gathered all the people together and told them:

> *Only take heed, and guard your life diligently, lest you forget the things which your eyes have seen, and lest they depart from your (mind and) heart all the days of your life; teach them to your children, and your children's children; especially how on the day that you stood before the Lord your God in Horeb, the Lord said to me, Gather the people together to Me, and I will make them hear My words, that they may learn (reverently) to fear Me all the days they live upon the earth, and that they may teach their children.*
>
> *(Deuteronomy 4:9-10, TAB)*

"Don't forget," Moses warned them. "Take heed!" He also told them clearly what God required of them:

> *And now, Israel, what does the LORD your God require from you, but to fear the LORD your God, to walk in all His ways and love Him, and to serve the LORD your God with all your heart and with all your soul, and to keep the LORD's commandments and His statutes which I am commanding you today for your good? ...You shall fear the LORD your God; you shall serve Him and cling to Him, and you shall swear by His name.*
>
> *(Deuteronomy 10:12-13,20, NAS)*

Moses warned Israel that punishment would come upon them if they failed to obey God's commandments, yet soon after Israel entered Canaan, they turned their backs on God and disobeyed Him.

God requires the same thing from you that He did from Israel:

- Fear the Lord.
- Walk in all His ways.
- Love Him.
- Serve and cling to Him with all your heart and soul.
- Keep His commandments.
- Walk in the authority of His Name.

Make this declaration: "I will serve You and walk in Your ways. I will love You, serve You with all my heart, and keep Your commandments. I will walk in the authority of Your Name."

Fearing Giants More Than God

Scripture reading: Deuteronomy 31:24-30

The children of Israel rebelled and disobeyed God because they did not fear Him. Without the fear of God in their hearts motivating them, they could not obey God. God had purposed in His heart to have a people who would fear, love, serve, and obey Him all the days of their lives. The type of fear He planned for them to have was not one that would cause them to run or hide from Him in terror.

When God met with Israel on Mount Sinai and talked to them with an audible voice out of the fire, He wanted them to see His power and glory so they would have a reverential fear, a sense of awe at seeing His manifested Presence and hearing His voice.

When He punished them for their disobedience in the wilderness, He wanted them to learn to fear Him as a holy and righteous God who is not only long suffering, kind, and merciful, but who is also a God of wrath and judgment who will not allow the wicked to go unpunished. He wanted them to have a holy fear that would create a wholesome dread of displeasing Him.

When the children of Israel reached the border of the Promised Land and heard the negative report concerning the giants in the land, they were afraid. The entire congregation wept and cried aloud that night. They were so afraid of the giants that they decided to go back to Egypt.

Joshua and Caleb pleaded with them: *"… do not rebel against the LORD. And do not be afraid of the people of the land, because we will swallow them up. Their protection is gone, but the LORD is with us. Do not be afraid of them"* (Numbers 14:9-10, NIV).

Because of their fear, the people rebelled and disobeyed God. They feared the giants more than they feared God. They were more concerned about falling into the hands of their enemies than they were about displeasing Almighty God and reaping His wrath.

As a father dealing with a rebellious son, God punished them for their acts of disobedience and yet – even after they reaped all the judgments, curses, famines, diseases, captivity, and death they refused to yield under the mighty hand of God. Knowing the wrath of God and the punishment for their disobedience, they chose to rebel and follow their own way.

On the threshold of each new day, we have the same choice. You can fear the giants or you can walk in the reverential fear of God. What will be your decision?

Make this declaration: "I will not fear the giants. I will walk in the reverential fear of God."

Fearing Man Or Fearing God?

Scripture reading: Deuteronomy 5:24-33

The God we serve is a holy and righteous God. He has not changed. He hates sin and will not tolerate disobedience. Those who walk in disobedience are living under the wrath of God which will be poured out upon the disobedient when Christ returns.

Our God is a consuming fire (Deuteronomy 4:24). Sin cannot remain in His presence because it is consumed. God's dealings with the nation of Israel – their unbelief, rebellion, disobedience, His pleading, long suffering, wrath, and judgment that came upon them – have been given to us as examples and warnings.

The children of Israel did not fear God. Their hearts were hardened against Him. God told Moses:

> *O that there were such an heart in them, that they would fear me, and keep all my commandments always, that it might be well with them, and with their children forever!* *(Deuteronomy 5:29)*

Without this fear, they could not obey God.

Under the New Covenant God made with us, He has placed His fear in our hearts and has promised us:

> *They will be my people, and I will be their God. I will give them singleness of heart and action, so that they will always fear me for their own good and the good of their children after them. I will make an everlasting covenant with them: I will never stop doing good to them, and I will inspire them to fear me, so that they will never turn away from me.*
>
> *(Jeremiah 32:38-40, NIV)*

God has circumcised our hearts (Romans 2:29). He has removed hearts that were once hardened and rebellious against Him, given us a new heart, and placed His fear in it so that we have a desire and a will to walk in obedience to Him. He has placed His Spirit within us giving us the power to obey. It is this holy fear of God that must now motivate our lives to walk holy before Him.

As Jesus commissioned and sent the twelve disciples out, He commanded them: *"...do not fear those who kill the body, but are unable to kill the soul; but rather fear Him who is able to destroy both soul and body in hell"* (Matthew 10:28, NAS). We are commanded to fear God rather than man.

Examine your own heart today. Do you fear man more than you fear God? Is the favor of man more important than that of God?

Make this declaration: "I will not fear man more than I fear God. I will fear God!"

God Judges Sin

Scripture reading: 1 Peter 1:13-22

God has warned us of His wrath that is to come upon the unbelieving and disobedient. He has warned us of the eternal fate of a literal burning lake of fire that has been prepared and is waiting for those who continue to rebel and sin against Him. Knowing that there is a time of judgment coming, we are to work out our own salvation "with fear and trembling" (Philippians 2:12). Our thoughts and actions are to be governed by a holy fear.

Peter told the believers:

> *Wherefore gird up the loins of your mind, be sober, and hope to the end for the grace that is to be brought unto you at the revelation of Jesus Christ; as obedient children, not fashioning yourselves according to the former lusts in your ignorance: But as he which hath called you is holy, so be ye holy in all manner of conversation; Because it is written, Be ye holy; for I am holy. And if ye call on the Father, who without respect of persons judgeth according to every man's work, pass the time of your sojourning here in fear.*
>
> *(1 Peter 1:13-17)*

One of Satan's strategies today to lead Christians into rebellion and disobedience is to take away the fear of God and cause us to fear man more than we fear God. His strategy is to cause us to become hardened, indifferent, and develop a lax attitude toward sin. He wants us to abuse the grace of God and use it as a covering to follow after the lusts of our flesh.

Satan has deceived many Christians into thinking that because they are born again they will not stand before God in judgment. They live with one foot in the world, trying to get away with as much as they can without going too far into open sin. They think that God will somehow overlook their transgressions. The fear of God is not in their hearts.

There are Christians today who are walking after the lusts of their flesh – mingling with the world, using profanity, watching movies filled with lust, greed, and violence, participating in social drinking and other worldly pleasures – who are walking around claiming to be "the righteousness of God in Christ." They have been deceived by false teaching that when they were saved they were covered by Christ's righteousness and, regardless of what they do or say, when God looks at them He doesn't see their sin. All that He sees is Christ's righteousness.

Satan is a liar! When you were born again, you were cleansed and sanctified by the blood of Jesus. You were declared righteous through faith in Jesus Christ (Romans 3:22). You were restored into right standing with God. However, if you yield yourself to unrighteousness and disobey God, you stand condemned before God. Unless you repent and turn away from your sin, it will be judged. Examine your heart today.

Make this declaration: "I reject the world, its lusts, and its ungodly system. I am cleansed and sanctified by the blood of Jesus and declared righteous through faith in Jesus Christ."

God's Fear In Your Heart

Scripture reading: 1 Peter 4:1-7

Not all Christians today who claim to be righteous are righteous. John wrote, *"Little children, let no man deceive you: he that doeth righteousness is righteous, even as he is righteous* (1 John 3:7). We are made righteous through faith in Jesus Christ, but we must *"follow after righteousness"* (1 Timothy 6:11) by yielding ourselves to the Holy Spirit.

There are Christians today who have brought God down to their own level. They have a concept of God as an earthly Father – a "daddy God" – Who is loving and kind, Who overlooks their sins and weaknesses, and never punishes them when they do wrong.

It is true that God is loving. He is forgiving. He is merciful and long suffering. We have entered into a covenant relationship with Him where we no longer fear to come into His presence and can enter boldly into the throne of grace, whereby we cry out, "Abba, Father!" (Romans 8:15). But God is also a holy, pure, righteous, God of judgment and wrath. He hates sin. He hates rebellion and disobedience and He does not overlook it. In His great love and mercy He forgives and pardons, just as He did Israel. He has provided full and free salvation through the blood of Jesus. Those who rebel, disobey God, and fail to repent and turn from their sin will be judged.

Do not be deceived! Every man must give an account of himself before God (Romans 14:12). Paul told the Romans that God will render to every man according to his deeds. He said: *"...to those who by perseverance in doing good seek for glory and honor and immortality, eternal life; but to those who are selfishly ambitious and do not obey the truth, but obey unrighteousness, wrath and indignation"* (Romans 2:7-8, NAS).

What has happened to the fear of God that once caused Christians to walk holy and upright before Him, crucifying the lusts of the flesh instead of playing the harlot with the world? What has happened to the fear of God that caused the saints to stay all night on their faces before God, repenting and crying out for God's mercy for the sins of the people?

God wants to put His fear in our hearts. He has said, *"I will put my fear in their hearts, that they shall not depart from me"* (Jeremiah 32:40). This fear is not a fear that will cause you to live in dread of what God is going to do to you the minute you do something wrong.

It is a fear that will cause you to turn away from anything that might be displeasing to Him.

Satan wants to steal the fear of God out of your heart, so that you will walk in disobedience and be defeated. Don't let him do it!

Make this declaration: "Satan cannot steal the fear of God out of my heart. God is putting His fear in my heart."

January 16

Rekindling The Fear

Scripture reading: Isaiah 6:1-13

The fear of God has gone out of His house. There are approximately four hundred references in the Bible concerning the fear of God, yet how many pastors, teachers and evangelists are there today who are teaching and warning people to live their lives in the fear of God? How many sermons do you hear from holy men of God concerning the wrath and judgment of God that is coming upon the earth? Not very many.

Do you want to know why? Because people don't want to hear it. They don't want to hear about crucifying the flesh, self-denial, or holiness. When they hear preaching under the power and anointing of the Holy Spirit that exposes the disobedience and sin in their lives, they say the person is preaching legalism or condemnation upon them. They would rather hear how to be successful and prosperous and enjoy the blessings of God. They would rather hear something that would allow them to live their lives following after their own wills, living their lives to please their selfish desires, fulfilling the lusts of their flesh.

There are many Christians today who are walking in disobedience to God and His Word. The fear of God is missing. They are afraid to live their lives according to what God has revealed to them in the Word because they fear what other people will think of them. They are afraid people will think they are foolish or fanatic.

There are Christians to whom God has spoken to step out into some area of ministry, but they have rebelled and are walking in disobedience because they are afraid of what their friends or family will say.

When God speaks to you – regardless of how insignificant it may seem, regardless of how foolish it may sound, regardless of the consequences and how impossible it may seem – do it in the fear of the Lord.

Because the nation of Israel did not have a heart that feared God, they disobeyed and forfeited their inheritance. They were defeated and destroyed. God's people today need to rekindle the fear of God in their hearts. We need to have a fresh revelation of God in all His holiness as Isaiah did when he fell on his face before God and said, *"Woe is me! for I am undone; because I am a man of unclean lips, and I dwell in the midst of a people of unclean lips"* (Isaiah 6:5). Once we see Him in the beauty and greatness of His holiness, the fear of God will be rekindled in our hearts, motivating us to walk in love and one-hundred percent obedience to Him all the days of our lives.

Make this declaration: "When God speaks to me – regardless of how insignificant it may seem, regardless of how foolish it may sound, regardless of the consequences and how impossible it may seem – I will do it in the fear of the Lord."

GUARANTEED FAILURE

Scripture reading: Numbers 14:36-45

One of the greatest strategies Satan has used since the beginning of time is to tempt believers to reject God's voice and follow after their own self-will. This strategy is one of the most dangerous because many people are so deceived that they cannot determine when they have fallen into the snare of the enemy.

Israel was defeated because they rejected the voice of God and followed their own self-will. Instead of going forward to take possession of the land as God had commanded them, they rebelled. They refused to listen. They refused to follow God into the Promised Land, but chose instead to follow their own will and go back to Egypt. This was the height of their rebellion against God. They chose their will above God's will. They rejected Him as the supreme authority of their lives.

Moses and Aaron recognized the full significance of this act of rebellion and fell on their faces in fear before God. Joshua and Caleb, in great anguish, tore their clothes and pleaded with them not to rebel against God. The children of Israel refused to listen. Once they made decision to follow their own way, nothing anyone could possibly say or do could change their minds.

The people were ready to stone Caleb and Joshua. In their rebellion against God's authority they wanted to ease their consciences by shutting out those who represented God's will, and who were willing to follow Him in obedience. As they were getting ready to stone them, the glory of God appeared before all the people at the Tent of Meeting. God was ready to destroy the entire congregation. Moses interceded and God forgave their rebellion, but He did not let them go unpunished. He told Moses that all those twenty years of age and older would not be able to enter into the Promised Land, but would die in the wilderness.

God told Moses to *"turn back tomorrow and set out toward the desert along the route to the Red Sea"* (Numbers 14:25, NIV). The next morning they came to Moses. *"We have sinned, they said. We will go up to the place the LORD Promised"* (Numbers 14:40, NIV). Once again they were rejecting the Word of the LORD and choosing to go their own way. When God told them to go forward, they turned back. Now that He commanded them to turn back, they decided to go forward. Moses said, *"Why are you disobeying the LORD's command? This will not succeed!" – and it didn't* (Numbers 14:41-43, NIV).

If you are disobeying God's command, you will not succeed. You will fail...guaranteed.

Make this declaration: "God has planned no failures for me. I will obey His commands and succeed in all I do."

January 18

Total Obedience

Scripture reading: 1 Samuel 15:1-9

Moses told Israel that because of their rebellion in refusing to enter the Promised Land God would not go to battle with them, but in rebellion Israel persisted and went forward, depending upon their own strength to deliver them from their enemies. God had positioned them for victory but through their self-will they chose defeat.

Satan's strategy is to cause you to reject God's Word and follow your own way. He will deceive you into compromising what God has told you to do and make you think you have obeyed God when you are actually following your self-will.

In the Old Testament account of King Saul, God had anointed him as king to rule over the people and commanded him to attack the Amalekites and totally destroy everything that belonged to them including women, children, cattle and sheep, camels and donkeys. Yet, Saul followed his own will and disobeyed. Instead of destroying everything as God had commanded, he spared the best of the sheep, the cattle, and lambs.

God told Samuel, *"I am grieved that I have made Saul king, because he has turned away from me and has not carried out my instructions..."* (1 Samuel 15:11, NIV). When we refuse to allow our lives to be governed by God and His Word and choose to live as we please, we have actually rejected God and turned away from Him.

Early the next morning Samuel went to see Saul. When he reached him, Saul said, *"The LORD bless you! I have carried out the LORD's instructions"* (1 Samuel 15:13, NIV). Saul was deceived. He thought he had obeyed God.

Samuel asked: *"What then is this bleating of sheep in my ears? What is this lowing of cattle that I hear?"* (1 Samuel 15:14, NIV).

Saul justified his actions. He answered, *"The soldiers brought them from the Amalekites; they spared the best of the sheep and cattle to sacrifice to the LORD your God, but we totally destroyed the rest"* (1 Samuel 15:15, NIV). Saul did not accept responsibility for his disobedience. He was in authority over the soldiers and had the ability to command them to do whatever he desired. The soldiers did not make him disobey God. No man or woman has the power to make you disobey God. You disobey by an act of your will.

God doesn't want just ninety-five percent obedience. A man who is ninety-five percent faithful to his wife is one-hundred percent unfaithful to her. Are you walking in total, one-hundred percent obedience today?

Make this declaration: "I will govern my life by the principles and promises of God. I will walk in one-hundred percent obedience."

A New Generation

Scripture reading: 1 Samuel 15:10-31

In the battle with the Amalekites, instead of obeying God King Saul decided to do things his way. He compromised the instructions God had given him and spared the best of the animals to offer as a sacrifice to God. But God was not pleased with his partial obedience.

This is where Satan deceives many Christians into thinking they are acceptable to God. They are submissive and obedient to God in some areas of their lives, but refuse to be obedient in other areas, and choose instead to follow their own way.

Samuel asked him, *"Why did you not obey the LORD? Why did you pounce on the plunder and do evil in the eyes of the LORD?"* (1 Samuel 15:19, NIV).

"But I did obey the LORD," Saul insisted (1 Samuel 15:20, NIV). He was still unable to see that he had rebelled against God by following his own will instead of fully obeying what God had told him to do.

Samuel said, *"To obey is better than sacrifice, and to heed is better than the fat of rams. For rebellion is like the sin of divination, and arrogance like the evil of idolatry. Because you have rejected the word of the LORD, he has rejected you as king"* (1 Samuel 15:22-23, NIV).

Saul finally acknowledged his sin. He said, *"I have sinned. I violated the LORD's command and your instructions. I was afraid of the people so I gave in to them"* (1 Samuel 15:24, NIV).

Saul acknowledged his sin, but did not repent. He continued to follow his self-will and was defeated. Because he rejected God, followed his self-will, and disobeyed God, King Saul forfeited his kingdom. The same thing happened to Israel the first time they came to Kadeshbarnea. They rejected God, acted in self-will, disobeyed, and forfeited their Promised Land .

King Saul rejected God and lost the kingdom. The people of Israel rejected God and lost their Promised Land . But God raised up a new generation who entered in to claim God's promises.

God is raising up a new generation – a new breed of spiritual warriors – who are prepared for war and will cross over to claim the promises of God. You are part of that new generation. You will not turn back. You will not be denied. Even if you failed in times past, God is saying to you, "This is a new day." This is your time to take the kingdom. This is your time to claim the promises of God.

Will you take the first step today? If so, make this simple declaration…

Make this declaration: "This is a new day! I will not turn back. I will not be denied. I will claim all the promises God has for me."

The River Of God

Scripture reading: Ezekiel 47:1-10

One of the truths God has taught us in this ministry is that He moves in cycles. A cycle is a period of time with a specific beginning and ending point. Right now, we are in God's end-time, harvest-time cycle and its ultimate conclusion is the culmination of the harvest and the return of our Lord Jesus Christ.

God has revealed to me that as part of this end-time cycle, the River of God's Spirit is rising throughout the whole world and it will crest at unprecedented levels. This divine, supernatural river is a new manifestation and demonstration of God's power that cannot be measured. It is this manifestation of His presence that will empower us to enter in and claim the spiritual territory God has promised us. Israel passed through a river to claim their land. We, too, will pass through a river – the river of God's manifested glory.

The Prophet Ezekiel saw a vision of this River of God and described it as a river that could not be passed over which issued into the desert and the sea, and everywhere the waters touched, life sprang forth (Ezekiel 47:5-9). As we are entering the new millennium, the River of God is intensifying as never before. It will not be a trickle. It will not be a puddle. It will not be a stagnant pool of the stale, aging waters of tradition. It will be a new, intensified downpour! Get ready for a deluge. A cascade. A torrent.

One of the major truths God has taught us in this ministry is that all truth is parallel – what is happening in the natural world reflects what is being manifested in the spiritual world. A few years ago, through the world wide weather phenomenon called El Nino, we witnessed floods where there had never before been floods. Small streams were turned into raging waters. Tidal waves of tremendous dimensions completely deluged places like Papua New Guinea. Right now, in the spirit world, God is preparing to parallel this natural phenomena. There is a River of God – a new, intensified tidal wave of His manifested power – that is about to hit this world that will be totally unprecedented:

> *And it shall come to pass afterward, that I will pour out my spirit upon all flesh; and your sons and your daughters shall prophesy, your old men shall dream dreams, your young men shall see visions.*
>
> *(Joel 2:28)*

God's Spirit will be poured out upon all flesh – your sons, your daughters, the young and the old. Everywhere this River of God flows, new life will spring forth. Everyone who is touched by its waters will experience supernatural healing – a total regeneration of body, mind, soul, and spirit.

Plunge in today. Let that river flow into every area of your life and ministry.

Make this declaration: "I receive the pouring out of God's river of supernatural power into my spirit today!"

Let The River Flow

Scripture reading: John 4:1-14

When this great River of God begins to flow over us, the focus will not be on the channel it flows through – the movement, denomination, or human vessel – but on the water itself! That is because the flow of this divine river will be so powerful, so unlimited, and so God-directed that no one will be able to take credit for it. It will not be a work of man, but a work of the Holy Spirit!

In Ezekiel's vision, he saw five depths of water in the river (Ezekiel 47:2-5). There was a trickle, ankle-deep water, knee-deep water, loin-deep, then it became a river to swim in, so deep it could not be passed over. These various levels have been interpreted many ways by Bible scholars. Some believe it represents different stages of any one outpouring of the Holy Spirit. Others believe it to be stages of spiritual maturity.

One thing is for sure, if we follow the spiritual analogy of the natural parallel of a river, it takes the deeper water for a river to flow to the sea. In scriptures, the seas are symbolic of the nations – so this means that the river that flows from God is to flow through us – the temple – to the nations of the world.

The direction of the flow of this river is also significant:

> *...These waters issue out toward the east country, and go down into the desert, and go into the sea: which being brought forth into the sea, the waters shall be healed.*
>
> *(Ezekiel 47:8)*

Isaiah 14:17 states that Satan caused the world to become a wilderness. Here we see the River of God flowing into the wilderness bringing life. The sea mentioned in this passage refers to the Dead Sea, which symbolizes the Dead Sea of fallen humanity.

The living water from God is being released into the temple of God – the Church, you and me – to flow into the Dead Sea of sinful humanity and into the wilderness of this world and bring life. And ...

> *...whithersoever the rivers shall come, shall live and there shall be a very great multitude of fish, because these waters shall come thither: for they shall be healed; and every thing shall live whither the river cometh.*
>
> *(Ezekiel 47:9)*

Everywhere the river flows – everything and everyone it touches – will live! On the banks of the river are fishermen, who are pulling in tremendous catches of fish. These are the souls of men and women.

God wants you to claim your inheritance so you can be a blessing to the nations of the world. God wants the river of His manifested presence to flow through you into the wilderness of this world! Let the river flow!

Make this declaration: "The River of God is being released in me. I will be a blessing to the nations. God's manifested presence will flow through me into the wilderness of this world."

January 22

We Must Catch The Vision

Scripture reading: 2 Kings 4:1-7

This River of God's manifested power will not flow into a spiritual void. It won't reach those who are too busy with their own concerns to receive it – those who limit God by their own unbelief. This River of God will flow over those who – like the widow woman in Elisha's time – are prepared to receive it!

> *Now there cried a certain woman of the wives of the sons of the prophets unto Elisha, saying, Thy servant my husband is dead; and thou knowest that thy servant did fear the LORD: and the creditor is come to take unto him my two sons to be bondmen. And Elisha said unto her, What shall I do for thee? Tell me, what hast thou in the house? And she said, Thine handmaid hath not any thing in the house, save a pot of oil. Then he said, Go, borrow thee vessels abroad of all thy neighbours, even empty vessels; borrow not a few. And when thou art come in, thou shalt shut the door upon thee and upon thy sons, and shalt pour out into all those vessels...*
>
> *(2 Kings 4:1-4)*

This widow woman experienced an unprecedented, supernatural flow of oil because she heard and acted upon the word of the prophet. She caught the vision – she could see the oil multiplying, the flow intensifying, her vessel being filled, her debt being canceled – and she prepared herself to receive the River of God. To prepare ourselves to receive the River of God, we must renew our spiritual focus. Like the widow woman, we must catch the vision of what God is preparing to do.

The Bible states in Proverbs 29:18, *"Where there is no vision, the people perish."* All over the world there are born-again believers who are perishing. No – they are not perishing in sin. These believers have accepted salvation through Jesus Christ. They attend church services, they read the Bible, and may even be leaders in their churches.

But they are dying spiritually. Their lives are routine. There is no spiritual excitement. They have no goal or purpose. They have no spiritual vision, no focus.

- If you feel empty spiritually...
- If you long to be used of God, but do not understand your part in His plan...
- If you feel there is something more to Christianity than the routine into which your life has fallen...

Then you need to let God instill in you – or renew in you – spiritual vision.

When I speak of "vision," I am not talking about natural eyesight or what is visibly seen in a trance-like vision. I am referring to the process where God is raising up spiritually dead men and women with new purpose, meaning, and vitality to fulfill their God-given destiny.

That is what He wants to do for you – right now, today!

Make this declaration: "God is raising me with new purpose, meaning, and vitality to fulfill my destiny."

Developing A Vision

Scripture reading: Habakkuk 2:1-5

If we are to ever claim our spiritual inheritance, we must have vision. Joshua and Caleb saw Canaan and never lost that vision until they entered in to claim the promises of God. This vision involves seeing beyond the natural world into the spiritual world. It is understanding the divine purpose of God and recognizing your part in His plan. Spiritual vision provides direction. It provides challenge and structure for life. Without it, people become spiritually dead.

If you are to be a participator instead of a mere spectator in this new flow of God's power, then you must allow Him to birth spiritual vision in you. The natural birth process which brings a human baby into the world is similar to the process of the birth of a vision in the spirit world. You will experience the following stages: Conception, development, travail, transition, and – finally – the birth.

Conception: "Conception" means to create. A spiritual vision is created in your spirit by God, just as a baby is conceived in its mother's womb. Conception in the natural world occurs through intimacy, and the same is true in the spirit world. You must develop a new level of intimacy through prayer and the Word to prepare you to receive the vision.

In these early stages – just as an expectant mother with morning sickness – we experience some difficulties. But each obstacle becomes an opportunity for the demonstration of God's power as we prepare to receive the manifestation of the vision which has been instilled in us.

Development: When a woman is impregnated, an embryo is formed and from these basic cells of life all other features emerge. When God plants a vision within you, the basic vision – like the human cell of life – remains the same, but the features continually change.

The focus will always remain the same, but the details of the vision will continually develop. Like a growing embryo, its features will not be the same as it was yesterday, last week, or last month. This development will be a stretching experience – as it is in the natural world within the body of a mother – but the embryo must continue to develop or it will die.

Just as a mother who carries a child within her body, keep your focus on the vision. While a baby is developing, a pregnant mother will deny herself certain things for the good of the child. She watches her diet and monitors her activities. The same is true spiritually. If you are to receive the vision God wants to instill in you, you must deny yourself. You must monitor your spiritual diet and guard against activities that can abort your vision. You may have to set aside your own plans and ambitions, your own schedules and preconceived ideas.

Are you ready to get spiritually pregnant?

Make this declaration: "I am ready to get spiritually pregnant. The dreams and visions of God are being conceived in my spiritual womb."

January 24

Birthing Your Vision

Scripture reading: Isaiah 66:5-11

We are meditating on the stages of development in spiritual vision. Yesterday we reflected on conception and development. The next steps are travail and transition, then comes the birth of the vision.

Travail: In the King James version, Ecclesiastes 5:3 states, *"For a dream cometh through the multitude of business."* The meaning of the word "multitude" here is "great." According to one Hebrew definition, "business" means travail or difficulty. So a dream or vision comes through "great travail" or difficulty.

In the natural birth process, the time of travail parallels the spiritual travail which is necessary to birth vision. Natural travail is a time of intense, concentrated effort to birth a child which is commonly called "labor." As in the delivery of a baby, spiritual vision is born by similarly intense mental, physical, and spiritual concentration.

But just as in the natural birth process, it takes greater effort to intensify. One of the things that is so important to a vision is focus. Like a mother giving birth, you must remain focused. You must remain concentrated on what God says, His promises and His prophecies.

As a mother in travail hides herself from public gaze, so those in travail spiritually must get alone with God and prepare themselves to receive the vision God has for their life and ministry.

Transition: In the natural birth process there is a time during labor known as the time of transition. It is the most difficult time of travail right before the birth canal is opened to permit the birth of the child. When God births spiritual vision, there is also a time of transition. Transition means change. The spiritual vision God wants to instill in your life will require change. It will call for new commitment and dedication. During this time of transition while you are preparing to receive the new flow of God's power, you will experience pressure in every area of your life. Everything within you may cry out for relief from the spiritual birth pangs of what God is bringing forth.

Birth: Travail is a difficult experience, but it leads to the birth which brings joy:

> *A woman when she is in travail hath sorrow, because her hour is come: but as soon as she is delivered of the child, she remembereth no more the anguish, for joy that a man is born into the world.*
>
> *(John 16:21)*

Let God birth new visions in your spiritual womb this year.

Make this declaration: "Thank You, God, for the new thing You are birthing in me."

Two Undeniable Proofs

Scripture reading: Hebrews 6:13-20

There are two undeniable proofs that guarantee to us the fulfillment of every promise God has made to us. This guarantee is your security. It provides assurance that your vision will become reality. This guarantee assures that you cross over to claim your spiritual inheritance.

Here are the two assurances we are given that God's promises will be fulfilled:

First: God swears by Himself.

> *For when God made promise to Abraham, because he could swear by no greater, he sware by himself, Saying, Surely blessing I will bless thee, and multiplying I will multiply thee.* (Hebrews 6:13-14)

When God swore by Himself, in essence He was saying, "I will cease to be God if I do not fulfill My promise!" The promises of God are safe, secure, and sealed because they are made on the basis of Himself, that is, His unfailing record.

Second: He seals His promises with an oath.

In addition to swearing by Himself, God seals His promises with an oath:

> *For men verily swear by the greater: and an oath for confirmation is to them an end of all strife. Wherein God, willing more abundantly to shew unto the heirs of promise the immutability of his counsel, confirmed it by an oath…* (Hebrews 6:16 17)

You can trust God's promises because they are sealed by His sacred oath. He swears by it, and He does not lie! He is not a man that He should lie. It is impossible for God to lie.

God made promises to Abraham and his heirs – which you and I are – then He sealed the promises with two proofs: Himself and His oath…

> *That by two immutable things, in which it was impossible for God to lie, we might have a strong consolation, who have fled for refuge to lay hold upon the hope set before us.* (Hebrews 6:18)

God provided two immutable proofs to end all doubt, fear, and uncertainty. The synonyms that the thesaurus gives for the word "immutable" are "abiding, changeless, consistent, constant, enduring, ceaseless, continual, endless, permanent, stable, unchanging, unalterable, and eternal."

God's promises are abiding. They are changeless and consistent. They are constant, enduring, ceaseless, continual, lasting, permanent, stable, unchanging, unalterable – they are eternal! Another synonym given for "immutable" is "dateless." Unlike many so-called guarantees, there is no expiration date. There are no exclusions, no loop-holes.

Make this declaration: "God has sworn by Himself and sealed His Word with an oath. His promises to me will be fulfilled."

Seeking First The Kingdom

Scripture reading: Daniel 1:1-21

The Prophet Daniel was also a man of vision. He not only received visions and interpreted visions, but he was a man of purpose who knew how to order his life and set objectives to fulfill his destiny.

As a young man, Daniel rose to preeminence among the children of Israel (Daniel 1:3-6) and then gained preeminence among the wise of men of Babylon (Daniel 1:17-20). He was eventually given power over all of Babylon (Daniel 2:48-49; 5:29) and received similar power over the Medo-Persian empire (Daniel 6:1-3,28).

What was the key that enabled Daniel to achieve positions of power and influence in the midst of an ungodly society? It was because he was a man of purpose. That is one of the first things we learn about Daniel as a young man:

> *But Daniel purposed in his heart that he would not defile himself with the portion of the king's meat, nor with the wine which he drank: therefore he requested of the prince of the eunuchs that he might not defile himself.*
>
> *(Daniel 1:8)*

As a young boy in a strange land, Daniel was faced with an opportunity to violate God's law by eating the king's food. Despite his youth and the tremendous pressure to conform, Daniel *"purposed in his heart"* to uphold the law of God, no matter the cost. It was this strength of purpose that gave him the power to remain faithful when his allegiance to God was challenged by a law forbidding prayer (Daniel 6) and when he was elevated to a position of leadership (Daniel chapters 2 and 6). Because Daniel was a man of purpose, he was granted favor with key leaders who were used by God to enable him to fulfill his destiny (Daniel 1:9).

Every child of God needs to be a person of purpose and direction. A strong sense of purpose not only fosters the respect and admiration of others, it is essential to fulfilling your ministry. Jesus made it clear what our purpose should be:

> *But seek ye first the kingdom of God, and his righteousness; and all these things shall be added unto you.*
>
> *(Matthew 6:33)*

If you seek God's will first – His Kingdom and purposes – then all other things will be supernaturally provided. You will have the strength you need to accomplish God's plan. You will have the abilities required and the necessary finances. You will have the supernatural power equal to the vision God gives you.

Make this declaration: "I will seek God's Kingdom and purposes first and all other things will be supernaturally provided. I will have the strength and finances I need to accomplish God's plan. I will have the supernatural power equal to the vision God gives me."

You Are Well Able

Scripture reading: Deuteronomy 1:19-33

There is a difference between receiving a promise of God and possessing it. Israel certainly found that out!

In order to see the fulfillment of God's promises you must act upon them. You must move into a position where you can take possession. You must seize them by force! Jesus declared:

> *And from the days of John the Baptist until now the kingdom of heaven suffereth violence, and the violent take it by force.*
>
> *(Matthew 11:12)*

Don't be complacent. Don't vacillate.

- Don't say, "If it's God's will, He'll heal me." We know the promise and provision of God for healing. You can either settle for the sickness or make up your mind that you will not settle for anything less than the fulfillment of God's promise in your life.
- Don't say, "I might not be able to fulfill my vision." Continuously declare that your God-given vision will become a reality.
- Don't say, "I hope my family will get saved." Boldly declare, *"The Lord is not slack concerning his promise, as some men count slackness; but is long-suffering to us-ward, not willing that any should perish, but that all should come to repentance"* (2 Peter 3:9).

When God brought Israel to the border of the Promised Land, only two young men – Joshua and Caleb – declared, "We are well able." Their spiritual vision gave them the ability to focus on the promise God had made to Israel. They believed it and wanted to act upon it. But the majority of the people didn't have spiritual vision, so they did not enter in and claim their Promised Land. Instead, they returned to the wilderness to wander for another 40 years.

Don't wander in the wilderness for 40 years! Let God give you spiritual vision that will enable you to see beyond the circumstances of your impossibility:

> *Behold, the Lord thy God hath set the land before thee: go up and possess it, as the Lord God of thy fathers hath said unto thee; fear not, neither be discouraged.*
>
> *(Deuteronomy 1:21)*

You are at the edge of your Promised Land. Activate your faith. Don't fear. Don't be discouraged. Focus your vision, don't turn to the right or left. March forward in assurance to claim the promises of God. You are well able!

Make this declaration: "I am well able! I will march forward and claim all God has for me. I will not turn to the right or left!"

Eyes Of Faith To The Future

Scripture reading: Hebrews 11:8-19

You can be assured that every promise God has made to you will come to pass. He has sworn it by Himself and by His oath. Let's read this promise in the Amplified Version of the Bible:

> *Men indeed swear by a greater than themselves, and with them in all disputes the oath taken for confirmation is final [ending strife]. Accordingly God also, in His desire to show more convincingly and beyond doubt to those who were to inherit the promise the unchangeableness of His purpose and plan, intervened [mediated[with an oath. This was so that, but two unchangeable things [His promise and His oath] in which it is impossible for God every to prove false or deceive us, we who have fled [to Him] for refuge might have mighty indwelling strength and strong encouragement to grasp and hold fast the hope appointed for us and set before us.*
>
> *(Hebrews 6:16-18, TAB)*

God doesn't want the standard of how people think in the world to be the final word, so He goes beyond that. His desire is to show us convincingly and beyond a shadow of a doubt that His promises will be fulfilled to those who have fled to Him for refuge.

God has given you a vision. He has given you a ministry. He has made promises to you. Do not look to the greatness of the task before you, but look to the greatness of God. Be like the great Patriarch Abraham who *"...believed God even though such a promise just couldn't come to pass!"* (Romans 4:18, TLB).

Do not look to your own abilities to claim God's promises or fulfill your vision. In the natural, Abraham had no ability to fulfill God's plan, nevertheless...

> *Therefore sprang there even of one, and him as good as dead, so many as the stars of the sky in multitude, and as the sand which is by the sea shore innumerable.*
>
> *(Hebrews 11:12)*

When the vision was first given to Abraham, his past had been unproductive and his present was unproductive. But he did not accept the deadness of his own body nor that of Sarah's womb.

Perhaps your past has been spiritually unproductive and your present is still unfruitful, but God is instilling in you a new vision for His purposes. As Abraham, look with eyes of faith into the future knowing that *"He who has promised is able also to perform it."*

Make this declaration: "I will not look to the greatness of my task. I will not look to my own abilities. I will look with eyes of faith into the future knowing that He who has promised is able to perform it."

Have You Caught The Vision?

Scripture reading: Acts 9:1-22

One day, while riding down the Damascus Road, the Apostle Paul received a vision from God that changed his entire life – his purpose, his plans, his nature, even his name. In this vision, God told Paul...

> *...Rise, and stand upon thy feet: for I have appeared unto thee for this purpose, to make thee a minister and a witness both of these things which though hast seen, and of those things in the which I will appear unto thee; Delivering thee from the people, and from the Gentiles, unto whom now I send thee. To open their eyes, and to turn them from darkness to light, and from the power of Satan unto God, that they may receive forgiveness of sins, and inheritance among them which are sanctified by faith that is in me.* *(Acts 26:16-18)*

God gave Paul a vision of the souls of multitudes of Gentile people. (The term Gentiles refers to everyone who is not a Jew, so it means Paul was called to minister to all other nations of the world.)

When God gives you a vision He also provides spiritual and practical strategies for fulfilling it. God gave Paul objectives – a step by step plan – for ministering to the Gentile nations in order to fulfill his vision. Paul was to...

...Open their spiritual eyes from darkness to light.

...Turn them from the power of Satan to God.

...Lead them to forgiveness of sins.

...Reveal their spiritual inheritance made possible by faith (Acts 26:16-18).

Paul's vision became the central focus of his life. It was the force that compelled him to endure hardship, persecution, discouragement, shipwreck, and physical abuse. It motivated his writing. It directed his travels. It dictated his life style.

Paul always remained true to his God-given vision. Despite discouragement, rejection, and persecution, he never lost his focus. God raised him up to be one of the greatest leaders in New Testament times and his ministry is still impacting multitudes through the tremendous books he authored under the anointing of the Holy Spirit. Near the end of his life, the Apostle Paul was able to reflect back over his ministry and declare, *"I was not disobedient unto the heavenly vision"* (Acts 26:19).

God is looking for men and women of vision. Joshua was able to lead Israel into the Promised Land because he had seen it 40 years earlier. He had caught the vision.

Have you caught the vision of all God wants to do in and through you?

Make this declaration: "I will remain true to the vision God gives me. I will not be disobedient unto the heavenly vision."

Three Simple Keys

Scripture reading: Genesis 15:1-6

Every promise of God is based upon His unfailing record. He has always fulfilled His Word. He has never failed because...

> *God is not a man, that he should lie; neither the son of man, that he should repent: hath he said, and shall he not do it? Or hath he spoken, and shall he not make it good?*
>
> *(Numbers 23:19)*

When you really understand this, there will be...

...No more doubt

...No more vacillating

...No more questioning

God never fails. He always fulfills His promises.

When Israel came to the border of their Promised Land the first time, they did not enter. Years later, when Israel once again returned to that border, Joshua led the people across the Jordan River to take possession.

There are three simple keys to the promises of God. If you want to claim God's promises, then you must settle these issues in your heart once and for all:

1. There is divine stability and constancy to every promise of God. His promises never fluctuate.
2. The promises are given to us by a God who cannot lie.
3. These promises are made in Christ who purchased them with His own blood. By His death, Christ made effective the covenant which contains all the promises of God.

Abraham believed God's promise to give him a son despite the fact he and his wife, Sarah, were beyond the child-bearing years. Like Abraham, you must keep the vision God has given you alive by faith. If you let the vision die, you will lose the promise.

Keep the vision of your healing alive.

Keep the vision of household salvation alive.

Keep the vision of your ministry alive.

The promises of God are not just passing words. When they are supernaturally birthed within you then you must keep them alive by faith. Never let go of the promises of God!

Abraham didn't just receive his promise from God and continue to act nonchalantly saying, *"If it comes to pass okay; if not...oh well!"* I think every day when he woke up he asked God, *"Where's my son?"* He reminded God of His promises. He waited. He was expecting. The promises – the vision he received – influenced every area of his life. He planned for the fulfillment of these promises. He positioned himself to receive his miracle.

Make this declaration: "There will be no more doubt, vacillating, or questioning in my life. God's promises never fluctuate, they are given by one who cannot lie, and they are sealed with the blood of Jesus Christ."

Refuse To Be Defeated

Scripture reading: Psalm 78:1-72

Satan's strategy is to keep you out of your Promised Land, to prevent you from inheriting all the blessings God has promised you...

- He will cause you to doubt God and His Word.
- He will tempt you to murmur against God.
- He will cause you to turn away from God.
- He will cause you to refuse to hear God's voice.
- He will cause you to reject God's voice and reject His Word.
- He will steal the fear of God out of your heart.
- He will cause you to fear man more than you fear God.
- He will cause you to reject God's will and follow your self-will.
- He will deceive you into compromising what God has spoken to you.
- He will try to steal your vision.

During this past month, God has brought you to a new level of spiritual destiny. He has positioned you for victory. He has placed His Spirit within you, giving you the will, desire, and power to obey. God has set a new powerful spiritual dimension before you and is saying, "Go up and possess it!"

Your spiritual destiny is in your hands. You have a choice...

...To obey God and, by faith, take full possession of your spiritual inheritance and enter this new spiritual dimension, or ...

...To rebel and disobey God and, through your unbelief, forfeit your inheritance and live in defeat.

Go forward! Take offensive action and defeat Satan's strategy to lead you into disobedience and rebellion against God.

- Know the voice of God.
- Hear the voice of God.
- Obey the voice of God.
- Walk in the manifestation of God's glory.
- Believe God's promises.

Behind you is a life of struggle, failure and defeat. Before you lies a new spiritual territory:

- A place of peace and security.
- A place of blessing and provision.
- A place of power and victory.
- A place of joy and contentment.

Go forward in the power of Almighty God and take possession of your spiritual inheritance! Refuse to be defeated!

Make this declaration: "I will know the voice of God. I will hear and obey it. I will walk in the manifestation of God's glory and believe His promises."

February

During Brother Cerullo's early ministry, giant tractor-trailers carried the huge 120' by 240' tent and equipment for Brother Cerullo's crusades from city to city.

Morris Cerullo Miracle Crusade, 1956, Erie, Pennsylvania: Thousands packed Brother Cerullo's tent meeting, night after night, to receive God's miracle healing and salvation.

FEBRUARY
OVERCOMING OBSTACLES

What are the impossibilities, hindrances, and problems preventing you from taking hold of the promises God has given you? What are the obstacles hindering you from receiving the answers to your prayers, the things you desperately need, and the total victory God has provided for you?

God has shown me seven major obstacles which block the fulfillment of God's promises and provision for your life. These obstacles are barriers – walls that are standing in your way hindering you from taking possession of God's promises, just as the walls of Jericho prevented Israel from taking their land.

Like the children of Israel who marched around the Jericho walls seven times, God has directed me during the next few months to lead you in seven marches around seven major walls in your life. Day by day, we are going to identify these seven obstacles – the walls – which block you from receiving the fulfillment of God's promises and walking in the total victory He has planned for you. Then, using the strategies God has given us, we are going to see these walls come down in Jesus' Name!

The seven major obstacles God has shown me that are holding you back from taking the victories you need are like the massive walls of Jericho separating the Israelites from taking possession of the Promised Land. These walls separated the children of Israel from the fulfillment of God's promises. They were a formidable barrier which stood between them and their Promised Land. Behind Israel was the Jordan River and forty years of wandering in the wilderness. Before them was Jericho, the first major stronghold they must conquer before taking possession of the land.

Jericho was a strong, fortified city with two massive walls surrounding it. The outer wall was six feet thick and the inner wall twelve feet thick. Both walls were thirty feet high, fifteen feet apart, and connected together with houses built across them. The total combined distance across the two walls was thirty-three feet. These walls formed a formidable barrier. In the natural, they were impossible to penetrate. Joshua and the children of Israel did not have the weapons nor military might to tear them down and there was absolutely no possible way for them to break through the walls.

God directed Israel to march around the walls of Jericho once a day for six days. The warriors, armed for battle, were to go before the Ark of the Covenant. Seven priests marched in front of the Ark sounding seven trumpets. During this time they were commanded not to speak a word. No sound was to be heard beside the trumpets.

The Ark of the Covenant was the central focus of this entire procession. The outcome of the battle was dependent upon one thing: The God of battles was with them. Without His presence and divine intervention, there could be no victory. The Ark was the place where

God's manifested presence dwelt above the Mercy Seat. God wanted the people to know it was His divine presence in their midst that would make them victorious.

On the seventh day the people were to march around the city seven times. On the seventh time around, the priests were to sound a long, loud blast of the trumpet. When the people heard this signal, they were to shout and the Lord promised that when they did this, the walls would fall down. The walls of Jericho collapsed exactly as God promised because Joshua and the children of Israel acted upon the strategy God gave them.

During the next few months, you will learn strategies that will break down every wall in your life. As we begin this march, if you act upon each God-given strategy you learn, the "Jericho walls" in your life will fall.

- Get ready to see your walls of impossibilities come crashing down!
- Get ready to see the walls hindering you from taking hold of the fulfillment of the promises God has given you crumble!
- Get ready to see the walls blocking you from receiving the answers you need from God broken down!

This month we are marching against the walls of fear, unbelief, and confusion. In the next few months we will march against disobedience and unforgiveness. We will also learn how to win the battle for the mind and the tongue.

Are you ready for our supernatural march? Then put on your spiritual boots and let's get going!

The Obstacle Of Fear

Scripture reading: Isaiah 43:1-19

As we begin our march against the obstacles standing in the way of our possessing God's promises, remember that just as the Ark of the Covenant was with the children of Israel as they marched around Jericho, the presence, power, and anointing of God are with you!

He is with you and will give you total victory!

The number one obstacle blocking Christians today from taking hold of God's promises and walking in the power He has planned is fear. God's word to you today, dear friend, is "Fear not!" God is saying:

> *...Fear not: for I have redeemed thee, I have called thee by thy name; thou art mine. When thou passest through the waters, I will be with thee; and through the rivers, they shall not overflow thee: when thou walkest through the fire, thou shalt not be burned, neither shall the flame kindle upon thee. For I am the* LORD *thy God, the Holy One of Israel...*
>
> *(Isaiah 43:1-3)*

The first obstacle the children of Israel faced as they entered the Promised Land was fear. God had promised upon His oath to give them the land, but when the ten men sent to spy out the land saw giants in the land and large, fortified cities, they became fainthearted. A spirit of fear and intimidation gripped their hearts. Instead of rejecting fear and getting their eyes upon God who had rolled back the Red Sea and miraculously delivered them out of Egypt, they allowed their minds to be controlled by a spirit of fear.

When these men returned to give their report, they caused this spirit to spread until all of Israel was fearful. The people heard the bad report and a spirit of fear entered their hearts and minds until they were controlled and dominated by it. None of that fearful, unbelieving generation entered the Promised Land. Only Joshua and Caleb, who focused their faith upon God and His mighty power, were able to enter the Promised Land and take possession of God's promises. They said, *"we are well able to conquer it!"* (Numbers 13:30, TAB) and *"the Lord is with us; fear them not"* (Numbers 14:9, TAB).

There are hundreds of different manifestations of fear. We are living in a day when fear stalks the streets of our cities and people are bound by fear for their safety. Other major fears people face include those of failure, rejection, pain, sickness, danger, want, growing old, man, and death.

Fear and perplexity are part of the signs Jesus said would be upon the earth before His coming. He said men's hearts would fail them because of fear and of the things which are coming upon the earth (Luke 21:26). As believers, however, we are to look up and rejoice. Our redemption draws nigh!

Make this declaration: "I am beginning my march on fear. I will not fear the circumstances around me. I will look up and rejoice. My redemption draws near!"

February 2

Recognizing Fear For What It Is

Scripture reading: Psalm 27

When you face a serious disease, financial difficulties, a dangerous situation, or some unexpected problem, immediately a spirit of fear will try to gain a stronghold in your mind. If you fail to take authority over it, it will control and dominate your heart and mind until you are no longer trusting and believing God.

When God gives you a promise and directs you to take a step of faith and act on that promise, a wall of fear will try to block you. God has spoken, "Now is your time!" Now is your time to receive the fulfillment of the promises God has given you – to take hold of God's promises, that wall of fear must be demolished.

When God directs you to step out in faith on His Word to witness, pray for the sick, or launch out into any area of ministry, a spirit of fear will come against your mind to hinder you from doing the will of God. During the next few days I want to give you five strategies that will help you conquer the obstacle of fear:

Strategy number one: Recognize fear for what it is. Although we are surrounded on every hand by fear, God does not intend for us to live our lives bound, tormented, or oppressed by a spirit of fear. God does not want you to be fearful concerning the circumstances you are facing. He wants you to be set free from every fear that has you bound and to know how to reject and take authority over every fear that tries to gain a stronghold in your life.

You must recognize that fear is a spirit Satan is using to torment, control, dominate, oppress, and bind eight out of ten people today – including born-again Christians. Satan's objective is to build a wall of fear in your mind…

- To immobilize your faith.
- To stop you from taking possession of God's promises.
- To cause you to disobey God.
- To stop you from doing the works of God.
- To stop you from operating in the gifts of the Holy Spirit.

You must learn to recognize fear for what it is – sin. Fear is the opposite of faith and God's Word declares, *"Whatsoever is not of faith is sin"* (Romans 14:23). At the very root of fear is a lack of trust and unbelief. If you are fearful, you are not trusting God. Trust in God is a bold confidence and total reliance upon Him that He will fulfill His Word.

Fear draws your focus off God and onto your circumstances. When an individual becomes fearful it is because he no longer has his eyes upon the unlimited power of God and is no longer believing and trusting Him to deliver him or meet his needs.

Where is your focus today? On your fears or on your God? Why not confess your fears to Him right now and trust Him to meet your need?

Make this declaration: "I recognize fear for what it is – a sin. It immobilizes my faith, hinders me from possessing God's promises, causes me to disobey God, and prevents me from doing the works of God and operating in the Gifts of the Spirit. I reject fear!"

Taking Authority Over Fear

Scripture reading: Isaiah 51:11-15

In order to break down the wall of fear in your life, you must identify the fears that are attacking your mind as the work of the demonic spirit of fear, bind it, and take authority over it in Jesus' Name. If you allow it to operate, fear will spiritually immobilize you until you are unable to release your faith and receive what you need from God.

A spirit of fear will torment your mind until you no longer have any peace. Your mind will be in a constant state of turmoil. If you fail to resist and take authority over it, it will gain entrance into your mind until you become obsessed – your thoughts, emotions, and will come under its control.

The spirit of fear will try to bind you with a fear of man which is a trap *"The fear of man bringeth a snare: but whoso putteth his trust in the LORD shall be safe"* (Proverbs 29:25).

If your heart and mind are continually filled with a fear of what other people think of you – fear of ridicule, rejection, scorn, and fear of what they might do to you – you will be so bound you will be unable to walk in the freedom and victory Christ has made possible for you. Fear of man will keep you from doing the work God has given you to do. Once you allow it to take hold in your life, instead of living your life to please God you will be living it to please man.

There is nothing man can say or do that can destroy us or hinder the fulfillment of God's promises in our lives. Nothing! God said,

> *I, even I, am he who comforts you. Who are you that you fear mortal men, the sons of men, who are but grass, that you forget the Lord your Maker, who stretched out the heavens and laid the foundations of the earth, that you live in constant terror every day because of the wrath of the oppressor, who is bent on destruction? For where is the wrath of the oppressor?*
>
> *(Isaiah 51:12-13, NIV)*

God has not given you a spirit of fear. As a child of God, there is no room in your life for it. God intends you to be fearless. He has made it possible for you to live your life free from fear. When you face sickness or disease, financial problems, family problems, trials, tribulations, or circumstances which seem impossible and you just don't know what to do, He expects you not to fear!

Fear is contrary to the nature of a child of God. You have not received a spirit of fear, but of power. The Apostle Paul said, *"God hath not given us the spirit of fear, but of power, and of love, and of a sound mind"* (2 Timothy 1:7). If you have fear, it is because you have allowed a spirit of fear to attack your mind and you need to take authority over it, in the Name of Jesus!

Make this declaration: "The fear of man brings a snare. God has not given me a spirit of fear, but of power, and love, and a sound mind."

Dealing With The Spirit Of Fear

Scripture reading: Isaiah 12:1-6

God does not intend you to ever be bound by a spirit of fear. He does not intend that your mind be tormented, oppressed, or obsessed by a spirit of fear. He has given you a spirit of power. The fearlessness and boldness which God has planned to be manifested in your life is imparted to you through the power of the Holy Spirit.

During the past few days we have meditated on the first strategy to overcome fear which is to recognize fear for what it is. Are you ready to receive the next strategy into your spirit?

Strategy number two: Identify the fears that have gained a stronghold in your mind and deal with them in the spirit by casting them out in Jesus' Name. You cannot overcome your fears in your natural strength. You must recognize you are in a spiritual battle and know that through the power of the Holy Spirit you will be set free. What you are facing is a spirit of fear that is trying to bind and torment your mind until you are unable to reach out by faith and take hold of God's promises.

Beloved, what are the fears attacking your mind? Name them, one by one, and through the power and authority of the Holy Spirit begin to cast them out, in the Name of Jesus! Rebuke the spirit of fear. Use today's declaration and say,"You foul, tormenting spirit of fear, I rebuke you in the Name of Jesus and command you to loose your hold on my mind!" Then name every fear binding you and take authority over it.

Learn to recognize any trace of the spirit of fear the moment it attacks your mind and resist it. Refuse to allow fear to have any place in you. Bring every thought into obedience to the Word. That is where the third strategy against fear comes in...

Strategy number three: Resist every attempt of Satan to fill your mind with fear. Satan will try to use a spirit of fear to attack your heart and mind, steal your peace and joy, and paralyze your faith so that you will be unable to take hold of God's promises and receive what you need from Him.

Don't allow your mind to be filled with fearful thoughts. Reject them and fill your mind with the Word. To walk in freedom over fear you must resist. James said: *"Submit Yourselves therefore to God. Resist the devil, and he will flee from you"* (James 4:7). Peter exhorts us to resist the devil, *"steadfast in the faith"* (1 Peter 5:9). The word "resist" is translated from the Greek word, *"anthistemi,"* which means "to oppose and to withstand." The moment the spirit of fear tries to attack your mind, you must not give in to it, but resist, oppose, and withstand it in the Name of Jesus!

Make this declaration: "You foul, tormenting spirit of fear, I rebuke you in the Name of Jesus and command you to loose your hold on my mind!"

Focus On God

Scripture reading: Psalm 3

In order to conquer fear you must first, recognize it for what it is; second, identify the fears that have gained a stronghold in your mind and deal with them in the spirit by casting them out in Jesus' Name; and third, resist every attempt of Satan to fill your mind with fear.

Now here is the fourth powerful strategy against fear:

Strategy number four: You must get your eyes focused upon God and release your trust in Him. What are the fears you are facing? Fear of sickness? Fear of not having sufficient income to meet your family's needs? Fear of failure? Fear of danger? Fear of becoming homeless? Fear of growing old? Fear of losing your job? Fear of death?

God has entered into a covenant with you. He has given you promises which provide for all your needs. He has promised to deliver you out of the hands of all your enemies, to protect you, heal your body, strengthen you, and supply all your needs.

The fearlessness God wants you to have is based upon your faith and trust in Him and His faithfulness to His promises to you. As you release your faith and trust God to fulfill His promises, there is no fear! You are covered and protected by His mighty hand. God has said, *"I have put My words in your mouth and: have covered you in the shadow of My hand..."* (Isaiah 51:16, TAB). He has promised never to leave you, to fight for you, to go with you through the waters of adversity and fiery tribulation and give you victory over all your enemies. God has promised:

> *"...I will not in any way fail or give you up nor leave you without support. (I will) not, (I will) not, (I will) not in any degree leave you helpless nor forsake nor let (you) down, (relax My hold on you) Assuredly not!"*
>
> *(Hebrews 13:5, TAB).*

As a result of this promise, you can boldly and fearlessly say:

> *"The Lord is my Helper, I will not be seized with alarm – I will not fear or dread or be terrified. What can man do to me?."* *(Hebrews 13:6, TAB)*

When fear of any kind attacks your mind, you must reject it and use your trust in God as a mighty weapon to destroy fear. King David, a powerful, fearless warrior who faced many dangerous situations and formidable enemies said: *"I will not be afraid of ten thousands of people, that have set themselves against me"* (Psalm 3:6). He said: *"The LORD is on my side; I will not fear. What can man do unto me?"* (Psalm 118:6).

The key to his fearlessness was not based upon the fact that he was never afraid. There were times he was fearful, but in those times as an act of his will he learned to reject fear and trust God. Focus on God today and release your trust in Him.

Make this declaration: "God has not given me a spirit of fear, so I will not fear! I will reject fear and release my trust in God!

THE GOD OF YOUR CIRCUMSTANCES

Scripture reading: Psalm 56

The Psalmist David said:

> *What time I am afraid, I will trust in thee. In God I will praise his word, in God I have put my trust, I will not fear what flesh can do unto me.*
> *(Psalm 56:3-4)*

David made a conscious decision to cast out fear and trust God. As he released his trust in God, he won the battle against fear. In other words, David was saying: "When I feel fear overtaking me, by an act of my will – regardless of the circumstances – I will put my trust in God and I will not fear!"

David refused fear. Repeatedly in the Psalms he declared, "I will not fear"…

> *Yea, though I walk through the valley of the shadow of death, I will fear no evil: for thou art with me; thy rod and thy staff they comfort me.*
> *(Psalm 23:4)*

> *The LORD is my light and my salvation; whom shall I fear? the LORD is the strength of my life; of whom shall I be afraid?* *(Psalm 27:1)*

> *Though an host should encamp against me, my heart shall not fear: though war should rise against me, in this will I be confident.*
> *(Psalm 27:3)*

> *Therefore will not we fear, though the earth be removed, and though the mountains be carried into the midst of the sea;* *(Psalm 46:2)*

As you begin your march to break down the wall of fear in your life, this is part of the strategy God has given you to cause that wall of fear to come crashing down! Use your trust in God to counterattack the fears that come against your mind. What are the fears that are blocking you from taking hold of God's promises to you? Forget your circumstances and get your eyes focused on God!

You maybe thinking, "Brother Cerullo, you just don't know the circumstances I am facing. How can I not be afraid?" While it is true that I may not know your exact circumstances, I do know the God of your circumstances. That is why I can encourage you to choose to trust God instead of allowing fear to paralyze you. Trust is a bold confidence and total reliance upon God. Look at your circumstances and – like David – say to the Lord, "I will not fear! I will put my trust in You, the God of my circumstances!"

Make this declaration: "When I am afraid, I will trust in God. In God I have put my trust. I will not fear what flesh can do unto me."

Fix Your Mind On God

Scripture reading: Psalm 112:1-10

We have learned that in order to conquer fear you must recognize it for what it is; identify the fears that have gained a stronghold in your mind and deal with them in the spirit by casting them out in Jesus' Name; resist every attempt of Satan to fill your mind with fear; and focus upon God and release your trust in Him. Now here is the fifth powerful strategy against fear:

Strategy number five: Fix your mind and keep it fixed upon God. God has promised to guard, protect, and keep you in perfect and constant peace as long as you have your mind fixed upon Him:

> *Thou wilt keep him in perfect peace, whose mind is stayed on thee: because he trusteth in thee.*
>
> *(Isaiah 26:3)*

The word "stayed," in this verse is translated from a Hebrew word which means "to lean upon or take hold, establish, be steadfast." When your heart and mind are fixed upon God you will not be afraid because God will cover your mind with His perfect peace! David said concerning the righteous, *"He shall not be afraid of evil tidings: his heart is fixed, trusting in the Lord. His heart is established, he shall not be afraid"* (Psalm 112:7-8). Releasing your trust is the key that will enable you to fix your mind. Is your mind fixed, fully persuaded, and convinced, that God will fulfill His promise to you? Are you trusting God in the circumstances you are facing?

To keep your mind fixed upon God, you must trust Him at all times, under all circumstances, believing He is with you to sustain you and meet your need. Regardless of the circumstances you face today, fix your mind, keep your thoughts focused upon God's faithfulness to fulfill His promises.

To keep your mind fixed upon God, you must guard what you allow into your heart and mind. Fear gained an entrance into the minds of the children of Israel through what they saw and heard. When the spies saw the giants and the fortified cities, a spirit of fear attacked their minds. Instead of rejecting the fear and fixing their hearts and minds upon God Who had rolled back the Red Sea and miraculously delivered them out of Egypt, they allowed their minds to be controlled by a spirit of fear. When the people of Israel heard the fear-filled words of the report, a spirit of fear entered their minds until they were filled with unbelief which led to rebellion and disobedience. Refuse to listen to fear-filled words. Don't speak fear-filled words and don't focus your eyes upon your circumstances. Fix your mind on God.

Make this declaration: "I am fully persuaded and convinced that God will fulfill His promises to me. Today, my mind is fixed on Him instead of my circumstances."

Five Strategies Against Fear

Scripture reading: Isaiah 41:10-13

When Jairus heard the report that his daughter had died, Jesus told him, *"Fear not: believe only, and she shall be made whole"* (Luke 8:50). If he had kept his heart and mind fixed upon the outward circumstances of his dead daughter instead of fixing his heart and mind on Jesus and His power to heal, he would have been fearful and would not have been able to release his faith and receive the miracle he needed.

You can deal a death blow to the spirit of fear when you use the spiritual arsenal God has given you and fill your mind with His Word. God has declared:

> *Fear thou not, for I am with thee: be not dismayed; for I am thy God I will strengthen thee; yea, I will help thee; yea, I will uphold thee with the right hand of my righteousness.*
>
> *(Isaiah 41:10)*

You can fix your mind upon God and release your trust in Him because God has promised you:

> *For I the Lord thy God will hold thy right hand, saying unto thee, fear not, I will help thee.*
>
> *(Isaiah 41:13)*

When you keep your mind fixed upon God and fill your mind with His Word, God covers your mind with His peace. It becomes a powerful, supernatural shield which guards your mind and keeps it in constant peace and free from all fear.

When your mind is fixed upon God, you will be:

- Set free from a fear of man!
- Set free from a fear of death!
- Set free from a fear of lack or want!
- Set free from a fear of danger or calamity!
- Set free from a fear of sickness!
- Set free from all bondage to fear!

As we close out our meditations on the first obstacle that we must defeat, to claim God's promises, let us review the five powerful strategies we have learned that will conquer fear:

Strategy one: You must recognize fear for what it is – sin: Romans 14:23.

Strategy two: You must identify the fears that have gained a stronghold in your mind and deal with them in the spirit by casting them out in Jesus' Name: 2 Corinthians 10:4; John 14:14

Strategy three: You must resist every attempt of Satan to fill your mind with fear: 2 Timothy 1:7; James 4:7.

Strategy four: You must focus upon God and release your trust in Him: Hebrews 13:5.

Strategy five: You must fix your mind and keep it fixed upon God: Isaiah 26:3

Make this declaration: "I will not fear! Spirit of fear, I am dealing you a death blow. I command you to go in Jesus' Name!"

THE OBSTACLE OF UNBELIEF

Scripture reading: Matthew 7:7-11

During this month's devotions we are examining the obstacles that will try to prevent us from claiming God's promises. We have talked about the first obstacle, fear. Now let us move on to the second obstacle which is unbelief.

Beloved, God has shown me that there are many Christians who are under siege by the enemy. They have been under such heavy attack physically and financially, within their families and on their jobs, that their faith has begun to waver. A struggle is going on within them. They believe God is able to deliver them. They believe God can heal their bodies, provide the finances they need, and supply their needs. They believe that His Word and His promises are true. Yet, because they have not yet seen an outward manifestation of the fulfillment of His promises or the answers to their prayers, they have become discouraged and are beginning to question, murmur, and complain.

- "How long must I wait before God intervenes in my circumstances?"
- "I've been fasting and praying, I've been claiming God's promises. Why can't I get the breakthrough I need?"
- "Why is God allowing these circumstances in my life?"
- "How long must I wait before God intervenes in my circumstances?"
- "Has God forgotten me?"

You may be going through a time when you are being tested. You don't understand why God has allowed certain circumstances to come into your life. You have become fearful and worried. In your heart a battle is going on to try to produce enough faith to believe God for the fulfillment of His promises to you. You may be struggling to have faith to believe God for the healing of your body, for a financial need, for the restoration of your marriage, or for some other desperate need and your prayers have seemingly gone unanswered. You don't know what to do and your faith is beginning to waver.

One of the greatest obstacles that will block you from taking hold of the things you need, from receiving answers to your prayers, and the fulfillment of God's promises, is unbelief. Unbelief is like the massive walls of Jericho separating and holding the children of Israel back from entering the Promised Land. It separates and prevents you from taking hold of the promises of God. Unbelief blocks and hinders you from walking in total victory in the circumstances you face. It robs you of the blessings God is ready to pour into your life. That is why during the next few days we are going on a march to destroy the wall of unbelief in your life.

Make this declaration: "I am beginning a march to destroy the wall of unbelief in my life. It will come down through the power of God!"

February 10

Shut Out Of The Land

Scripture reading: Hebrews 3:6-19

When most Christians face desperate situations in their lives:
... terminal illness.
... loss of job.
... mounting financial problems, where there seems no possible solution.
... seemingly impossible situations and problems within their marriage and with family members...

They face the seemingly insurmountable obstacle of unbelief. They pray and cry out to God. They bind and loose and do everything they know to do. They repeat God's promises over and over, make positive confessions, and exercise the power of positive thinking, hoping somehow to produce enough faith to receive the answer to their prayers. But, when they don't receive the answer they want when they expect it, they begin to vacillate in their faith.

The question is, why aren't we seeing more answers to our prayers? God has planned that the Church of Jesus Christ operate in the same manifestation of faith that Jesus operated under. Through the Holy Spirit, He has given you power and authority over Satan to heal the sick, cast out demons and raise the dead. Yet, in the majority of our churches there is something desperately wrong. We are not seeing a manifestation of what we profess to believe.

Why aren't we seeing a greater manifestation of God's healing power where the blind receive their sight, the crippled get out of their wheelchairs and walk, cancers are dissolved, and all manner of diseases are healed? We say we believe God's promises and believe He is able to perform miracles, yet, we are not experiencing a manifestation of them in our lives.

What has hindered the Body of Christ from walking in the powerful supernatural dimension of faith God has planned for us? Why are there so many weak and stumbling Christians in our churches today? What is missing? What is holding us back?

The major obstacle blocking the Church from entering the fullness of God's power and blocking you from receiving what you need from God is unbelief. It was fear and unbelief that kept an entire generation of Israel from entering the Promised Land:

> *But with whom was he grieved forty years? Was it not with them that had sinned, whose carcases fell in the wilderness? And to whom sware he that they should not enter into his rest, but to them that believed not? So we see that they could not enter in because of unbelief.*
> *(Hebrews 3:17-19)*

In the Amplified Version, verse 19 is translated, "unbelief had shut them out." Don't let your unbelief shut you out of God's Promised Land!

Make this declaration: "I refuse to be shut out of God's promises by unbelief."

Limited By Unbelief

Scripture reading: Mark 6:1-6

At Kadesh-Barnea God directed Israel to go in and take possession of the land, but when ten of the twelve spies spying out the land gave an evil report the peoples' hearts were filled with fear and unbelief:

Their unbelief led to rebellion against God. Unbelief turned their hearts away from Him. They forgot His promises and rejected His Word. As a result of their unbelief, they were defeated. This entire generation of unbelievers wandered in the wilderness 40 years where they died. Not one of them entered the Promised Land.

Just as unbelief shut that generation out of the Promised Land, it will shut you out and keep you from taking possession of the promises God has given you. Unbelief shuts off the supernatural life flow of God. One of the saddest examples, in the Word, of people shut off from experiencing the mighty manifestation of God's miracle power through Jesus is found in the Gospel of Mark: *"And he could there do no mighty work, save that he laid his hands upon a few sick folk, and healed them. And he marveled because of their unbelief"* (Mark 6:5-6).

In Nazareth, where Jesus lived and grew up, He could not do any mighty miracles among the people as He had in other cities throughout Galilee. In Capernaum there was such a powerful manifestation of God's miracle power flowing through Jesus in healing the sick and casting out devils that the entire city gathered together outside the door of the house where He was staying:

> *And all the city was gathered together at the door. And he healed many that were sick of divers diseases, and cast out many devils; and suffered not the devils to speak, because they knew him.*
>
> *(Mark 1:33-34)*

A great multitude from Galilee, Judea, Jerusalem beyond Jordan and Tyre and Sidon followed after Him to be healed:

> *For he had healed many; insomuch that they had pressed upon him for to touch him, as many as had plagues. And unclean spirits, when they saw him, fell down before him, and cried, saying, Thou art the Son of God.*
>
> *(Mark 3:10-11)*

He had healed a paralyzed man. He healed a leper. He cast out a legion of demons from the demon possessed man of Gadara. He healed the woman who had an issue of blood for twelve years and raised Jairus' daughter from the dead. Yet, when Jesus came to Nazareth He could not do any mighty miracles. It was not because His power was limited. His power had not changed. He did not withhold His power from the people because of any unwillingness on His part. The only thing that limited Him was their unbelief.

Are you limiting the work of God in your life through unbelief?

Make this declaration: "I won't let unbelief limit God's work in my life or hinder me from receiving His promises."

Take The Mask Off

Scripture reading: Acts 10:34-43

Jesus came to earth for a specific purpose: *"...For this purpose was the son of God manifested to destroy the works of the devil" (1 John 3:8)*. God anointed Jesus with the Holy Ghost and power and He *"...went about doing good, and healing all that were oppressed of the devil, for God was with him"* (Acts 10:38).

Jesus came to set people free, heal the sick, cleanse the leper, and cast out devils but their unbelief limited what He could do. It shut off the people from the flow of the manifestation of God's miraculous power that multitudes had experienced in the other cities.

Within the Church today we are seeing a mighty manifestation of God's power being released. In our crusades and meetings we are seeing people with cancer and all forms of diseases being healed. Men and women are being delivered and set free from years of addiction to drugs and alcohol. Multiplied thousands are being saved. In every meeting there are hundreds who are receiving a fresh anointing of God's power being released in their lives to work the works of God. But, what grieves my heart is that in the midst of this great outpouring, there are many Christians who have limited God's power to work on their behalf and have shut off the flow of God's miracle-working power in their lives through their unbelief.

The power of God has not changed. It is the same today as it was when Jesus walked upon this earth two thousand years ago. God is ready to meet the needs of His people. He is ready to meet whatever need you may be facing today. He is ready to fulfill His promises to you, but He is limited through your unbelief.

God has directed me over the next few days to lead you on a march to tear down the wall of unbelief. God has said, "Now is your time!" The time has come for you to destroy the wall of unbelief preventing you from entering into the full manifestation of God's power and activating the supernatural faith of God to take hold of His promises!

To tear down the wall of unbelief you must be willing to humble yourself. Don't allow pride to hinder you from acknowledging any unbelief which may have entered your heart. As we begin this march, you must allow God to reveal any areas of unbelief in your life, repent, and get rid of it. It is time to take the mask off!

Make this declaration: "I am receiving the word of God's prophet: Now is my time! I am tearing down the wall of unbelief."

Unbelief Shuts Off The Flow

Scripture reading: 1 Peter 1:3-9

God has not planned for you to live a life of fear, doubt, discouragement or unbelief! He has not planned any defeats for you! He has made it possible for you to have the same supernatural faith operating in your life, the same manifestation of faith that Jesus operated under during His earthly ministry.

God has planned for you to have a supernatural faith activated in your life that is so strong that no trace of unbelief will be able to remain. This faith will enable you to overcome every obstacle, every circumstance, and every attack of the enemy and you will not waver or falter when affliction or persecution come for the sake of the Gospel. It is His faith within you that will enable you to defeat Satan and overcome the world: *"And this is the victory that overcometh the world, even our faith"* (1 John 5:4).

Having the faith of God does not mean that you will not face trying circumstances. There is not one verse in the Word of God that teaches that if you have faith you will never be sick, never have need of anything, or never have to endure affliction and suffering. In all the trials, afflictions, and testings the Apostle Paul faced, he said, *"But none of these things move me..."* (Acts 20:24). Not one time did unbelief overwhelm him or cause him to weaken in faith.

Do you know why he was able to say that? He was not trusting in the strength of his own faith or ability to produce faith. He had a supernatural faith within him that did not waver.

Regardless of the circumstances you are now facing or will face in the future, there is nothing you cannot overcome when the supernatural faith of God is activated within you. It will enable you to overcome all fear, doubt, and unbelief that will try to gain a stronghold in your life.

You may not feel as if you have any faith at all. The problems you face may seem too much for you to bear, but God wants to prove Himself strong on your behalf. During the trial of your faith, He wants you to grow stronger, to take hold of the supernatural faith He has given you and activate it. Your faith will be tried in the fire and is more precious than gold (1 Peter 1:6-7).

There is only one way you will be able to break down the wall of unbelief and tap into the supernatural power of God you need in the circumstances you are facing. You must come boldly before God in faith without wavering. The moment you become fearful or begin to waver, unbelief gains a stronghold in your heart. God is ready to meet your need, but His power is limited to work on your behalf through your unbelief. Your unbelief shuts off the flow of God's power in your life.

Make this declaration: "God has not planned for me to live a life of fear, doubt, discouragement or unbelief! He has not planned any defeats for me! I can have the same supernatural faith operating in my life that Jesus did during His earthly ministry."

YOU CAN RECEIVE WHAT YOU ASK

Scripture reading: James 1:5-8

Those who waver or doubt will not receive anything from God. The word "waver" in these verses is translated from the Greek word for doubt, *diakrino*. It refers not to a weakness of faith, but a lack of it.

When we come to God, we must come to Him in faith, not with a faith that wavers or fluctuates according to our circumstances, not with our own natural ability to believe, but with the supernatural faith that is imparted by God into our hearts.

One of Satan's strategies is to keep you wavering, to keep you trusting in the power of your faith instead of activating and exercising the faith of God that is in and through Jesus Christ. Satan knows that if he can keep you wavering, he will keep you from experiencing the fulfillment of God's promises to you.

- He will bring symptoms of sickness and disease upon your body in an attempt to make your faith waver so that you will not receive your healing.
- He will try to use circumstances to make you doubt the salvation of your loved ones or His provision for supplying your needs.
- He will try to make you worry concerning the basic necessities of life – how to feed and clothe your family, how to pay your bills, how to buy the things you need.

As long as you are wavering and allowing doubt to creep in, you will not be able to receive what you need from God.

Satan doesn't want you to enter into a relationship with God where you really know Him, where you are daily trusting and depending upon Him, where you come into the full knowledge that His faith is in you, and you are drawing upon that faith as your life Source.

Jesus said:

> *I say unto you, That whosoever shall say unto this mountain, Be thou removed, and be cast into the sea, and shall not doubt in his heart but shall believe that those things which he saith shall come to pass; he shall have whatsoever he saith.*
>
> *(Mark 11:23)*

The Greek word for believe is *pisteuo*, which refers to more than mental assent, and means a continual reliance upon and trust in God. Jesus was saying that if we have the supernatural faith of God and do not doubt or waver in our hearts but believe – continually relying upon and trusting in God – we will receive whatever we ask!

Make this declaration: "God, I am doing what you said. I am asking in faith for your wisdom. I believe I will receive!"

Where Are Your Eyes Today?

Scripture reading: Matthew 14:22-33

When the manifestation of faith is truly in our lives, it cannot be explained. It doesn't make any difference what we are facing, how big it is, or what it is – there is an absence of doubt. There's not the slightest tinge of fear, concern, or doubt.

To break through the wall of unbelief, you must get your eyes off your circumstances and focus your faith upon God. As long as you continue to focus your eyes on your physical condition, your pain, the impossibilities of your condition, your financial problems, or the problems in your marriage, you will open the door to fear and doubt. As fear sets in, your faith will begin to waver and you will begin to doubt or question God. "I wonder if God has really heard me?" You will question His promises and wonder if He will do what He said.

Look at what happened to Peter when he stepped out of the boat into the stormy sea to walk on the water to Jesus. Peter said to Jesus:

> *"Lord, if it is You, command me to come to You on the water." And He said, "Come!" And Peter got out of the boat; and walked on the water and came toward Jesus.*
> *(Matthew 14:28-29, NAS)*

Peter, knowing that Jesus calling him to come to Him, stepped out of the boat and began walking on the water despite the waves churning and crashing around him. As long as Peter kept his eyes focused upon Christ, he was able to walk on water in the midst of the storm. The moment he began to look at his circumstances, however, fear entered his heart and he began to waver and doubt: *"But seeing the wind, he became afraid, and beginning to sink, he cried out, saying, "Lord, save me!"* (Matthew 14:30, NAS). What he saw with his natural eyes caused his faith to waver.

As Peter was sinking, Jesus reached out His hand and saved him. As he pulled Peter out of the water, He asked Peter, *"O you of little faith, why did you doubt?"* (Matthew 14:31, NAS). Jesus asked, "Why did you doubt, Peter? I was standing before you to enable you to do the impossible. My power is unlimited, Peter. Why did you doubt?"

There are many Christians today like Peter. They receive a promise from God and launch out in faith upon that promise. As long as they keep their faith focused upon Christ and act upon His promise to them, they are able to do the impossible – to walk in faith and victory even in the midst of the fiercest storms, trials, or testings. But the moment they get their eyes upon their circumstances, they open themselves up to fear and doubt and their faith begins to waver. As a result, they fail to receive the fulfillment of God's promises and the answers to their prayers and they are defeated.

Where are your eyes today? On Jesus or the storm of your circumstances?

Make this declaration: "I am taking my eyes off the storm of my circumstances today and fixing them on God. I will walk in faith and victory."

SHUT OUT UNBELIEF

Scripture reading: Matthew 12:33-37

When Peter begin to sink in the storm because he took his eyes off Jesus, the Lord rebuked him and asked, "Why did you doubt?" Peter had seen Jesus heal all manner of diseases, cast out demons, open blind eyes, multiply the five loaves and two fish to feed 5,000, and raise Jairus' daughter from the dead. Yet, when he took his eyes off Christ and began to focus on his circumstances, he began to doubt.

Beloved, is your faith beginning to waver concerning the promises God has given you? Have you allowed your eyes to become focused upon your circumstances until your heart has become filled with fear and unbelief and you cannot get the breakthrough you need? Christ is saying to you, "Why do you doubt?"

There is no reason to ever doubt or question! Jesus' power and presence are just as real in the midst of His people today as it was when He walked upon the earth 2,000 years ago. It's time for no more doubt, no more wavering. We are breaking down the walls of unbelief.

In Hebrews 3, the Apostle Paul reminds us that the children of Israel failed to enter the Promised Land through their unbelief and then warns: *"Take heed, brethren, lest there be in any of you an evil heart of unbelief, in departing from the living God"* (Hebrews 3:12). One of the greatest dangers you will ever face is that unbelief would gain entrance into your heart. Paul said, "Take heed!" You must be constantly on guard and watching so there will be no opening for unbelief to enter your life. Shut out all unbelief and refuse to allow it into your heart!

Once unbelief enters your heart, it will affect your thoughts, the words coming out of your mouth and your actions. Instead of releasing faith to believe God for the things you need, you will be living in doubt and defeat. Jesus said, *"...For out of the abundance of the heart the mouth speaketh"* (Matthew 12:34). If you allow even the slightest trace of unbelief into your heart, you will eventually begin to murmur, complain, and speak words of doubt, fear and unbelief. If doubt remains, it will eventually lead to disobedience and rebellion against God.

God intends you and me to operate under the same power, the same authority, the same anointing, and the same manifestation of faith that Jesus had! But how is that going to be possible? The supernatural faith manifested in the lives of the disciples and believers in the Early Church was not something they were able to produce through their own efforts. They didn't try to work it up. They received a divine infusion of the supernatural faith of God on the Day of Pentecost when Jesus came to dwell in them by the Spirit. Faith came with Jesus. He is the Source, He is the Author and Finisher of your faith. (Hebrews 12:1-2). Jesus is synonymous with faith. There is no manifestation of faith, no reality of faith, without Jesus!

Shut out unbelief! Don't allow your heart and mind to be filled with thoughts of doubt and unbelief. Reject them and keep your heart and mind fixed upon God and His Word. Meditate upon His promises until every fibre of your being is literally charged with faith!

Make this declaration: "I am shutting the door on unbelief. I am rejecting it in the Name of Jesus. I will keep my heart and mind fixed upon God and His Word. I will meditate on His promises."

You Have His Faith

Scripture reading: Acts 3:1-16

When you received Jesus, you received His faith. To break through the wall of unbelief and have His supernatural faith manifested in your life, you must come to Him. There is no other Source, no other way.

Peter was among those who ran in fear as the soldiers came to the Garden of Gethsemane to arrest Jesus. As Jesus was being tried, it was Peter who denied Him three times. After Christ's resurrection, Peter was with the other disciples hiding behind closed doors when Jesus appeared and rebuked them for their unbelief (Mark 16:14). Yet we later see Peter – this same man whose faith had wavered, who had been filled with fear and doubt – operating in the supernatural faith of God to speak the word to a lame man and command him to rise and walk in the Name of Jesus!

> *Then Peter said, "Silver and gold have I none, but such as I have give I thee. In the name of Jesus Christ of Nazareth rise up and walk. And he took him by the right hand, and lifted him up: and immediately his feet and ankle bones received strength."*
> *(Acts 3:6-7)*

It was this same man who spoke the word of faith and raised the dead:

> *But Peter put them all forth, and kneeled down, and prayed; and turning him to the body said, "Tabitha, arise." And she opened her eyes: and when she saw Peter, she sat up.*
> *(Acts 9:40)*

There was such supernatural faith manifested through Peter's life that they placed the sick in the streets so that just his shadow would fall on them and they would be healed:

> *Insomuch that they brought forth the sick into the streets, and laid them on beds and couches, that at the least the shadow of Peter passing by might overshadow some of them.*
> *(Acts 5:15)*

A transformation took place within Peter and the other disciples that enabled them to manifest the same supernatural faith as Jesus. They had an experience where the supernatural faith of Jesus Christ was released within them.

Jesus had told them to go to Jerusalem and wait until they were baptized with the Holy Ghost and power (Acts 1:4-5). He told them, *"But ye shall receive power, after that the Holy Ghost is come upon you"* (Acts 1:8). He was declaring, "When I get back to My Father, the Holy Ghost is going to come upon you. And once the power, the anointing, and the same experience I had comes upon you, you will be witnesses, you will have my faith, and you will be able to do the same works!"

Make this declaration: "God's power is coming upon me. I have the anointing to do the works of God."

Breaking Down The Wall Of Unbelief

Scripture reading: Galatians 2:16-22

Faith is not a natural life force. If you think by mental assent, by reaffirmations of slogans or sayings, or even quoting the Scripture, you can receive this mountain-moving faith, you are sadly mistaken. When you receive a breakthrough in your understanding – you know that faith is a life source and that Jesus is the Source of faith – then your struggling will cease. You will have enough faith to believe God for the things you need, receive the fulfillment of His promises, and to do the works of God.

He that hath an ear, let him hear! The Spirit is calling you. He is calling me. He is calling us to move into an experience where His powerful faith is flowing through us. The manifestation of the power and anointing – the faith of God – was so strong in the lives of the apostles and disciples in the Early Church that everywhere they went, God's power was released and people were healed and made whole.

What was the key to the release of this supernatural manifestation of faith? It was the result of their willingness to be crucified with Christ and lay down everything at His feet in total surrender. They no longer lived after the dictates of their flesh. They were no longer motivated by the desires of their carnal minds.

The Apostle Paul was not dependent upon his own natural faith. He lived his life in the power of the supernatural faith of Jesus Christ Who lived within him. He said, *"I am crucified with Christ, nevertheless I live...And the life where I now live...I live by the faith of the Son of God, who loved me and gave himself for me" (Galatians 2:20).*

Faith is a life to be lived. Paul lived his life by the faith of Jesus Christ which was in him. He drew his very life from this faith. He was strengthened and sustained by it. That is why his faith never faltered and never failed. It was Christ's faith flowing through him.

The proportion of faith that is released in your life will be according to the proportion you are crucified with Christ.

The Spirit of God is calling us to a fresh dedication of ourselves, a dying to self which will allow a full manifestation of Christ's power, anointing, and the supernatural faith of God to be released within us.

It is time.

No more doubting.

No more wavering.

We are breaking down the wall of unbelief!

Make this declaration: "I will not doubt. I will not waver. The wall of unbelief is coming down."

DEFEATING UNBELIEF

Scripture reading: Mark 11:22-26

Today and tomorrow, I want to give you five strategies that will destroy the wall of unbelief so you can receive the fulfillment of God's promises in your life.

Strategy number one: Repent of any unbelief. You must first recognize and acknowledge any areas in your life where you may have allowed unbelief to enter into your heart. Confess these to God and ask Him to remove every trace of unbelief from your heart and to release His supernatural faith within you.

Strategy number two: Shut out unbelief from your heart and mind. You must guard your heart and refuse to allow unbelief to gain a stronghold in your life. Worry or anxiety is a form of unbelief and you must refuse to allow it any place in your life. You must not allow your mind to be filled with worry over the circumstances or problems you face. Repent of every worried or anxious thought and activate your faith by filling your heart and mind with the Word.

Strategy number three: Activate the supernatural faith of God. Christ, who lives within you is the Author and Finisher of your faith. He alone can impart the faith you need. Don't look to your own faith or depend upon your ability to believe. When Christ releases His supernatural faith within you, you will not need to struggle to believe. There will be a deep inner knowing that what you ask will come to pass. Move to a higher level of consecration where you are willing to die to yourself. Remember, the proportion of faith that is released in your life will be according to the proportion you are crucified with Christ.

Strategy number four: Appropriate God's supernatural faith. God wants you to live your life by His faith. Paul said, *"...the life which I now live...I live by the faith of the Son of God..."* (Galatians 2:20). He told the Romans, *"...The just shall live by faith"* (Romans 1:17). You must no longer regulate your life according to your five senses, but according to the faith of God. God wants you to step out in faith and minister to the desperate needs of those around you, praying for the sick and loosing those who are bound by Satan.

As you wage war against Satan, release the supernatural faith of God. It is a powerful shield which will quench the fiery darts of the enemy (Ephesians 6:16). Act upon God's promises and His Rhema Word to you. The Rhema Word is the Word spoken by God to your heart.

When we receive it, we are not receiving words of man; we are receiving thoughts of God. Within the Rhema lies the faith. When the promise comes, when the personal words come directly from God to you, do not struggle for the faith. The faith is already there with the promise. All that is necessary is for you to act upon the personal word or promise from God. Faith is a fact, but faith is an act!

Make this declaration: "I am receiving the Rhema Word of God. The faith is already within me to receive the promises of God. I am going to act upon that faith today!"

Five Powerful Strategies

Scripture reading: Luke 17:11-19

Here is the next strategy for breaking through the obstacle of unbelief:

Strategy number five: Don't waver. The ten lepers who came to Jesus were healed as they acted in obedience to Christ's command, *"... Go show yourselves unto the priests..."* (Luke 17:14). They were healed as they went. The man with the withered hand was made whole as he acted in obedience to Christ's command, *"...Stretch forth thine hand..."* (Mark 3:5). The disciples caught a miraculous draught of fishes as they acted in obedience to Christ's command, *"...Launch out into the deep, and let down your nets for a draught!"* (Luke 5:4). The impotent man at the pool of Bethesda, who had been sick thirty-eight years, was healed as he acted in obedience to Christ's command, *"...Rise, take up thy bed, and walk"* (John 5:8). The blind man was healed as he acted upon Christ's command, *"...Go, wash in the pool of Siloam..."* (John 9:7).

God is speaking to you today that now is your time! As you receive His Rhema Word to you, the faith is already there with the promise. All that is necessary is for you to act upon the promise.

What are the promises God has given you that you have been holding on to? What are the circumstances where you desperately need an answer from God? It is time to break through the walls of unbelief and claim those promises. Whatever your needs are right now, God has placed within you His supernatural faith which will enable you to take hold of those things – the healing, finances, restoration of your marriage, whatever you need or desire from God – and bring them into reality. As you release your faith and act upon God's promises, you will break through the wall of unbelief to take possession of the things He has promised you.

Before we conclude today's devotional, let's review once more the five strategies to break through the obstacle of unbelief:

- Strategy number one: Repent of any unbelief.
- Strategy number two: Shut out unbelief from your heart and mind.
- Strategy number three: Activate the supernatural faith of God.
- Strategy number four: Appropriate God's supernatural faith.
- Strategy number five: Don't waver.

From this moment on, let every struggle in your life to produce faith, to believe God's promises, to trust Him for the things you need, cease in the Name of Jesus. Receive the revelation that God has placed His supernatural faith within you. Draw your faith from Jesus, Who is the Author and Finisher of your faith. Take the title deed God has placed in your hands and take possession of the things you need.

Make this declaration: "I repent of any unbelief. I shut unbelief out of my heart and mind. I activate and appropriate God's supernatural faith!"

Confronting Confusion

Scripture reading: 1 Corinthians 2:9-14

This month we are eliminating some of the obstacles that prevent us from claiming God's promises. We have talked about fear and doubt and learned powerful strategies to confront these walls in our lives. The next obstacle we will face is confusion.

Are you facing circumstances, problems, and crises in your life that are causing inner turmoil? Are there things in your life that are causing you to be uncertain, fearful, worried, and perplexed? Have things happened in your life that you just don't understand and you don't know what to do? It seems you have exhausted every possible means of determining the solution to your problems. You know what God's Word says. You have prayed, struggled, reasoned, and still are unsure of what action to take, what to say, or what direction to go.

Instead of having God's peace and knowing what to do, you are facing a wall of confusion.

Confusion will rob you of your peace and joy. It will hinder you from fulfilling the will of God and from taking possession of God's promises and receiving what you need and desire from Him. Satan's strategy is to bring you to a point in your life where you are confused and perplexed, where your mind is in a constant state of turmoil, and you are faced with problems and circumstances that are so complex you just don't know what to do.

Your mind is the battlefield where the battle of confusion must be fought and won. This is one of the reasons why it is so important for you to have a breakthrough where you are no longer depending upon your natural mind with your limited abilities or human understanding. God wants you to have a one-hundred percent victorious mind of Christ but as long as you live your life according to the limitations of your carnal, natural mind, you will be unable to take possession of God's promises. The Apostle Paul said:

> *But the natural man receiveth not the things of the Spirit of God: for they are foolishness unto him: neither can he know them, because they are spiritually discerned.*
>
> *(1 Corinthians 2:14)*

As long as you are still living according to your natural mind and are controlled by it, you cannot understand the things of God. You are unable to know the truth, you cannot discern God's voice or know His will, and you open yourself up to confusion, frustration, and defeat.

We are entering a period of time that is unparalleled in the history of the world and the Church. This is, without a doubt, the greatest hour the Church of Jesus Christ has ever faced!

In the midst of the great turmoil and upheaval that permeates every phase of our society, God has planned for the Body of Christ to be strong, immovable and fearless, regardless of what we may see happening all around us.

We must confront and conquer the obstacle of confusion.

Make this declaration: "Today, I begin a march on the wall of confusion. It will come down through the power of God."

February 22

God Is Not The Author Of Confusion

Scripture reading: Luke 21:25-36

We are living in a time of great perplexity and confusion which Jesus said would be one of the end-time signs of His coming (Luke 21:25-28). Jesus said there would be *"distress of nations with perplexity."* The word "perplexity" translated from the Greek means "at one's wit's end, at a loss of how to proceed."

Today we are witnessing this confusion and perplexity in the nations of the world as we see political and economic upheaval on an unprecedented scale. The nations are facing crisis situations – financial collapse, skyrocketing budget deficits, political turmoil and violence, governments being overthrown, the AIDS epidemic, famine, hopelessness – with no solutions!

In the midst of all the confusion and turmoil, we are going to have clarity of vision and purpose:

- We are going to know God's will...
- We are going to know how the enemy is attacking us and how to overcome...
- We are going to know how to tear down Satan's strongholds and take possession of all that belongs to us!

You don't have to be limited or bound by your own natural mind with its limited wisdom that is subject to confusion, frustration, and defeat. You can tap into God's wisdom and walk in power and victory as you act upon the supernatural wisdom He imparts to you. His wisdom is pure, full of peace, free from doubt, and unwavering.

As children of the Living God, we are not bound or limited by the confusion and turmoil in the world today. We have a choice! We can have God's wisdom and clarity of vision and purpose or we can depend upon man's sensual, natural wisdom and bring confusion upon ourselves.

Today God has positioned you for victory, and by His Spirit He is telling you to go forward to take possession of the land and all He has provided for you. He is telling you to drive out the enemy – Satan and all his evil principalities – and take hold of the healing, deliverance, and supernatural provision you need for your life. God is telling you to go forward to take the power and authority He has given you to take your city and nation for God!

You cannot allow yourself to become confused, fearful, or frustrated because God is not the author of confusion. When you act according to the truth He has revealed in His Word and obey His voice when He speaks and directs you, there will be no confusion.

The words "confused" or "frustrated" should never come out of the mouth of a Christian! You must not allow yourself ever to become confused, but you must take hold of the one-hundred percent victorious mind of Christ.

Make this declaration: "I am going to know God's will. I am going to know how the enemy is attacking me and how to overcome. I am going to know how to tear down Satan's strongholds and take possession of all that belongs to me!"

The Choice Is Yours

Scripture reading: 1 Corinthians 14:26-40

You must learn how to draw upon the supernatural wisdom of God for direction and guidance in the circumstances you face in every area of your life. God is not the author of confusion...Satan is! Instead of confusion, fear, and frustration, God wants you to have His wisdom operating in your life so you know His will and purpose.

God wants you to have His wisdom in the major decisions you must make. When you face circumstances that, in the natural, seem impossible and you don't know which way to turn, He does not want you to become confused or afraid. He wants you to draw upon His wisdom.

God is a God of order, design, and peace. Confusion does not come from Him. The Apostle Paul, in writing to the Corinthians concerning order in the church, told them, *"For God is not the author of confusion, but of peace, as in all churches of the saints"* (1 Corinthians 14:33). Whenever there is confusion, strife, contention, or turmoil, you can be assured it is not from God.

The word "confusion" translated from the original Greek in this verse means "instability, a state of disorder, disturbance, confusion, turmoil." There is no doubt – Satan is the source from which confusion originates. It is his objective to attack your mind and bring it to a state of instability, disorder, and turmoil. He will attack you with doubt, fear, and unbelief until your mind is clouded with confusion and indecision. You will become unstable, fluctuating back and forth, not knowing what to do. James said, *"A double minded man is unstable in all his ways" (James 1:8).*

You can have the supernatural wisdom which comes from God, enabling you to walk in faith and victory or you can have the worldly, natural wisdom which results in confusion, frustration, and defeat. The choice is yours. The Apostle James compared these two types of wisdom. He said:

> *But if you have bitter jealousy (envy) and contention (rivalry, selfish ambition) in your hearts, do not pride yourselves on it and thus be in defiance of and false to the Truth. This (superficial) wisdom is not such as comes down from above, but is earthly, unspiritual (animal), even devilish (demoniacal). For wherever there is jealousy (envy) and contention (rivalry and selfish ambition), there will also be confusion (unrest, disharmony, rebellion) and all sorts of evil and vile practices. But the wisdom from above is first of all pure (undefiled); then is it peace-loving, courteous (considerate, gentle.) (It is willing to) yield to reason, full of compassion and good fruits; it is wholehearted and straightforward, impartial and unfeigned free from doubts, wavering and insincerity.*
>
> *(James 3:14-17, TAB)*

Is this wisdom operating in your life today?

Make this declaration: "I bind confusion and release the spirit of wisdom to operate in my life today."

February 24

Destroying The Wall Of Confusion

Scripture reading: Romans 8:1-11

Man's natural mind is at enmity and in rebellion against God. Paul told the Romans, *"...the sinful mind is hostile to God. It does not submit to God's law, nor can it do so"* (Romans 8:7, NIV).

The wisdom that comes from God is not a product of the natural mind. It does not come as a result of research or study. Neither is it a product of this world. It does not come from the outstanding scientists or leading intellects of our day. It is pure and untainted from any outside influence. God's wisdom brings peace and is full of mercy.

The wisdom from God is supernatural. It comes from the supreme mind of God and reveals His will, thoughts, and purposes. You do not ever need to be brought to a point of confusion where your mind is in a constant state of turmoil, full of fear and doubt.

Regardless of the circumstances you face, you do not need to be confused, indecisive, or double minded. As you live according to the wisdom He imparts, the end result is peace instead of confusion. Instead of your mind being in turmoil, filled with fear, doubt and indecision, it will be covered with His peace (Isaiah 26:3).

By His Spirit within, you can draw from God's wisdom and face every circumstance knowing that you will not be defeated or brought to confusion. Paul told the Corinthians:

> *For who has known or understood the mind (the counsels and purposes) of the Lord so as to guide and instruct (Him) and give Him knowledge? But we have the mind of Christ, the Messiah, and do hold the thoughts (feelings and purposes) of His heart.* *(1 Corinthians 2:16, TAB)*

God does not intend you ever to be confused or fearful. He has ordained that you live your life according to His divine wisdom. God has placed His Spirit within you so you can have the mind of Christ and know and walk in His divine wisdom.

By God's Spirit within you, He reveals His mind, His divine counsel, His will, His thoughts and purposes so you may know and partake of all He has prepared for you. In 1 Corinthians 2:7-10, Paul told the Corinthians that the spirit of God within searches the deep things of God and reveals them. John said, *"But ye have an unction from the Holy One, and ye know all things."* The Holy Spirit within you is the Source from which you draw God's wisdom. He teaches and reveals truth. John said, *"But the anointing which ye have received of him abideth in you, and ye need not that any man teach you: but as the same anointing teacheth you of all things, and is truth and is no lie, and even as it hath taught you, ye shall abide in him"* (1 John 2:27).

The truth is in you. The anointing is in you. Release the truth and anointing to defeat the walls of confusion in your life today.

Make this declaration: "The truth is already within me. The anointing is in me. I release the truth and anointing to defeat the walls of confusion in my life today."

Rejecting God's Will

Scripture reading: 2 Chronicles 32:1-8

- Is there confusion in your life?
- Are you confused about God's direction?
- Are you at a point of frustration, where you are facing circumstances for which there doesn't seem to be a solution?
- Are you at a point where you are facing problems which have you distressed, frustrated, and confused?

There are five major factors which bring confusion into your life:

1. Sin: Romans 14:23
2. Leaning to the arm of the flesh: 1 Corinthians 2:14
3. Lack of knowledge of the truth: James 3:14-17
4. Lack of hearing God's voice clearly: 1 Corinthians 14:33
5. Lack of clarity of direction – not knowing God's will: James 1:8

When Israel was at the border of their Promised Land, God told them to not be dismayed (Deuteronomy 1:21, TAB). The word "dismayed" is translated from a Hebrew word which means "to break down either by violence or confusion and fear." God was telling them not to be confused or fearful. Instead of acting in faith, however, Israel rebelled. Their sin prevented them from entering in and taking possession of all God had promised them. Because of sin, they brought confusion upon themselves and wandered in the wilderness for forty years.

Jeremiah, the weeping prophet, cried out concerning the nation of Israel:

> *We lie down in our shame, and our confusion covereth us: for we have sinned against the LORD our God, we and our fathers, from our youth even unto this day, and have not obeyed the voice of the LORD our God."*
> *(Jeremiah 3:25)*

The thing which caused them to sin and led to their defeat was that they reacted to their circumstances from a position of confusion and frustration instead of upon the Word of the Lord. They acted according to their natural minds. They forgot God's miraculous power to deliver them and set their eyes on the strength and might of their enemies.

Their confusion was not the result of not hearing God's voice or a lack of clarity of direction. God had spoken and clearly revealed His will to them. It was clearly the result of shutting their ears to God's voice, rejecting His will, and substituting their own wills.

What about you? Are you rejecting God's will and substituting your own will? Examine your heart today.

Make this declaration: "I will not walk in sin. I will not lean on the arm of the flesh. There will be no lack of knowledge of the truth in my life. I will be able to hear God's voice clearly and know His will."

February 26

ELIMINATING CONFUSION

Scripture reading: James 3:13-18

During the next few days, we want to meditate on five strategies that will eliminate confusion in your life.

Strategy number one: Reject every trace of confusion and act on God's promises.

God has brought you to a place of spiritual destiny. He has given you many promises but if you allow confusion to gain a stronghold in your mind it will block you from taking possession of these promises.

When we look at the rebellion of the children of Israel, we wonder, "Why would they rebel against God and refuse to go into the Promised Land and take possession of it?" God had just supernaturally delivered them out of the hands of the Egyptians. They had seen Him part the Red Sea and had walked across on dry ground with a wall of water on the right and on the left of them. How could they refuse to believe His promises and rebel against Him?

But today we are guilty of doing the same thing when we fail to believe God's promises and act on them! God has clearly revealed His will to us through His Word. He has given us His promises of salvation, healing, deliverance, supernatural provision, and victory over all the power of the enemy. He has entered into an everlasting covenant whereby all our needs are met in Him. So, the first strategy to break through the wall of confusion is to reject every trace of confusion and act on God's promises. When you fail to do this, you are rejecting His Word and are in rebellion.

Strategy number two: Wait upon God and draw upon His wisdom. Your strength when you face desperate circumstances for which there is no possible solution is not to run here and there trying to work things out according to your limited understanding. You can either chose to let your mind be thrown into turmoil or to take hold of the mind of Christ and let God's peace envelope you. Instead of leaning to your own understanding – looking to man for a solution, to physicians, to banks for loans, to your own resources – you must draw upon God's supernatural wisdom and discernment.

God has placed His Spirit within you so you can know and walk in His divine wisdom. You will never be confused, indecisive or double-minded when you act according to the wisdom God gives you. The supernatural wisdom He imparts is from God. It is full of peace, free from doubt, and unwavering (James 3:17).

Make this declaration: "I reject confusion in the Name of Jesus. I will wait upon God and draw from His wisdom."

God Will Direct You

Scripture reading: John 8:31-36

If you are to eliminate confusion you must first, reject every trace of confusion and act on God's promises and, second, wait upon God and draw upon His wisdom. Here is the next strategy for confronting the obstacle of confusion:

Strategy number three: Reject Satan's lies and act on the truth. In the circumstances you face, Satan's strategy is to feed you misinformation – lies. His objective is to keep you from knowing and acting upon the truth. If you do not know the truth, it is easy to be sidetracked and led into rebellion against God.

Jesus told His disciples, *"If ye continue in my word, then are ye my disciples indeed, and ye shall know the truth and the truth shall set you free"* (John 8:31-32). It is knowing the truth that will set you free and enable you to break through any confusion in your mind. The truth will drive out Satan's lies, set your mind free, and enable you to be victorious over all the power of the enemy.

In the circumstances you are facing, don't listen to Satan's lies. Reject them. Satan is the father of lies and there is absolutely no truth in him. *"Resist the devil, and he will flee from you"* (James 4:7). Use the truth of God's Word to resist him and drive his lies out of your mind. Regardless of how impossible, how bleak, or how discouraging the circumstances are, you will not become confused or afraid when your mind is filled with truth and you are acting on it.

Saturate your mind with the Word until it is part of your being, where every aspect of your mind is not only filled with the Word, but renewed and controlled by it. The truth will set you free from confusion!

Strategy number four: Act knowing you have heard from God. If ever there was a time when we need to hear the voice of God clearly, it is today. As we face the challenges and opportunities God has placed before us in the new millennium, we must be able to hear God's voice clearly and distinctly. There can be no confusion in our minds concerning what we are to do. We must know. We must hear from God!

I don't know about you, but I don't want to take one step or make one plan until I have waited before the Lord and have heard His voice. God never intended His people to flounder around or grope in the dark to know His direction and leading. He has planned for you to come into His presence and wait upon Him until you hear His voice clearly. He has planned for you to seek Him and draw upon His wisdom. As you come before Him in seasons of prayer, He will direct you. You will hear His voice and you will know the way you are to go.

You have God's promise that He will direct you: *"Trust in the LORD with all thine heart, and lean not unto thine own understanding. In all thy ways acknowledge him, and he shall direct thy paths"* (Proverbs 3:5-6).

Make this declaration: "I reject Satan's lies. I will act upon the truth today, knowing I have heard from God."

LED BY THE SPIRIT

Scripture reading: Exodus 3:17-22

If you are to eliminate confusion you must first, reject every trace of confusion and act on God's promises; second, wait upon God and draw upon His wisdom; third, reject Satan's lies and act on the truth; and fourth, act, knowing you have heard from God. Here is the fifth and final strategy that will enable you to conquer confusion:

Strategy number five: Face your circumstances knowing God's will. Lack of clarity in direction will cause confusion which enters in when we fail to hear God's voice, know His will, and take the right direction.

God has always planned for His people to have clear direction and know His will. After He delivered the children of Israel out of Egyptian bondage, He did not leave them out in the wilderness on their own to find their way to the Promised Land. His presence went before them to lead and guide them:

> *And the LORD went before them by day in a pillar of a cloud to lead them the way; and by night in a pillar of fire, to give them light; to go by day and night. He took not away the pillar of the cloud by day, nor the pillar of fire by night, from before the people.* *(Exodus 13:21-22)*

There was no confusion among the great multitude of more than two million people in knowing God's direction. They were not divided or disorganized with a lot of different groups, each going a different way. God supernaturally led them and they all moved together in unity.

When the cloud lifted off the tabernacle, they journeyed. But as long as it remained upon the tabernacle, they did not move. They didn't move one inch until God directed them to go forward (Numbers 9:17-22).

There is no reason why we should ever be confused or frustrated concerning God's will or direction in our lives. God has made the way clear. He has revealed His will to us through His Word. We do not ever need to be in doubt. By His Spirit within us, we can know God's will and direction in every circumstance we face.

God's Presence – His Spirit – dwells within our spirits and as we yield and submit ourselves to Him, He will guide and direct us. As sons of God, we are led by His Spirit. *"For as many as are led by the Spirit of God, they are the sons of God"* (Romans 8:14).

The reason so many Christians are confused and frustrated concerning God's will and direction for their lives, is because they are not depending upon the Spirit of God to lead them. They are not staying in God's presence and His Word long enough to find His will. They are not spending seasons of prayer and fasting, waiting upon God and seeking to know His will. They are looking for someone else to give them direction. They want someone to tell them what God's will is for their life.

Confusion comes when we fail to wait upon the Spirit of God for His direction. We must be led, directed, and controlled by God's Spirit.

Make this declaration: "I will face my circumstances today from a position of knowing. I will be led, directed and controlled by God's Spirit."

Become Spiritually Aggressive

Scripture reading: Exodus 23:20-33

We must be led, directed, and controlled by the Spirit of God. Just as the children of Israel did not move a forward until the cloud moved, we must not move until the Spirit of God directs us to move.

Instead of being led around by the hand and depending on others to lead us, it is time to grow up. We must learn to stay in God's presence until we develop a new sensitivity to the Spirit of God so when He speaks, we move. When He tells us to turn to the right, we will turn to the right. When He directs us to go to the left, we will go to the left. When He says sit still, we will sit still, waiting upon Him. When He tells us to go forward, we will go forward.

You must stay in the presence of God and in His Word until the Spirit tells you, "This is the way, walk in it." Do not act on your circumstances until the Spirit gives you direction. Seek God for His timing and His will. Regardless of how long your wait or what anyone else may say, don't move until the Spirit of God says you may move.

If you are at a point of confusion and frustration, wanting to know God's direction and will for your life today, break out of your confusion. God will make His will known to you by His Spirit. Do not become anxious or fearful. As you begin, in earnest, to seek Him with your whole heart, He will make the path clear to you. You won't need to stumble or grope or knock down any doors. He will lead you by His Spirit.

Beloved, it is not God's will for you ever to be brought to a position of confusion, frustration, and defeat. Those who keep their trust in God will not be brought to confusion. King David prayed: *"In thee, O Lord, do I put my trust. let me never be put to confusion"* (Psalm 71:1).

Turn your confusion upon the enemy and put him to flight. Take the fear, unbelief, and confusion Satan has tried to bring into your life and use it to bring confusion upon the enemy. God promised Israel, *"I will send my terror before you and will throw into confusion all the people to whom you shall come, and I will make your foes turn from you (in flight)"* (Exodus 23:27, TAB).

During this past month, you have learned strategies to combat fear, unbelief, and confusion. You will no longer be able to sit back while Satan destroys your home, finances, and attacks your body. You will not sit back and wait for him to attack you. You will rise up in the Spirit and put your enemies to flight because just ahead of you lies the Promised Land with all of its promises and provisions. If you want to claim it, you must become spiritually aggressive!

Make this declaration: "The walls that have prevented me from advancing are being knocked down through the Name of Jesus. Ahead of me lies the Promised Land. I will claim it!"

March

Prayer is one of the vital keys to winning the major spiritual battles – Satan's battle for your mind. You will learn the power of prayer and many other strategies during the coming month that will equip you with a one-hundred percent victorious mind.

March
Winning The Battle For Your Mind

God has shown me seven major obstacles which block the fulfillment of God's promises and provision for your life. These obstacles are barriers – walls that are standing in your way hindering you from taking possession of God's promises, just as the walls of Jericho prevented Israel from taking their land.

During February, we marched against the obstacles of fear, unbelief, and confusion. This month we will spend the entire month focusing on one major obstacle – the major battlefield where most spiritual attacks occur – your mind.

As Joshua entered into the promised land, God told him not to fear or be dismayed. As we are entering into the new millennium, God's message to us has not changed. God has not given us a spirit of fear, rather He has given us a strategy for victory.

- Do not fear when you see signs of the end times coming to pass.
- Do not fear when economic crises occur.
- Do not be dismayed when wars break out around the globe.
- Do not dread disaster.

God dealt with Joshua to give him the proper mind set for leading Israel into Canaan – one free of fear and dismay. During this month, God is going to deal with you and give you a renewed mind which will enable victorious Christian living in the new millennium. You will learn how to program your mind for victory, arm yourself with a militant mind, win the war over depression, and have a one-hundred percent victorious mind.

When Joshua and the children of Israel marched against Jericho, they were not leaning to the arm of the flesh nor depending upon their own wisdom or strength. The key to their victory was they were totally dependent upon God and His power to deliver them. They acted upon the strategy God gave them, in complete trust and confidence He would do as He had promised.

The key to your victory is in your total dependence upon God and His power. You must act upon the strategies God has given you in complete trust and confidence that He will break down every wall.

The battlefield where Satan is waging war against the Church today is in our hearts and minds – at the very center of our beings – where our entire moral and mental abilities are located. He is attacking us at our most vulnerable point where our thoughts originate, where we receive God's Word and have faith to believe His promises, where our wills, emotions, desires and our mental abilities are located.

Are you ready to receive a renewed, one-hundred percent totally victorious mind? Then let's begin…

The Main Battlefield

Scripture reading: Mark 7:15-23

Where is the main battlefield where the war against Satan and his principalities must be fought and won? Satan's objective is to attack at the very center of man's existence – within his innermost being – his heart and his mind.

Jesus, in explaining to His disciples the source from which man is defiled and sin is conceived, said:

> *For from within, (that is) out of the heart of man, come base and wicked thoughts: sexual immorality, stealing, murder, adultery, Coveting (a greedy desire to have more wealth), dangerous and destructive wickedness, deceit; unrestrained (indecent) conduct; an evil eye (envy), slander (evil speaking, malicious misrepresentation, abusiveness); pride – (that is) the sin of an uplifted heart against God and man; foolishness (folly, lack of sense, recklessness, thoughtlessness). All these evil (purposes and desires) come from within, and they make the man unclean and render him unhallowed.*
>
> *(Mark 7:21-23, TAB)*

He told the Pharisees:

> *O generation of vipers, how can ye, being evil, speak good things? for out of the abundance of the heart the mouth speaketh. A good man out of the good treasure of the heart bringeth forth good things: and an evil man out of the evil treasure bringeth forth evil things.* *(Matthew 12:34-35)*

In these verses, we see that man's heart is his innermost being where either good or evil thoughts reside. A man's words and actions are determined by what is in his heart. If there are thoughts of hatred, anger, jealousy, or pride in an individual's heart, they will be manifested in the words that come out of his mouth. Whatever is stored in a man's heart within his innermost being will eventually be manifested in his actions.

If you allow your heart to be filled with thoughts of doubt, fear, and unbelief regarding the circumstances you are facing, it will be manifested through the words you speak and you will not be able to receive what you need from God because of the unbelief in your heart.

The heart is the center of man's existence. It is in the heart man receives the Word, is convicted by the Holy Spirit, and is saved (Romans 10:8-10). It is in the heart – within man's innermost being – where he has faith to believe God's Word and His promises or harbors doubt and unbelief that hinders him from receiving from God what he needs. Jesus said, *"Have faith in God. Truly I say to you, whoever says to this mountain, 'Be taken up and cast into the sea,' and does not doubt in his heart, but believes that what he says is going to happen it shall be granted him"* (Mark 11:22-23, NAS).

Make this declaration: "I will learn to fight the battle of the mind effectively during this month. I will reprogram my mind through God's Word."

Where Sin Is Conceived

Scripture reading: Romans 7:13-25

It is in man's heart – his innermost being – where sin is conceived. It was in Satan's heart that sin originated. His "heart" was lifted up because of his beauty (Ezekiel 28:17). It was in his "heart" he exalted himself above God (Isaiah 14:13-14). The sin of Ananias and Sapphira originated in the heart (Acts 5:3).

The Greek word for heart is *kardia,* which includes man's entire mental and moral activity – the inner man. It is recognized as the place where the emotions, desires, perceptions, thoughts, understanding, reasoning powers, conscience, intentions, the will, and imagination reside.

What is in a man's heart and mind determines what he is. *"For as he thinketh in his heart, so is he"* (Proverbs 23:7). In this verse, we see that the word "heart" is used to refer to the place where man's thoughts originate.

Because the heart and mind are so closely related, we must also look at what God's Word says concerning our minds to have a clear understanding of this battlefield. The Greek word used to refer to the mind is *nous,* which is used to describe the mental functions of perception, understanding, knowing, feeling, judging, and determining. The mind is comprised of the will, emotions, thoughts, and imagination.

Our minds are Satan's major point of attack and it is here where the war must be fought and won. It is in your mind where Satan is unleashing his all-out war. The war we are fighting is not against flesh and blood. It is not a mental battle fought through the power of positive thinking, "head knowledge," psychological techniques, or our limited mental abilities. It is a spiritual battle in our minds, which must be fought and won by the power of the Holy Spirit working within us!

All of the strategies Satan has developed and is using against us today are directed toward our minds. Satan doesn't want Christians to have a free, clean, healthy, renewed, victorious mind. His strategy is to attack and corrupt your mind, to hinder you from being transformed into Christ's image, to prevent you from living your life in victory. His strategy is to bombard your mind with fiery darts of unbelief, fear, worry, confusion, temptation, depression, oppression, and deception. He will try to fill your mind with evil desires and lusts of the flesh. He will try to fill your mind so full of the cares of this world and yourself, that you will have no time for Christ and the Word.

Satan is going to do everything within his power to fill your mind with anxiety, fear, defeat, and discouragement concerning the circumstances in your life until your mind is so weakened that you become depressed – frustrated and filled with a sense of hopelessness and despair. But Satan's power over your mind has been broken!

Make this declaration: "Satan, your power is being broken over my mind. Frustration, hopelessness, despair, and depression will go in the Name of Jesus."

Total Provision For Victory

Scripture reading: Romans 6:4-23

God has made total provision so you can be victorious in every battle, just as He did for the nation of Israel as they prepared to enter their promised land. God has provided you with everything you need to defeat the enemy. Jesus has already destroyed the works of Satan and broken his power over you.

You are facing an enemy that has already been defeated. Two thousand years ago when Jesus cried, "It is finished!" the war against Satan was won. Through His precious blood that was shed, we are totally delivered out of Satan's hands. He has no claims over us whatsoever! *"For sin shall not have dominion over you..."* (Romans 6:14).

We are no longer living in the kingdom of darkness under Satan's dominion as "prince of this world." We are joint-heirs with Christ in the Kingdom of God. Satan has no power over us. He is defeated and God is going to crush Satan under our feet!

God does not expect you to face Satan and the evil powers and principalities of wickedness in your own strength. He has given you power and authority over all the power of Satan through the Holy Spirit within you. He has given you powerful spiritual armor that makes you invulnerable and invincible against all of Satan's attacks. He has provided you with weapons of warfare, enabling you to pull down every stronghold of the enemy (2 Corinthians 10:4-5).

God has given the power to *you* to pull down Satan's strongholds and totally demolish them so they are nonexistent, but your victory over Satan is not automatic. Simply because you have been born again and are a child of God with the Spirit of God dwelling in you, does not mean that you are automatically going to be victorious over the temptations, lies, and attacks of Satan. Victory is yours only as you put on the armor, use your spiritual weapons of warfare, and resist Satan.

You must take aggressive action and press the battle in the Spirit, using the power and authority God has given you, if you are to win the battles you face against Satan. The reason there are so many Christians today living defeated lives is because they have failed to take an offensive position and fight against Satan and his evil principalities.

In recent years, the Body of Christ has come into a greater awareness of this spiritual warfare we are fighting against Satan. Most Christians today know a battle is raging, but they don't understand it. During the coming days of this month we will learn where the major battlefield is, how Satan attacks, and how to defeat him.

Make this declaration: "I am facing an enemy that is already defeated. God has given me the power to pull down his strongholds and totally demolish them so they are nonexistent."

Going To The Root Cause

Scripture reading: Matthew 12:33-37

Through my message on the *New Anointing* that I have preached around the world, multiplied thousands have learned that we are not fighting against flesh and blood. Our battles are not with family members, our husband or wife, employer, fellow workers, or any individual. Our battles are not with sickness and disease – cancer, high blood pressure, heart trouble, or diabetes. Our battles are not the problems we are facing on our jobs or in our finances.

We must go beyond the surface to the root cause. In prayer we must go deep into the spirit realm, locate the real enemy – the evil spirits that are at work – and bind them in the Name of Jesus.

Spiritual warfare is not just a theory. It is a reality! Although we cannot see it with our natural eyes, this war we are fighting against Satan is just as real as the wars we see taking place in the natural world that are being fought with tanks, airplanes, missiles, and guns.

This war where we are wrestling, struggling, and fighting in one-on-one combat is not taking place in the natural realm. We see the effects of Satan's attacks in the natural world as he attacks our bodies, our finances, our families, and our circumstances. We see the effects of his attacks in the lives of our loved ones in rebellious children, lost family members, and those who are bound by drugs or alcohol. But until you begin to recognize where this spiritual battle is taking place and take offensive action against Satan, you will continue to live in defeat.

Where is the battlefield where the war against Satan and his principalities must be fought and won? Satan is attacking us at our most vulnerable point. His objective is to attack at the very center of man's existence, within his innermost being.

Jesus, in explaining to His disciples the source from which man is defiled and sin is conceived, said that sinful conduct comes out of the heart. He told the Pharisees:

> *0 generation of vipers, how can ye, being evil, speak good things? for out of the abundance of the heart the mouth speaketh. A good man out of the good treasure the heart bringeth forth good things: and an evil man out of the treasure bringeth forth evil things.* *(Matthew 12:34-35)*

In these verses, we see that man's heart is his innermost being, where either good or evil thoughts reside. A man's words and actions are determined by what is in his heart. If there are thoughts of hatred, anger, jealousy, or pride in an individual's heart they will be manifested in the words that come out of his mouth. Whatever is stored in a man's heart, within his innermost being, will eventually be manifested in his actions.

What is in your heart? What is in your innermost being?

Make this declaration: "Dear God, I open my heart to You today. Reveal to me what is in my innermost being. Help me go to the root cause."

As You Think In Your Heart

Scripture reading: Acts 5:1-11

If you allow your heart to be filled with thoughts of doubt, fear, and unbelief regarding the circumstances you are facing, it will be manifested through the words you speak and you will not be able to receive what you need from God because of the unbelief in your heart.

The heart is the center of man's existence. It is in the heart man receives the Word, is convicted by the Holy Spirit, and is saved (Romans 10:8-10). It is in the heart – within man's innermost being – where he either has faith to believe God's Word and His promises, or harbors doubt and unbelief that hinders him from receiving what he needs from God. Jesus said:

> *Have faith in God. "Truly I say to you, whoever says to this mountain, "Be taken up and cast into the sea," and does not doubt in his heart, but believes that what he says is going to happen, it shall be granted him.*
> *(Mark 11:22-23, NAS)*

What is in your heart and mind determines what you are. Jesus taught the multitudes that it is possible to sin in the heart, the innermost being, even before the sin is manifested outwardly. He said:

> *But I say unto you, That whosoever looketh on a woman to lust after her hath committed adultery with her already in his heart.*
> *(Matthew 5:28)*

It is in man's heart – his innermost being – where sin is conceived. It was in Satan's heart that sin originated. His heart was lifted up because of his beauty (Ezekiel 28:17). It was in his heart that he exalted himself above God. (Isaiah 14:13-14).

When Ananias and Sapphira lied about the sale of their land and kept back part of the money for themselves, Peter asked, *"But Peter said, Ananias, why hath Satan filled thine heart to lie to the Holy Ghost, and to keep back part of the price of the land?"* (Acts 5:3).

The Greek word for heart is *kardia* which includes man's entire mental and moral activity, the inner man. It is recognized as the place where the emotions, desires, perceptions, thoughts, understanding, reasoning powers, conscience, intentions, the will, and imagination reside.

Peter asked Ananias, "How is it that Satan has so filled your heart?" In other words, he was asking, "Why have you allowed Satan to so fill your innermost being, your heart, mind, your thoughts that you lied to the Holy Spirit? The thoughts Ananias and Sapphira allowed Satan to plant in their hearts took root, sin was conceived in their hearts, and sin was manifested through their actions.

What is in a man's heart and mind determines what he is, *"For as he thinketh in his heart, so is he…"* (Proverbs 23:7). In this verse we see that the word "heart" is used to refer to the place where man's thoughts originate. How's your thinking today?

Make this declaration: "What is in my heart determines who I am. I will think today on only things that edify. I will bring every thought under the control of the Holy Spirit."

March 6

The Major Point Of Attack

Scripture reading: 1 Corinthians 2:1-16

If a man's heart and mind are filled continually with evil thoughts, lusts of the flesh, carnal desires, and wicked imaginations, he will be evil. If his heart and mind are continually filled with fear, negative thoughts, and his own abilities he will be cowardly, weak, and defeated. On the other hand, if a man's heart and mind are filled with good thoughts – God's Word, faith, and the power of the Holy Spirit – he will be strong and victorious.

Since the heart is the center of man's existence, the innermost being where all the issues of life flow forth, it is vitally important that we make every possible effort to continually protect it from the enemy. We are warned: *"Keep your heart with all vigilance and above all that you guard, for out of it flow the springs of life"* (Proverbs 4:23, TAB).

The war against Satan must be fought and won in your heart and mind. The battlefield where Satan is waging war against the Church today is in our hearts and minds, at the very center of our beings, where our entire moral and mental abilities are located. He is attacking us at our most vulnerable point where our thoughts originate, where we receive God's Word and have faith to believe His promises, where our wills, emotions, desires, and our mental abilities are located.

Because the heart and mind are so closely related, we must also look at what God's Word says concerning our minds to have a clear understanding of this battlefield. The Greek word used to refer to the mind is *nous*, which is used to describe the mental functions of perception, understanding, knowing, feeling, judging, and determining. The mind is comprised of the will, emotions, thoughts, and imagination.

Our minds are Satan's major point of attack! And it is here where the war must be fought and won. It is in your mind where Satan is unleashing his all-out war. The war we are fighting is not against flesh and blood. It is not a battle fought with our mental abilities, through the power of positive thinking, head knowledge, psychological techniques, or our limited natural abilities. It is a spiritual battle in our minds which must be fought and won by the power of the Holy Spirit working within us!

All of the strategies Satan has developed and is using against us today are directed toward our minds. Satan doesn't want Christians to have a free, clean, healthy, renewed, victorious mind. His strategy is to attack and corrupt your mind, to hinder you from being transformed into Christ's image, and prevent you from living your life in victory. His strategy is to bombard your mind with "fiery darts" of unbelief, fear, worry, confusion, temptation depression, oppression, and deception.

Make this declaration: "Satan, I reject unbelief, fear, worry, confusion, temptation, depression, oppression, and deception. I will have a free, clean, health, renewed, victorious mind."

A New Mind

Scripture reading: Colossians 1:21-23

Satan will try to fill your mind with evil desires of the flesh. He will try to fill your mind so full of the cares of this world and yourself that you will have no time for Christ and the Word.

Satan is going to do everything within his power to fill your mind with anxiety, fear, defeat, and discouragement concerning the circumstances in your life until your mind is so weakened that you become depressed, frustrated, hopeless, and discouraged.

Satan is going to try to oppress and weigh your mind down with such heavy burdens and pressures, until you are unable to break through the thick cloud of depression over you to reach out in faith and receive what you need from God. But Satan's power over your mind has been broken!

Paul referred to the battle that takes place in our minds. He told the Romans:

> *For in my inner being I delight in God's law; but I see another law at work in the members of my body, waging war against the law of my mind and making me a prisoner of the law of sin at work within my members.*
> *(Romans 7:22-23, NIV)*

In these verses, Paul was speaking of the carnal desires of flesh which war against the mind (thoughts, understanding, feelings, will, emotions), and try to make it a slave to the sin that lies within the old carnal nature.

Paul continues by explaining that through Christ we have been set free from this sinful nature with its sinful desires, thoughts, and attitudes, so that we are now able to serve God with our minds and hearts (Romans 7:24-25). Then, in Romans 8:2-3, Paul explains how sin has been deprived of its power over us. The power of sin which once ruled over our innermost beings – our hearts and minds – has been broken, and we are no longer slaves to the sinful desires of our carnal nature.

Satan has been defeated! Jesus destroyed the works of the devil and set us free from the power of sin. Sin no longer has dominion over us. We were once alienated, enemies in our minds toward God. We were bound by our own stubborn, selfish wills to live our lives to please the flesh. But now, through Jesus, our hearts and our minds – our innermost beings – have been one-hundred percent cleansed, one-hundred percent set free from the power of sin. Our hearts and minds have been one-hundred percent renewed and are continuing to be renewed daily. We have been reconciled to God (Colossians 1:21-22).

Our minds, with their fleshly desires, emotions and wills, which were once under Satan's control, have been set free from his power. God has placed His Spirit within us and we have been born again. Our old sinful nature has not been transformed...we have a new nature. All things have become new. We have a new heart, a new mind, with renewed attitudes, will, and desires.

Make this declaration: "I have been set free from Satan's power. My fleshly desires, emotions, and will which were once under Satan's control are now under God's control."

March 8

A Victorious Mind

Scripture reading: John 10:7-11

Jesus paid the price so we can have a one-hundred percent victorious mind. Satan no longer has power over us. He does not have power to make us sin. He cannot control our minds, thoughts, wills and emotions. He has no power whatsoever over us except what we allow him to have. Yet, eighty or ninety percent of Christians today are living in defeat because they do not recognize that their minds are Satan's battlefield, or they fail to resist and take offensive action against him.

Satan is creating havoc in the minds of many Christians and the vast majority do not even recognize what is happening to them! Their minds are in a constant state of turmoil. They are continually being bombarded with fear, worry, doubt, and confusion, yet they do not understand Satan is attacking and a battle is raging.

There are Christians who are being tempted with carnal desires and lusts of the flesh. A fierce battle is raging within them. Satan has a vice-like grip on their minds and they are unable to be free of his tormenting spirits. They are unable to resist temptations and are yielding to them. As a result, their minds are filled with fear, guilt, and condemnation and they are defeated.

The majority of Christians today are in a weakened state of mind and are victims of depression. They are gripped with a sense of hopelessness, frustration, and despair. They are spiritually immobilized, unable to release their faith, to think clearly, to cope with the problems and circumstances they are facing. In this depressed state of mind, they feel the only way out is to escape, to run from their problems. In some severe cases of depression, their minds are in such turmoil they begin to think about suicide.

There are Christians sitting in our pews today who are oppressed by Satan. They are tormented, frustrated, and confused. Yet they are unable to take authority over Satan's forces and walk in Christ's victory because they do not know where and how Satan is attacking or how to defeat him.

There is no reason why God's people today should live in defeat. There is no reason why they should be confused, worried, fearful, anxious, depressed, or oppressed in their minds. Jesus did not pay the supreme price and give His life as a sacrifice to free us from the power of sin so that we would live in victory over the enemy fifty percent, seventy percent, or even ninety-five percent of the time.

He did not give us the power of the Holy Spirit so we would be overcome by the enemy and walk around discouraged and defeated in our minds. Jesus did not come to earth, destroy the works of the devil, and free us from Satan's power so that we would be fearful, weak, and powerless. Jesus paid the price so that you and I can have a one-hundred percent victorious mind one-hundred percent of the time!

Make this declaration: "Satan has no power over me except what I permit him to have. Therefore, I will not allow him to have any power at all!"

Satan's Strategy Of Depression

Scripture reading: 1 Samuel 30:1-6

Perplexity! Stress! Fear! Worry! Hopelessness! Despair! Discouragement! Depression! Defeat! These negative forces permeate our environment and affect our society today. But, in the midst of these negative forces, we do not have to become victims of depression!

Regardless of the things which are coming upon this earth, regardless of the circumstances, trials, and attacks of Satan upon our minds and bodies, we do not have to become fearful, worried, or depressed as the rest of the world, who are living in darkness.

You must recognize depression for what it really is – a major strategy Satan is going to use to defeat you. God has planned for your heart and mind to be filled with His peace which will keep you even during your greatest trial, temptation, or attack and will enable you to be victorious.

More than likely, you have been depressed at least one time during your life. You may feel discouraged or depressed right now concerning the circumstances you are facing. If you are depressed, I am believing God to set you free.

In the massive attack Satan is now launching against the Church, God revealed to me Satan is using depression as a major strategy against Christians in an attempt to stop the progress we have been making in pushing back the forces of darkness in the nations of the world and winning souls into the Kingdom of God. I am warning you now, so you will be prepared and know how to take aggressive action to resist and defeat his attacks upon your mind.

There are multiplied thousands of Christians today who are depressed and oppressed by Satan because they have failed to recognize depression as an attack of Satan and do not know how to fight and win the war over depression. They are being spiritually immobilized.

Many do not recognize when they are depressed. Those who do may try to ignore it. Most Christians consider depression a natural reaction when they face heavy trials or attacks of the enemy, such as long-term sickness, marital problems, death of a loved one, loss of a job, or extreme financial difficulties.

There are ministers and Christian leaders who are experiencing "spiritual burnout." They have been carrying such a heavy workload, pushing themselves twelve to fourteen hours a day, seven days a week, giving themselves unselfishly in meeting the needs of others by preaching the Word, praying for the sick, counseling, conducting meetings and seminars. Yet, they have failed to take time to spend in God's presence being strengthened and refreshed by His Spirit. They have been so involved in fulfilling the ministry God has given them that they have failed to be renewed daily in the spirit of their minds.

Depression is a major tool of the enemy – but you do not have to succumb to it. God has promised you a one-hundred percent victorious mind. Take authority over depression today!

Make this declaration: "I take authority over depression. It will not operate in my life today."

Set Your Mind For Victory

Scripture reading: Psalm 69:1-36

There are Christians who are depressed and have a spirit of defeat oppressing their minds because of past failures. Satan is bringing up past experiences where they have sinned, and is using it against them so they will become depressed and willing to give up.

There are Christians who are depressed because of guilt over habits, temptations, and sin they have been unable to overcome. They have struggled to be free from certain habits and temptations but, when they continue to fail, their hearts and minds are filled with guilt.

Many Christians today are frustrated, worried, and depressed over their financial problems. They have overdue bills and other commitments they are unable to pay because their income is insufficient to meet their basic needs.

There are Christians deeply distraught and depressed concerning their marriage relationships. There are those who have unsaved mates and there is no peace or harmony in their home. There are others whose mates have left them for someone else or have asked for a divorce. Others feel their mate no longer loves and appreciates them.

On the job, there are thousands of Christians who are depressed. Faced with a heavy work load, pressures, demands, and deadlines they are unable to cope, feel defeated, and worn-out mentally and physically.

Many Christians today are depressed because of pain or sickness. Research indicates eighty-three percent of patients with chronic pain are depressed. When they become sick, experience pain or a long-term illness, their minds are often filled with doubt, worry, and self-pity.

Depression is not a normal reaction for Christians to problems and adverse circumstances. Depression is more than a sign of the times. It is more than a leading cause of major illness. It's time that we go to the root cause and recognize depression for what it really is. Depression is a spirit that Satan is using to attack the minds of Christians today!

We must not be afraid to recognize it and call it by name. We are not helpless victims of depression. Regardless of the great distress and perplexity – the stress, fear, and hopelessness that surrounds us – we do not have to be depressed. We can fight and win the battle against Satan's attack on our minds through depression.

Set your mind for victory. Refuse to allow Satan, anyone, or anything to program your mind for failure, depression, and defeat. From this day forward, determine that by the power of the Holy Spirit you will not allow your mind to become depressed by Satan.

Make this declaration: "I am setting my mind for victory. I refuse to allow Satan, anyone, or anything to program my mind for failure, depression, and defeat.

God's Guarantee

Scripture reading: Acts 16:19-33

Satan is going to attack Christians in their minds with such intensity that they will become victims of depression and experience one defeat after another unless they recognize depression for what it is and take aggressive action against it.

Satan wants you to accept depression as if there is nothing you can do about it. He wants you to be depressed and stay depressed, so that you will be unable to fulfill God's purpose and plan for your life.

To overcome depression and walk in victory over it, you must learn to recognize it the moment a spirit of depression attacks your mind. You must recognize that the thoughts which you allow to remain in your heart and mind – not the circumstances – are causing your depression. None of the circumstances you face or will ever face in your lifetime can cause you to be depressed or defeated. You do not have to ever be depressed again! God has promised you:

> *No temptation (trial) has seized you except what is common to man. And God is faithful; he will not let you be tempted (tried) beyond what you can bear. But when you are tempted (tried), he will also provide a way out so that you can stand up under it.*
>
> *(1 Corinthians 10:13, NIV)*

This is God's guarantee to you against depression. God will not allow any circumstances, trial, temptation or attack of the enemy to come into your life that you are unable to bear and, during every circumstance you face, He will make you victorious!

The Apostle Paul had every possible reason to be discouraged and defeated. Think about the circumstances he faced. He was beaten many times, stoned, shipwrecked, and imprisoned. Yet, he was not depressed. He said:

> *"We are troubled on every side – yet not distressed; we are perplexed, but not in despair; Persecuted, but not forsaken; cast down, but not destroyed..."*
>
> *(2 Corinthians 4:8-9)*

In the prison cell where he and Silas had been thrown for preaching the Gospel, Paul had every reason to be depressed (Acts 16). His back was bruised and bleeding. He was in chains. He was, no doubt, hungry and physically exhausted. He had been unjustly thrown in jail. But, he was not depressed. He was not fearful, worried, angry, or filled with self-pity concerning his circumstances. Instead of griping, complaining, and feeling sorry for themselves, Paul and Silas sang praises to God and won the victory over depression. As they praised and worshiped God, He supernaturally intervened in their circumstances.

Make this declaration: "There is no temptation I am experiencing but what is common to man. God is faithful. He will not let me be tempted beyond what I can bear. He will provide a way of escape."

Satan's Strategy Of Oppression

Scripture reading: Psalm 77:1-20

God has shown me that Satan is oppressing many of God's people today, just as he oppressed the Syrophoenician woman's daughter in New Testament times. Don't be deceived, demon "oppression" and "possession" are just as real today and as prevalent as during Jesus' ministry two thousand years ago!

It's time for the Church of Jesus Christ to recognize the Satanic oppression coming against God's people, confront Satan and his demon forces, and take authority over them just as Jesus did two thousand years ago.

Satan is illegally exercising dominion over Christians – tormenting their minds, afflicting their bodies, oppressing their loved ones, oppressing them in their finances and in every way possible.

Christians are being harassed, tormented, and troubled by Satan and his demon spirits. Satan's objective is to wear them down physically and mentally until they become so oppressed they feel hopeless and give up.

When a Christian is oppressed it is an outward manifestation that Satan is attacking the mind or the body and sometimes both. A Christian can be oppressed mentally or physically and not be possessed or totally under the control of unclean spirits.

Demons are spirit beings. They have no bodies. Their objective is to invade and indwell human minds and bodies so they will have greater control over the individual. They first attack and oppress the mind in an attempt to wear down the individual's resistance where he will accept their evil thoughts and suggestions.

Demons of fear, doubt, unbelief, worry, and depression will attack and torment a person's mind until they become so oppressed they are unable to appropriate by faith the promises of God in their circumstances. In this state of oppression it often seems there is such a heaviness in the spirit that there is a thick spiritual darkness like a cloud surrounding the individual that cannot be penetrated. The person may have "head knowledge" but because of the heavy oppression by these spirits, be unable to act in faith on God's promises.

If an individual fails to resist these attacks and does not reject the thoughts and suggestions the unclean spirits are trying to plant in his mind, his emotions will become controlled by these spirits. Eventually his will becomes so weakened he yields to the oppressing spirits until his mind is in a constant state of torment.

Make this declaration: "I will resist Satan's attacks. I will reject the thoughts and suggestions of unclean spirits. My emotions will not be controlled or tormented by demonic forces."

Recognizing Demonic Forces

Scripture reading: Luke 13:11-17

Demon spirits often oppress people by causing a physical condition in their bodies and keeping it there. This was the case of the woman with a spirit of infirmity who had been bound and oppressed by it for eighteen years. She was bowed together and unable to lift herself up (Luke 13:10-11). This woman was not "possessed," she was oppressed by an evil spirit. Jesus discerned that a spirit of infirmity was causing her condition.

It was for this purpose He had come! God had anointed and empowered Jesus with the Holy Spirit to heal and set free all those who were oppressed by the devil. He had come to destroy the works of the devil!

> *And when Jesus saw her, He called her to Him and said to her, Woman, you are released from your infirmity! Then He laid (His) hands on her and instantly she was made straight, and she recognized and thanked and praised God.*
>
> *(Luke 13:12 13, TAB)*

When the leader of the synagogue reprimanded Jesus for healing on the Sabbath, Jesus said, *"And ought not this woman, a daughter of Abraham whom Satan has kept bound for eighteen years, be loosed from this bond on the Sabbath day?"* (Luke 13:16, TAB).

One day they brought a man to Jesus who had a blind and dumb spirit. The demon oppressing this man had afflicted his eyes and ability to speak. After Jesus took authority over it and cast it out of the man, he was able to both see and talk. Jesus loosed the man from Satan's power and set him free. *"Then a blind and dumb man, under the power of a demon, brought to Jesus and He cured him, so that the blind and dumb man both spoke and saw"* (Matthew 12:22, TAB). This man was not "possessed" – fully under control of the demon, spirit, soul, and body.

All sickness and disease is from Satan. God never intended man to know sickness and disease. It came as a result of man's fall. However, there are many Christians today who have sicknesses and diseases that are a direct result of satanic oppression where a spirit has hold on their bodies and is causing a particular affliction.

It is time for us to recognize how Satan and his demonic forces are attacking, afflicting, and oppressing God's people today! This is the first major step we must take before we take authority over the spirits oppressing us, break their hold, and be loosed from their oppression in our lives.

Make this declaration: "Upon the authority of God's Word, I declare that my eyes will be open to recognize how Satan and his demonic forces are attacking, afflicting, and oppressing God's people today."

March 14

Rise Up In New Strength

Scripture reading: Luke 4:14-30

Millions of Christians are oppressed because of spiritual ignorance concerning Satan's strategies and methods of attacking and oppressing them. There has been a lack of revelation, not only regarding the work of the enemy, but also concerning the power and authority we have today over all the power of Satan and his evil principalities.

Not understanding the need to detect and resist the evil spirits oppressing and exercising authority over them, many have become victims. Instead of exercising dominion over Satan and his principalities, Satan and his demons are exercising authority and dominion over them.

Satan is oppressing God's people today in their minds, bodies, finances, on their jobs, and through circumstances. But God is exposing the strategies Satan is using in his attack against the Church and we are not ignoring his devices!

Jesus was anointed by the Holy Spirit and sent by God to deliver and set free all those oppressed by the devil. He has destroyed the works of the devil. We have been anointed by the Holy Spirit to stand in Christ's stead, confront the power of Satan, and set free all those who are oppressed by Satan.

It is time for you to rise up in new strength to exercise authority over the unclean spirits Satan will use to try to oppress you. There may be spirits oppressing your mind, continually bombarding you with lusts of the flesh, tempting you to sin. A spirit of fear may be tormenting your mind to such an extent you are bound by it and constantly living in fear. Spirits of worry and doubt may be attacking you until your mind is in a constant state of turmoil. Spirits of heaviness, hopelessness, frustration, or depression may be coming against you until you have reached a point in your life where you feel you cannot go on – where you are worn out, tired of resisting, tired of doing spiritual warfare.

But the bonds of oppression over God's people are going to break! Just as God delivered Israel out of the hands of her oppressors, He will deliver His people today out of the oppression of Satan and his demon forces. Christians who have been tormented in their minds by spirits of lust, adultery, fornication, and sexual perversion are going to be delivered and set free! Christians who have been under heavy oppression by the enemy – who feel as if they have been pushed down, spiritually repressed, who have no freedom in their spirit, no joy, no peace, who are unable to freely worship and praise God – are going to be set free! The dark cloud of oppression is going to lift and they are going to walk in new freedom and liberty in the Spirit.

The key to being set free from oppressing spirits and walking in victory over them is in knowing the power and authority that is yours over Satan and his demon forces and in using it! The moment you feel a spirit of oppression, don't hesitate or be fearful in dealing with the demons that are attacking you, your family, friends or loved ones. Take authority over them and cast them out, in the Name of Jesus.

Make this declaration: "I am anointed by the Holy Spirit to stand in Christ's stead, confront the power of Satan, and set free all those who are oppressed by Satan."

You Can Have A Strong Mind

Scripture reading: Ephesians 1:1-23

The reason the majority of Christians today are not walking in this dimension of power and victory is because they have not had a revelation of the powerful, one-hundred per cent victorious mind God has made possible for us to have. Many Christians are bound by limitations they have set in their minds. They do not think it is possible for them to have a totally hundred percent victorious mind. They have settled for less than what God has made possible for us to have.

God wants you to have the powerful, victorious mind of Christ in operation in your life. Paul told the Corinthians, *"But we have the mind of Christ, the Messiah, and do hold the thoughts (feelings and purposes) of His heart"* (1 Corinthians 2:16, TAB). God has planned that we be transformed into Christ's image (Romans 8:29) and this transformation takes place by the renewing of our minds (Romans 12:2 and 2 Corinthians 3:18).

As we fill our minds with the Word and a knowledge of Christ, we are constantly changed and transformed into His image. As our minds, thoughts, wills, emotions, desires, and purposes are renewed, we become governed by the Holy Spirit and begin to manifest Christ's life.

Paul told the Ephesians to be constantly renewed in the spirit of their minds (Ephesians 4:23-24). God has planned that our minds be filled with revelation knowledge so that we will be able to know and understand the riches of our inheritance and the power we possess as children of God.

Paul prayed that God would give the Ephesians the spirit of revelation in the knowledge of Christ, *"That the God Of our LORD Jesus Christ, the Father of glory, may give unto you the spirit of wisdom and revelation in the knowledge of him"* (Ephesians 1: 17). How? *"By having the eyes of your heart flooded with light…"* Why?

> *…so that you can know and understand the hope to which He has called you and how rich is His glorious inheritance in the saints – His set-apart ones. And (so that you can know and understand) what is the immeasurable and unlimited and surpassing greatness of His power in and for us who believe, as demonstrated in the working of His mighty strength.*
> *(Ephesians 1:18-19, TAB)*

God has planned that we have a strong mind with the ability to control every thought, imagination, desire, and emotion, and – through the spiritual weapons He has given us – bring them into obedience to Christ and conformity to the will of God. Don't settle for anything less.

Make this declaration: "God has planned that I have a strong mind with the ability to control every thought, imagination, desire, and emotion. Through the spiritual weapons He has given me, my mind will come into conformity to the will of God."

March 16

YOU ARE NOT A VICTIM

Scripture reading: James 4:7; 2 Timothy 1:7; Romans 12:1-2

God is calling us to have a sound, well-disciplined mind. There are many Christians today who feel they cannot control the thoughts, desires and emotions that fill their minds. They feel there is nothing they can do to stop themselves from the worry, temptations, depression, and opposition Satan brings against them.

Satan wants you to think you are a helpless victim and that there is nothing you can do to control the thoughts in your mind. Satan is a liar! He does not have the ability to control our minds. We do not have to accept the thoughts that he tries to plant in our minds. He cannot force us to think bad thoughts. He cannot force us to worry. He cannot force us to be fearful, discouraged, or depressed.

Satan's power over our minds has been broken. He only has the power to attack our minds when we fail to resist, allowing him to have access into our minds. *"...Resist the devil, and he will flee from you"* (James 4:7).

The extent to which we allow Satan to have access into our minds and to control our thoughts is the degree he has control over us. If we become fearful, confused, worried, or depressed, it is because we choose to be that way. I know that sounds hard to believe, but it is true. It is only when we fail to use our wills to resist the thoughts Satan tries to plant in our minds that he gains access and controls our thoughts.

There are Christians today who are allowing Satan to have a "hey-day" in their minds without putting up any resistance or spiritual effort to take their thoughts into captivity. Not only are they allowing every negative, doubting, fearful thought and every temptation to enter their minds, they are allowing them to remain and are dwelling upon them. Their excuse is, "I cannot help myself."

You can help yourself! God has made it possible for you to have a one-hundred percent free, one-hundred percent victorious mind, but you must be willing to take aggressive action. Recognize Satan's attack on your mind, resist the thoughts he tries to plant in your mind, take authority over them, and cast them out in the Name of Jesus.

Paul told Timothy: *"For God hath not given us the spirit of fear, but of power, and of love, and of a sound mind"* (2 Timothy 1:7). The Greek word for *sound* is *sophronismos*, which is a calling to soundness of mind or to self-control. It literally means discipline. In this verse, Paul was not simply making a statement to Timothy telling him that he had a sound, well-disciplined mind. The word fear in this verse means cowardice. He was encouraging Timothy not to be fearful and to develop a strong, sound, well-disciplined mind. He was calling Timothy to discipline his mind by taking authority over fear and set his mind on fulfilling the ministry to which God had called him.

Make this declaration: "I have a sound, strong, and well-disciplined mind, one that is free from the spirit of fear."*

*For complete teaching on overcoming fear, be sure to write to Morris Cerullo and request your copy of his book, *One Demon Spirit.*

The Vicious Cycle Of Defeat

Scripture reading: Ephesians 4:17-24

God has planned us to have a sound mind, one that is well-disciplined, one that is one-hundred percent free, one-hundred percent clean, one-hundred percent healthy, one-hundred percent renewed, and one-hundred percent victorious.

As we face the counterattacks of Satan, God is calling us to have a sound, well-disciplined mind that is controlled by the Holy Spirit. It is not ours automatically. To have a well-disciplined, victorious mind, we must:

1. Take aggressive action against Satan's attacks on our minds – recognize, resist and cast out all thoughts he tries to plant into our hearts and minds.
2. Continually bring every thought into captivity.
3. Continually be renewed in our minds by "putting off" the old carnal nature with its carnal mind and bringing our thoughts, wills, desires, and attitudes under the control of the Holy Spirit and into conformity with the will of God.
4. Fill our hearts and minds with God's Word and the knowledge of Christ.

Satan's objective in attacking us is to wear us down until we become spiritually weary and faint in our minds. He knows that if our minds are weak we will be weak and he will be able to gain access and defeat us.

In this massive counterattack Satan is launching against the Church, he is going to try to cause you to become so tired of your circumstances, your sickness, your struggling, and not having enough money to pay your bills that you will be ready to give up.

Becoming spiritually weary is one of the most dangerous positions you can be in. I am warning you now, so that you will not fall into Satan's trap. No doubt, you have fought heavy spiritual battles before, but they are not going to compare with the battles you are going to face during this counterattack. Satan is intensifying his attacks on the Church. He is going to try to literally wear you out, and – unless you are aware of what is happening and prepare yourself for the battle – you will not be able to stand. You will live in a cycle of defeat.

Satan will attack you through your circumstances so that you will become caught in this cycle of defeat. He will bombard your mind with fear, worry, and doubt until you become spiritually weary. At this point, if you faint in your mind, you will become discouraged, lose hope, and resign yourself to accept your circumstances. Your discouragement will lead to depression where you will be spiritually immobilized and caught in the vicious cycle of defeat.

Make this declaration: "I am taking aggressive action against Satan's attacks on my mind. I will continually bring every thought into captivity. I will renew my mind by putting off the old carnal nature and filling my heart with God's Word."

Take Authority Over Your Mind

Scripture reading: Hebrews 12:1-15

Paul warned New Testament believers not to become weary and faint in their minds, due to the great opposition and persecution they were facing. He told them:

> *...let us run with patience the race that is set before us, Looking unto Jesus the author and finisher of our faith; who for the joy that was set before him endured the cross, despising the shame, and is set down at the right hand of the throne of God. For consider him that endured such contradiction of sinners against himself, lest ye be wearied and faint in your minds.*
>
> *(Hebrews 12:1-3)*

The Greek word for *faint* means "to unloose, relax, and make weak." Satan wants to weaken you. He wants you to become so overwhelmed by your circumstances, so weary fighting the battles you have day after day in fulfilling the ministry God has given you, that you faint and relax in your mind.

He wants your mind to be spiritually tired and weak, so you will:

- let up on your intensified prayer time.
- cut down on the time you read and study the Word.
- stop resisting his attacks.
- stop taking control of your thoughts.
- stop casting out doubt and unbelief.
- stop releasing your faith.

You may be at this dangerous point right now, where you are spiritually weary and you have fainted in your mind. You may be so tired of the spiritual battles that you are ready to give up. You may be tired of resisting temptation. Your mind may be in constant turmoil. You may be depressed or severely oppressed by the enemy. Whatever strategy Satan may be using to attack your mind, don't give in! God has made it possible for you to walk in one-hundred percent victory over all the power of enemy. The moment you faint in your mind and stop resisting his attacks, you are surrendering yourself into his hands. You are defeated only when you fail to fight.

It is time to prepare your mind for aggressive action. To win the war against Satan and live in one-hundred percent victory over every attack, circumstance, and strategy Satan will use against you, you must have a strong, powerful, disciplined mind that is controlled by the Holy Spirit.

In his counterattack on the Church, Satan is going to attack your mind with such intensity and such force that you are left in a weakened, vulnerable condition. He doesn't want you to have a victorious, well-disciplined mind.

Make this declaration: "I will not become passive. I am preparing my mind for aggressive action against the enemy."

Strengthening The Loins Of Your Mind

Scripture reading: 1 Peter 1:1-13

Christians who have a passive mind are easy targets for Satan and are positioned for defeat. When their minds are in a state of passivity, their spiritual senses are dulled and inactive. They are vulnerable to every attack – deception, fear, temptation, oppression, depression – and eventually will lose control over their thoughts and their will.

It is time for God's people to rise up, to march forward in new strength with a one-hundred percent victorious mind! We must not be intimidated by the power of the enemy. We must not be fearful or cowardly. We must go out to face the enemy with new spiritual aggressiveness and determination to defeat him in every area of our lives.

The Apostle Peter, in writing to believers who were distressed because of the tremendous trials they were going through, said, *"Gird up the loins of your mind..."* (1 Peter 1:13). In essence, he was telling them to prepare their minds for aggressive action, to strengthen and brace their minds for the battles they were facing.

The first piece of armor the Roman soldier put on as he prepared for battle was his belt. The belt was a very significant part of the armor, tied around the soldier's waist, gathering in the short tunic he wore. It helped keep the breastplate in place and held the sword. The soldier's belt strengthened his loins and enabled him to fasten his armor close to him, whereby he was strengthened and better able to fight.

To tighten or gird up the belt indicated the soldier was ready for action. To loosen or slacken the belt was to go off duty. It is time for us to prepare for the battle by "girding up" and strengthening the loins of our minds. If we fail to do this, we're not going to be able to walk in the victory God has planned for us.

The *loins* refers to the chief seat of bodily strength. The body is knit to the loins. If the loins fail, the whole body fails. The loins we are to gird up and strengthen for battle are our hearts and minds. Our hearts and minds – our inner man – is the acting, ruling part of us. It is the center of our existence where all the issues of life flow forth. It is where our thoughts, emotions, wills, desires, understanding, reasoning powers, and intentions reside.

If our hearts and minds are strong and disciplined, if we know God's Word, if we have a clear understanding of His will, if our minds and wills are set to obey and do His will, if our thoughts are constantly being brought into obedience to Christ, then we will be strong and immovable even during times of great trial, temptation, or attack by the enemy!

Make this declaration: "I am girding up the loins of my heart and mind. I will be strong and immovable during times of trial, temptation, and attack by the enemy."

Walking In Victory

Scripture reading: Psalm 98:1-9

There are five powerful strategies which will make you victorious over Satan's attacks on your mind. The first strategy is to set your mind for victory rather than defeat. God has made it possible for you to have a one-hundred percent victorious mind, but you must discipline your mind through the power of the Holy Spirit within you and bring it into submission until it is fully governed and controlled by the Holy Spirit.

You can either set your mind for victory or for defeat. If you are continually thinking failure and defeat, you are going to be defeated. If you have your mind set and your thoughts continually fixed on the victory Christ has made possible for you, you will be victorious. *"For as he thinketh in his heart, so is he..."* (Proverbs 23:7).

In our San Diego World Evangelism headquarters, which we call the Mission Control Center, we have computers to help us produce publications to minister to you. What we program or enter into those computers is exactly what is going to come out. If the information and material we program into the computers is incorrect, that is what we are going to get. If what we program is one-hundred percent correct, it will be one-hundred percent accurate.

The same principle is true regarding our minds. The thoughts, desires, and imaginations which we allow to continually fill our minds are what are going to be expressed through our words and manifested in our lives.

As you prepare for the counterattack Satan is launching against the Church, it is time to strengthen your mind for battle! Begin today to take aggressive action against the enemy. Refuse to allow Satan to plant even one thought in your mind. If your mind is in turmoil, filled with fear or worry concerning the circumstances you are facing, don't allow your mind to be in a passive condition.

In the power of the Holy Spirit, take authority over the power of the enemy and reject every thought he tries to plant in your mind. Every time Satan tries to attack your mind, tell him:

> "Satan, you are a liar! You no longer have control over me. You have no control over my mind, thoughts, will, or emotions. God has not given me a spirit of fear, but of power, love and a sound, well-disciplined mind."

Gird up the loins of your mind, strengthen your mind for battle, and walk in the victory God has provided for you!

Make this declaration: "Satan, you are a liar! You no longer have control over me. You have no control over my mind, thoughts, will, or emotions. God has not given me a spirit of fear, but of power, love and a sound, well-disciplined mind."

Set Your Mind For Victory

Scripture reading: Colossians 3:1-16

Before Jesus came to this earth, He set His mind. He determined beforehand and set a predetermined course of action to do the will of God. Jesus knew He would be rejected, despised, and crucified, yet He set His mind to do the will of God regardless of what He had to endure. He did not allow anything to divert Him from completing the work He came to do. He set His mind on winning the war!

You must set your thoughts upon Christ, the Word, God's will, and His plan and purposes. In order to live in one-hundred percent victory in our minds and win the war against Satan, we too must set our minds that we are going to walk in the victory God has provided for us. We must take authority over our thoughts, desires, and intentions and focus them on Christ, the Word, and upon fulfilling God's will.

Paul told the Romans:

> *For they that are after the flesh do mind the things of the flesh; but they that are after the Spirit the things of the Spirit.* (Romans 8:5)

He told the Colossians:

> *If then you have been raised with Christ (to a new life, thus sharing His resurrection from the dead), aim at and seek the (rich, eternal treasures) that are above, where Christ is, seated at the right hand of God. And set your minds and keep them set on what is above – the higher things – not on the things that are on the earth. For (as far as this world is concerned) you have died, and your (new, real) life is hid with Christ in God.*
>
> (Colossians 3:1-3, TAB)

To "set your mind on things above" is to desire and seek after those things. You must see to it that your interests and desires are centered in Christ. "Setting the mind" means that a person fills his thought life and sets a thought pattern upon Christ, the Word, God's will, His plan and purposes. You must make a predetermined effort to program your thoughts and keep them centered upon Christ and things above instead of allowing them to be constantly filled with the cares of this life, your selfish desires, fear, unbelief, worry, doubt, or temptation.

Paul told the Philippians to think on things that were pure, lovely, gracious, excellent – things worthy of praise (Philippians 4:8, TAB). There are Christians today who have programmed themselves for failure and defeat by focusing on their circumstances instead of setting their minds for victory. You don't have to be one of them!

Make this declaration: "Today, I will think only on things that are true, worthy of reverence, honorable, seemly, just, pure, lovely, virtuous, and excellent."

Ready To Win The War

Scripture reading: Psalm 112:1-8

When our hearts and minds are firmly fixed and established upon God and His Word, there will be no fear. Regardless of the battles we may face, we will not be afraid, worried, or depressed:

> *He shall not be afraid of evil tidings: his heart is fixed, trusting in the Lord. His heart is established, he shall not be afraid, until he see his desire upon his enemies.*
>
> *(Psalm 112:7-8)*

As you respond in faith and act upon God's promise by fixing your mind upon Him, He will pour new strength into your innermost being – into your heart and mind, your thoughts, will, and emotions.

I am praying and believing God to pour His strength – His *dunamis* miracle power – into your mind so you can enter into a new dimension of power and victory greater than you have ever experienced! The prayer Paul prayed for the Ephesians is my prayer for you. I pray: *"That he would grant you, according to the riches of his glory, to be strengthened with might by his spirit in the inner man"* (Ephesians 3:16).

Set your mind for victory. Face your battles, problems, trials, and circumstances with a victorious mind that is one-hundred percent free from fear, one-hundred percent free from worry, and ready to win the war!

You are not a helpless victim that does not have any choice or control over your mind. You do not have to accept the thoughts Satan tries to place in your mind. He only has the power to attack your mind when you allow him to have access through your thoughts and imaginations.

Regardless of any of the circumstances you will ever face, you do not have to be defeated in your mind. In his counterattack on the Church, Satan is going to attack your mind through circumstances – and sometimes people – to try to cause fear, depression, discouragement, and defeat. Don't let him do it!

God has given you power over all of Satan's attacks on your mind. You must arm yourself with the mind of Christ. You must fill your thoughts with God's Word and His promises. Instead of negative thoughts of fear, doubt, and worry, fill your mind with thoughts of faith, praise, and thanksgiving to God for giving you victory. Replace your thoughts of anger, sadness, and worry with thoughts of love, joy, and peace. Set your mind for victory instead of defeat! Get ready to win the war!

Make this declaration: "I am not afraid of evil tidings. My heart is fixed, trusting in the Lord. My heart is established. I will not be afraid."

Resisting In The Power Of The Spirit

Scripture reading: 2 Corinthians 10:3-5

In addition to programming your mind for victory instead of defeat, you must also refuse to give Satan access to your mind. Satan wants you to have a passive mind and allow him to fill your mind with worry, doubt, fear, unbelief, defeat, failure, temptation and every type of negative thought.

When you are worried, discouraged, or depressed, he wants you to think you can't help yourself or that there is nothing you can do to control your thoughts. But you can arm yourself with a militant mind!

Unless a person is suffering from depression that is caused by physical problems, he or she is depressed because they choose to be depressed. If your mind is filled with illicit sexual desires or other carnal desires of the flesh, it is because you choose these thoughts to be there.

Through the power of the Holy Spirit, you have the ability to take authority and bring all your thoughts into captivity (2 Corinthians 10:5). If you have these types of thoughts in your mind, it is because you have not exercised your will to resist Satan, rebuke those thoughts, and reject them from your mind. Jesus has already defeated Satan and given you power over him, but that power is not activated until you resist Satan, his thoughts, lies, and attacks on your mind and body.

James told the believers, *"Submit yourselves therefore to God. Resist the devil, and he will flee from you"* (James 4:7). The Greek word for "resist" in this verse means "to set against; to withstand." You must rise up in the Spirit of God and withstand Satan and the evil powers and principalities that are coming against you. You must actively and aggressively oppose Satan by ejecting his thoughts from your mind.

The type of resisting James was talking about that forces Satan to flee is not a mental battle. It is not mind over matter or positive thinking. It is not just a matter of speaking words or saying, "I resist those thoughts in the Name of Jesus." You cannot resist and reject the thoughts Satan will place in your mind in the strength of your own natural mind and will. You are not using carnal weapons in this war against Satan:.

> *The weapons we fight with are not the weapons of the world. On the contrary, they have divine power to demolish strongholds. We demolish arguments and every pretension that sets itself up against the knowledge of God, and we take captive every thought to make it obedient to Christ.*
> *(2 Corinthians 10: 4-5, NIV)*

Don't kid yourself! Satan isn't afraid of you. He's not going to run from you. But he is afraid and will run from the One Who is in you, as you resist him in the power and anointing of the Holy Spirit.

Make this declaration: "Satan, you have no control over my mind. I command you, in the Name of Jesus, to loose your hold on my mind."

March 24

We've Got To Die

Scripture reading: 2 Peter 5:8-11

The Apostle Peter warned: *"Be sober, be vigilant, because your adversary the devil, as a roaring lion, walketh about, seeking whom he may devour: Whom resist stedfast in the faith…"* (1 Peter 5:8-9).

The moment you feel you are becoming fearful, worried, tempted, discouraged, or depressed, recognize it as an attack of Satan against your mind. When Satan attacks, you don't have ten minutes or even ten seconds. The instant you recognize Satan's attack on your mind, you must act. You must go deep into the spirit realm and resist the thoughts he is using against you, in the Name of Jesus.

There will be times when Satan's attack on your mind will be so strong it will be necessary for you to press the battle in the Spirit through seasons of prayer over a period of days, drawing upon His strength to resist and defeat Satan.

As you are resisting Satan steadfast in the faith, there will be times when you think you have won victory over worry, depression, or other attacks on your mind. Whatever you do, don't let your guard down for one moment! The moment you relax in your mind, thinking the battle is over, he will try to bring it back upon you or come against you using some other strategy. You must not let up. You must continue to press the battle in the Spirit, until you are walking in victory every day.

Like a soldier picking up his weapons before going out into battle, you must arm yourself with the "mind of Christ" – His will, His thoughts, His desires, His purposes. Peter told the believers in the Early Church:

> *So, since Christ suffered in the flesh (for us, for you), arm yourselves with the same thought and purpose (patiently to suffer rather than fail to please God). For whoever has suffered in the flesh (having the mind of Christ) has done with (intentional) sin – has stopped pleasing himself and the world, and pleases God. So that he can no longer spend the rest of his natural life living by (his) human appetites and desires, but (he lives) for what God wills.*
>
> *(1 Peter 4:1-2, TAB)*

We must take deliberate action to arm ourselves with the mind of Christ. It is not ours automatically. We must draw the line and settle forever that we will not allow sin to have any place in our lives whatsoever. To do this, we must be willing to put to death the lusts of our flesh including our own selfish desires, ambitions, and goals. We must strip ourselves – get rid of our carnal desires, thoughts, intentions – and put on the mind of Christ with His thoughts, desires, purposes, and intentions. Our wills must be totally surrendered to God and our lives totally consumed in fulfilling His will. We've got to die!

Make this declaration: "I am dead to sin, the world, and the flesh. I am alive to the thoughts, desires, purposes, and intentions of God."

Renewing The Spirit Of Your Mind

Scripture reading: Psalm 119:9-16

We must continually take authority over every aspect of our hearts and minds – our thoughts, desires, emotions, intentions – and bring them under control of the Holy Spirit.

This renewing of our minds to the mind of Christ must be a continual daily process. As we fill our minds with a knowledge of Christ – His will, thoughts, desires, and purposes – and bring our hearts and minds into submission to the Holy Spirit, we are being renewed *"in knowledge after the image of him that created him"* (Colossians 3:10).

There is absolutely no way a Christian can live in one-hundred percent victory over Satan and his strategies unless this continual renewal of the mind is taking place. To arm yourself with the mind of Christ, you must daily saturate your heart and mind until it is filled with the Word.

Paul said, *"Be renewed in the spirit of your mind"* (Ephesians 4:23). The renewal mentioned here is not referring simply to the natural mind with its natural abilities in retaining knowledge. He was not talking about filling our heads with "head knowledge" of the Word, Christ, and power He has given us over all the power of the enemy, or who we are in Christ.

The "spirit" of your mind refers to thoughts, attitudes, and desires which have been brought under the dominion and control of the Holy Spirit. It is the portion of your mind which brings you into fellowship with Christ and directs its energies toward pleasing God and fulfilling His will.

To be renewed "in the spirit of your mind" where you have armed yourself with the mind of Christ, you must fill your thoughts with God's Word and meditate on it continually – day and night – until the Word becomes an integral part of your life; until your thoughts, desires, intentions, and purposes are continually being transformed and molded into Christ's image.

As your mind is renewed daily and you go out into battle against Satan armed with the victorious mind of Christ you will not be defeated, regardless of the circumstances you face. Armed with the victorious mind of Christ, you will not be worried or fearful and you will not become weary or faint in your mind. You will be strong, immovable, and victorious over every attack, every obstacle, and every circumstance Satan may use against you.

God has revealed to me that Satan is now amassing his forces to come against the Church in a massive counterattack to stop the advance that we are making into his territory. He has already initiated his battle plan, but as we meditate on these truths together God is giving you advance knowledge – a spiritual breakthrough – so you will know what is happening, how Satan is attacking, where he is attacking and you will be able to activate your mind for battle.

Make this declaration: "My mind is being renewed in knowledge after the image of God who created me."

Preparing To Fight

Scripture reading: Psalm 98:1-9

God is calling the Church to prepare for battle. Satan is intensifying his attacks against God's people. To go out to battle without first activating and preparing your mind for the fight is like soldiers going out onto the battlefield without their weapons, unprepared to fight.

Before an army engages the enemy in battle, there must be a time of preparation. The weapons, tanks, and aircrafts, are checked to see if they are in good condition. The Commander-In-Chief and other officers meet to plan their battle strategies. The soldiers go through all sorts of maneuvers simulating and acting out the strategies they are going to use during battle. The army is placed on alert. They are prepared, physically, and mentally to confront and fight the enemy until the war is won.

As you face Satan's attacks on your mind, you must be prepared to fight and win the war. You must not be fearful or weak when the enemy attacks you. You must not be surprised, hide, feel sorry for yourself, whine, complain, or wonder why you are being attacked. We are on red alert! We have been warned about Satan's strategies. You must be spiritually and mentally prepared to fight and withstand Satan and his forces until victory is yours.

Paul told the Corinthians: *"Be on the alert, stand firm in the faith, act like men, be strong"* (1 Corinthians 16:13, NAS). He told the Ephesians: *"Finally, my brethren, be strong in the Lord, and in the power of his might"* (Ephesians 6:10).

When you go out to battle, you must be strong and face the enemy not, in your own strength, but in the power of Almighty God. You are going out to fight and win the war against Satan in the same mighty power that raised Christ from the dead (Ephesians 1:19-20). There is no room for defeat!

You must strengthen your mind for battle and prepare to fight, knowing that He is sending you forth in His strength, in the power of a strong, victorious mind. Paul said: *"Therefore put on God's complete armor, that you may be able to resist and stand your ground on the evil day (of danger), and having done all (the crisis demands), to stand (firmly in your place)"* (Ephesians 6:13, TAB).

A new strength is coming into God's people – into their hearts and minds, into their innermost beings. Receive this promise, by faith, and be strengthened in your mind. Be strong and move forward to take possession of the one-hundred percent victorious mind God has made possible for you to have.

Make this declaration: "I will be alert. I will stand firm and be strong. I will stand victorious in the day of trouble."

GUARDING YOUR HEART AND MIND

Scripture reading: Psalm 31:1-24

It is vitally important that you guard your heart and mind. Unless you consistently and faithfully do this, you're defeated before you even get out on the battlefield!

Your heart and mind are where all the issues of life flow forth. They are the center of your existence where your entire moral and mental activity originates; where your thoughts, will, emotions, desires and mental abilities are located.

Above everything else – your physical body, your business, career, finances, everything you are and have – you must diligently guard your heart and mind. They are your most vulnerable point and are where Satan is attacking with greater intensity than ever before.

Guarding your heart and mind is not assuming a passive position of sitting back waiting for Satan to attack. It is an offensive strategy requiring aggressive action! In the game of football, there are offensive guards and defensive guards. The offensive guard is aggressive. With great forcefulness and power he knocks players out of the way and blocks them to open the way for the runner to score a touchdown. The defensive guard's responsibility is to use every maneuver, every ounce of strength and determination he has, to stop the running back from the opposing team and prevent him from scoring a touchdown.

This is the type of aggressive action that is necessary to guard your heart and mind. To have a powerful, one-hundred percent victorious mind that will enable you to defeat all of Satan's strategies, attacks, deceptions, and weapons, you must keep a constant watch over your heart and mind.

We are exhorted: *"Keep your heart with all diligence and above all that you guard, for out of it flow the springs of life"* (Proverbs 4:23, TAB). To keep your heart and mind is to set a constant guard against all of Satan's attempts to gain access into your heart and mind. It requires aggressive action to exercise control over your will, thoughts, desires, emotions, and motives and bring them into submission to the Holy Spirit and keep them there.

It is not only necessary to guard against Satan's attacks and resist the thoughts, deceptions, and temptations he will try to use to defeat you, you must also guard against the carnal thoughts and desires of the flesh, the selfish inclination to follow your own will instead of setting your mind to seek after God's will. When these carnal attitudes and desires are detected, you must take aggressive action to rid yourself of them. You must continually guard your heart and mind under all circumstances, in all places, at all times with the greatest possible intensity. Why not start today?

Make this declaration: "I will guard my heart and mind with all diligence, under all circumstances, in all places and at all times with the greatest possible intensity."

Don't Do It

Scripture reading: Philippians 4:4-9

God has made it possible for you to have a powerful, victorious mind that is one-hundred percent free from fear and worry. Think about it. When you learn to guard your heart and mind, there is no reason you should ever be worried, depressed, confused, or defeated again!

The moment you recognize you are becoming anxious, your mind is troubled, afraid, or filled with worry and unbelief then you must take aggressive action. You must not allow your mind to dwell upon the problems or circumstances you are facing. You must recognize immediately when Satan begins to attack.

Like an offensive or defensive guard on a football team, you must exert spiritual energy to grab hold of every negative thought that tries to enter your mind. As you do this and begin to set your heart and mind upon God and His Word, He will guard your thoughts and keep your mind in perfect, constant peace: *"Thou wilt keep him in perfect peace, whose mind is stayed on thee: because he trusteth in thee"* (Isaiah 26:3).

Paul told the Philippians:

> *Do not be anxious about anything, but in everything, by prayer and petition, with thanksgiving present your requests to God. And the peace of God, which transcends all understanding, will guard your hearts and your minds in Christ Jesus.*
>
> *(Philippians 4:6-7, NIV)*

Jesus said not to worry about what you will eat, drink, or wear (Matthew 6:25). Paul said not to be anxious about anything. Yet, today the majority of Christians spend a great amount of time and energy worrying about the cares of this life and the circumstances they are facing until their minds are depressed and stressed by the enemy.

You may be thinking, "Brother Cerullo, you just don't know what I'm going through. I try not to worry, but I just can't help myself."

You can help yourself!

Your mind can be free from worry!

Your mind can be free from depression!

Jesus told us not to worry. He paid the price and made it possible for you to have a one-hundred percent victorious mind. Don't do it! It's as simple as that! Guard your heart and mind. Recognize thoughts of worry, doubt, fear, unbelief, failure, and defeat and stop them from entering your mind.

Make this declaration: "I will not be anxious about anything. In everything, by prayer and petition and with thanksgiving, I will present my requests to God and the peace of God, which transcends all understanding, will guard my heart and mind in Christ Jesus."

A New Dimension Of Power

Scripture reading: Matthew 6:25-34

When you are tempted to worry, instead of worrying, set your mind on God and His promises to you. Fill your heart and mind with thoughts of faith, praise, and thanksgiving even in the midst of the trials and adverse circumstances you are facing. Instead of worrying, begin to praise and give thanks unto God for the answer to your need.

You cannot do this in your own limited strength or in the power of your natural mind. Regardless of how strong your will may be, you cannot do this simply by willpower or by determining that you are no longer going to worry or allow yourself to be defeated. You must rise up in the power of the Holy Spirit within you. The battle must be fought and won in the Spirit.

When you recognize that your mind is in turmoil or you are feeling fearful, worried, discouraged, or defeated, take aggressive action. Get on your face before God in prayer and supplication. Go deep into the Spirit and begin to bind the power of the enemy. Draw upon God's strength and the power of the Holy Spirit within you to resist, oppose, and withstand Satan's attack on your mind. Exercise your authority over Satan by taking control over your thoughts and casting out every negative thought, worry, and doubt in the Name of Jesus. Command Satan to take his hands off your mind.

Say to him: "Satan, you have no control over my mind. I command you, in the Name of Jesus, take your hands off my mind! I resist you now in the power of Almighty God that is within me. I reject every negative thought, every fear, doubt, and worry from my mind. It is written, God has not given me a spirit of fear, but of power, love, and a sound mind. I claim the peace of God over my mind now, in the Name of Jesus."

Satan doesn't give up easily. He will try to bring back worry and doubt into your mind. Every time you sense that you are worried, discouraged, depressed or feel defeated, recognize it as an attack of Satan. Continue to rebuke and resist him in the Name of Jesus. As you continue to do this, he must leave and the thoughts of fear, worry, and unbelief must leave with him. He has no choice. As you actively resist him, he will flee!

Paul said, *"And the peace of God, which transcends all understanding, will guard your hearts and your minds in Christ Jesus"* (Philippians 4:7, NIV). The word "guard" in this verse is a military term which means "garrison." God has promised that as you refuse to worry and present your needs to Him in praise and thanksgiving, He will guard your heart – your thoughts, will, and emotions – as soldiers under military law guard a garrison (constantly watch over it, guard and protect it) so that it remains safe from enemy attack.

Make this declaration: "I am drawing upon God's strength and the power of the Holy Spirit to resist, oppose, and withstand Satan's attack on my mind."

March 30

Program Your Mind For Victory

Scripture reading: Psalm 103:1-22

Israel had a land promised to them. They had a promise of entering into God's rest, but they had to go out in power and authority and claim their inheritance. As you go out onto the battlefield to defeat Satan's attack on your mind, you will be able to take possession of the one-hundred percent victorious mind God has made possible for you to have.

Remember, however – in order to have a one-hundred percent victorious mind you must guard your heart and mind continually. You must recognize immediately when Satan begins to attack your mind with fear, anger, guilt, self-pity, and defeat and refuse to allow those thoughts to remain in your mind for even a minute.

For a moment, think about the circumstances you are facing right now. Have you allowed Satan to defeat you in your mind? Have you allowed anger, resentment, and self-pity to fill your heart and mind? Are you oppressed or depressed? Have you been deceived by Satan? You can be free right now from the spirit of depression over your mind and walk in new strength and power.

During this past month, I have shared with you five powerful strategies which will enable you to be victorious over all of Satan's strategies, attacks, deceptions, and weapons which he may deploy against your mind. As we close out this month's devotions, let's take a moment to review them:

- Offensive Strategy No. 1: Set your mind for victory!
- Offensive Strategy No. 2: Refuse to give Satan access to your mind!
- Offensive Strategy No. 3: Arm yourself with the mind of Christ!
- Offensive Strategy No. 4: Strengthen your mind for battle!
- Offensive Strategy No. 5: Guard your heart and mind!

Use these strategies in your life to defeat Satan's counterattack against you.

- Refuse to be intimidated.
- Become spiritually aggressive.
- Rise up in the new strength God is pouring into your innermost being, into your heart and mind.
- Face every circumstance, every problem, every challenge, every opportunity in the power of a one-hundred percent victorious mind.

Make this declaration: "I am setting my mind for victory. I am refusing to give Satan access to my mind. I am arming myself with the mind of Christ and strengthening my mind for battle. I will guard my heart and mind."

Now Is The Time

Scripture reading: Acts 10:34-43

From this day forward you can determine that you will no longer live your life according to your carnal, natural mind. In the circumstances you are facing, don't lean to your own understanding or depend upon the "arm of the flesh." Wait before the Lord until you know you have the mind of Christ. Once you know you have the mind of Christ in the situation you are facing, act boldly with full confidence and assurance knowing God will fulfill His Word.

When you know you have the mind of Christ, you won't be fearful or waver in your faith. You will be strong and victorious and you will be able to take hold of God's promises. When you know you have the mind of Christ, it will not be necessary for you to run here and there – from meeting to meeting – to try to find out what God is directing you to do.

When you act in accordance with the mind of Christ, there is no possibility for failure or defeat. You can live your life according to the one-hundred percent victorious mind of Christ. The Apostle Peter, in writing to believers who were distressed because of the tremendous trials they were going through, said, *"Gird up the loins of your mind..."* (1 Peter 1:13). In essence, he was telling them to prepare their minds for aggressive action, to strengthen and brace their minds for the battles they were facing.

The "loins" refers to the chief seat of bodily strength. The body is knit to the loins. If the loins fail, the whole body fails. If your mind is weak, you will live in defeat. Gird up, strengthen, and prepare your mind for aggressive action!

Resist, withstand, and oppose every attack of the enemy against your mind! Bring every thought into captivity! Be continually renewed in your mind!

As we conclude this month of powerful devotional readings concerning the mind, my prayer for you is...

- That you will break through the limitations of your carnal, natural mind and take hold of the supernatural...
- That you will take hold of the one-hundred percent victorious mind of Christ where you are walking in one-hundred percent victory in every area of your life...
- That you will not be content to settle for anything less than Christ has provided for you, but as you take hold of the one-hundred percent victorious mind of Christ you will take possession of all He has promised you!

You can do it...right now...today!

Make this declaration: "From this day forward, I will no longer live my life accord to my carnal, natural mind. I will have a one-hundred percent victorious mind!"

April

The platform at the Morris Cerullo Indonesia crusade was filled with discarded drugs and cigarettes following a miraculous deliverance service. This picture is a powerful testimony to the truth that God wants you to conquer the flesh in every area of your life – and this month's devotional readings will show you exactly how to do that!

April

Conquering The Flesh

In our devotions during the past few months we have been on a march to destroy the seven major obstacles which block the fulfillment of God's promises and provision for your life. These barriers – walls that are standing in your way hindering you from taking possession of God's promises – are as formidable as the walls of Jericho which prevented Israel from taking possession of their land.

During February, we marched against the obstacles of fear, unbelief, and confusion. Last month we spent the entire month discussing a major barrier – the battlefield where most spiritual attacks occur – your mind. This month's devotions focus on the last three remaining obstacles: Disobedience, the flesh, and the tongue.

Disobedience is sin and it cuts off communication and separates us from God. Without obedience, you can have no covenant relationship with God because disobedience breaks your fellowship with Him

In Joshua chapter 5, God told Joshua that Israel must be circumcised, a sign of dealing with the flesh, before advancing into their land to claim His promises:

> *And the Lord said unto Joshua, This day have I rolled away the reproach of Egypt from off you. Wherefore the name of the place is called Gilgal unto this day.*
>
> *(Joshua 5:9)*

Israel had to deal with the flesh in order to subdue their enemy and enter into God's rest. We, too, must deal with our flesh if we are to experience victory miracle living in the new millennium.

The Bible indicates that your tongue controls your destiny. You can change the entire course of your life by bringing your tongue under the control of the Holy Spirit. If you are living in discouragement, failure, or defeat in any area of your life, you can change it by changing how you speak, by restraining your tongue, and speaking forth life, blessing, and words of faith.

When Israel marched around Jericho on the seventh day, the Lord directed them to shout with a great shout. This was a shout of faith and victory because God had promised that the Jericho walls would fall flat and they would go in and take the victory that was theirs. Joshua told the people, *"Shout, for the Lord hath given you the city"* (Joshua 6:16).

This is your time to take possession of the promises you have been believing God for – healing, deliverance, family restoration, and financial provision are yours. During this month, the final walls that stand in your way – disobedience, the flesh, and the tongue – are going to crumble as you learn and apply powerful God-given strategies. So, go ahead – *"Shout; for the Lord hath given you the city"* (Joshua 6:16).

Disobedience Is Sin

Scripture reading: Joshua 6:1-20

The city of Jericho actually was Israel's from the very day they began their first march around its massive walls. They had God's promise and that was the same as if Jericho already belonged to them, but they were unable to take possession of it until they completed all God had directed them to do.

The powerful key to their great victory in taking the city of Jericho was not based upon Israel's natural abilities, their military strategies, their wisdom, or their strength. It was based upon their obedience. At any time during their seven marches around the walls of Jericho if they had failed to do what God directed them to do, they would have forfeited God's promise. The walls would not have fallen and they would not have taken possession of Jericho.

Disobedience will block you from taking possession of the promises God has given you. It will hinder your personal relationship with God and, unless it is dealt with, it will cause you to forfeit your spiritual inheritance.

Our entire relationship with God is based upon obedience. If there is no obedience, there can be no relationship. Our God, Who is altogether holy, pure, and undefiled, cannot be joined together in covenant relationship with those who are disobedient and rebellious. He will not do it!

Disobedience and rebellion cut off communication and separates us from God. When Adam and Eve rebelled against God in the Garden of Eden, their intimate relationship with Him was broken. They were driven from the garden and separated from God. When the children of Israel continued to rebel against God in spite of His warnings, He hid His face from them and their relationship was broken (Deuteronomy 31:17). God's covenant with Abraham was based upon obedience. By faith he obeyed and obtained the promise (Hebrews 11:8).

Disobedience is sin and if you fail to acknowledge it and repent, it will lead to defeat and destruction. The Church today is filled with Christians who are living defeated lives, unable to take possession of their inheritance as joint-heirs of the Kingdom of God because they are walking in disobedience to God and His Word. Many have refused to hear the Word of God, have closed their ears to hearing the truth, and are only willing to hear those things which "tickle" their ears. They are living in rebellion, following their own self-will. Others have yielded to the temptations Satan has placed before them and have compromised the Word of God to accommodate their own fleshly desires. As a result of their continual disobedience, they are living defeated lives because disobedience is sin.

Make this declaration: "I recognize disobedience for what it is – sin. I will not live a defeated life because of disobedience."

Overcoming Eight Strategies Of Satan

Scripture reading: Matthew 7:21-27

Satan would like you to think that God does not require the same type of obedience under the New Covenant as He did under the Old Covenant, but he is a liar. All acts of disobedience are sin, regardless of how small or large. It is through our obedience that we enter into a relationship where we become one with God.

Jesus said if we obey His teaching, He and the Father will come and live within us. If we obey, we will be saved and inherit eternal life. Jesus said, *"Verily, verily, I say unto you, If a man keep my saying, he shall never see death"* (John 8:51).

It is reverent submission and obedience to God that rewards us with the ultimate victory over Satan. Jesus said:

> *"Not every one who says to Me, Lord, Lord, will enter the kingdom of heaven, but he who does the will of My Father Who is in heaven. Many will say to Me on that day, Lord, Lord, have we not prophesied in Your name, and driven out demons in Your name, and done many mighty works in Your name? And then I will say to them openly (publicly), I never knew you; depart from Me, you who act wickedly - disregarding My commands."*
>
> *(Matthew 7:21-23, TAB)*

Paul told the Romans:

> *For it is not merely hearing the Law (read) that makes one righteous before God, but it is the doers of the Law who will be held guiltless and acquitted and justified.*
>
> *(Romans 2:13, TAB)*

There are eight basic strategies Satan employs to lead you step by step into rebellion and disobedience against God:

1. He will cause you to doubt God and His Word.
2. He will tempt you to murmur against God.
3. He will cause you to turn away from God.
4. He will cause you to reject God's voice and His Word.
5. He will steal the fear of God out of your heart.
6. He will cause you to fear man more than you fear God.
7. He will cause you to reject God's will and follow your own self-will.
8. He will deceive you into compromising what God has spoken to you.

If you allow even one area of disobedience to God or His Word to remain, Satan has a stronghold he will use to lead you further into disobedience.

Examine each of these eight areas in your own life as you wait in the presence of God today.

Make this declaration: "Through God's power, I will not doubt Him or His Word. I will not murmur against God, turn away from Him, or reject His voice, His Word, or His will. I will not fear man more than I fear God. I will not compromise."

Blessing Or Cursing?

Scripture reading: James 1:21-25

It is the obedient – those who obey the Word of God – who have the legal right to take possession of their inheritance and claim the covenant promises.

> *But-obey the message, be doers of the Word, and not merely listeners to it, betraying yourselves (into deception by reasoning contrary to the Truth). For if any one only listens to the Word without obeying it and being a doer of it, he is like a man who looks carefully at his (own) natural face in a mirror; For he thoughtfully observes himself, then goes off and promptly forgets what he was like. But he who looks carefully into the faultless Law, the (law) of liberty, and is faithful to it and perseveres in looking into it, being not a heedless listener who forgets, but an active doer (who obeys), he shall be blessed in his doing – in his life of obedience.* *(James 1:22-25, TAB)*

The disobedient are living under the wrath of God and unless they repent of their disobedience will be punished when Christ returns for His saints. Paul told the Thessalonians:

> *And to you who are troubled rest with us, when the Lord Jesus shall be revealed from heaven with his mighty angels, in flaming fire taking vengeance on them that know not God, and that obey not the gospel of our Lord Jesus Christ. Who shall be punished with everlasting destruction from the presence of the Lord, and from the glory of his power.*
> *(2 Thessalonians 1:7-9)*

God did not allow even one of those who had been disobedient to Him in the wilderness to enter the Promised Land. All the people twenty years and older died in the wilderness for their disobedience, just as God said they would. Only two men, Joshua and Caleb, along with the younger generation were allowed to take possession of their inheritance.

On the plains of Moab before Joshua led the people across Jordan to take possession of the Promised Land, God required the people to reconfirm the covenant He had made with their fathers on Mount Sinai. With all of Israel gathered together on the plains of Moab, Moses read all the laws and ordinances God had commanded them and reminded them of the covenant promises God had given them on Mount Sinai. He laid out the blessings that would overtake them in the land, if they were obedient to God (Deuteronomy 28:1-14). He also warned them of all the curses that would come upon them if they disobeyed (verses 15-68). That day they were given a choice to obey God and be blessed or disobey God and be cursed (Deuteronomy 30:15-20).

God sets before you this day blessing or cursing. Which will you choose?

Make this declaration: "I will be a doer of the Word and not a hearer only. I will choose to obey God and His Word and I will experience His blessing."

April 4

"I'M ONLY HUMAN"

Scripture reading: Psalm 53:1-6

Satan's major strategy to defeat you is to manipulate you into choosing to disobey God. Three of the major strategies he will use to lead you into disobedience are rebellion, self-will, and compromise.

Possessing God's promises was based upon whether or not Israel was obedient to God. They were given the choice: to obey God and be blessed or to disobey Him and be cursed. Their decision to obey or disobey was literally a life or death decision.

Joshua and the children of Israel crossed over the River Jordan and took possession of their inheritance. As long as they were obedient, God drove out their enemies and made them a strong and mighty nation. He multiplied and blessed them above the nations of the earth. He prospered them and gave them the spoils of their enemies. He established them as a holy people. They enjoyed an abundance of food, silver and gold, livestock, houses and land.

Instead of obeying and enjoy the blessings of God, however, Israel chose to disobey and all the curses that Moses warned them about came upon them as a result. The land which once flowed with milk and honey became a barren waste. They suffered from famine and disease. They were besieged and defeated by Assyrian and Chaldean armies, taken into Babylonian captivity for 70 years, and later ruled by the iron hand of Rome. During the Roman wars, 1,100,000 Jews died during the siege of Jerusalem under Titus by sword, pestilence, or famine and 97,000 were taken as prisoners to the slave market.

Israel was dispersed among the nations of the earth and became universally despised and oppressed. The nation which was to become as numerous as the sands of the sea dwindled to a small, insignificant number. All this was to bring them to the end of themselves – to show them their total inability to save themselves and their need for a Savior.

Satan's strategy today is to deceive you into thinking that God no longer requires this same one-hundred percent obedience under the New Covenant. Do not be deceived! God has not changed. He still requires one-hundred percent obedience. There are liberal preachers and teachers today who are leading Christians into disobedience and a lax attitude toward sin. They are teaching that since we no longer live under the law but under grace it is all right if we sin occasionally – "after all...we're only human." They teach about the love, long-suffering, and forgiveness of God, yet they fail to teach His hatred of sin and the wrath that will come upon all those who are walking in disobedience to His Word.

Now let me ask you...where do you stand? Are you walking in obedience to God and His Word, or are you excusing your sin by saying, "I'm only human"?

Make this declaration: "I will no longer use the excuse, 'I'm only human' because God's divine power is at work in me."

How Will You Escape?

Scripture reading: Hebrews 2:1-15

Without obedience, you can have no covenant relationship with God because disobedience breaks fellowship with Him. Jesus said, *"If you abide in My Word – hold fast to My teachings and live in accordance with them – you are truly My disciples"* (John 8:31, TAB). He said that if we obey His teachings, He and the Father will come in and live within us (John 14:23). He said that if we obey, we will never see death (John 8:51). John said, *"We know that we have come to know him if we obey his command"* (1 John 2:3, NIV).

Satan's strategy is to deceive you into thinking that it is impossible for any man to walk in one-hundred percent obedience to God. Do not listen to his lies. God has made it possible under the New Covenant. There is absolutely no excuse for Christians to allow any areas of disobedience to remain in their lives. God has given us the power through the Holy Spirit to walk in obedience to Him and keep His Word. Should we disobey and sin against Him, we have been granted permission to enter the Holy of Holies where we can come boldly into His presence and receive forgiveness.

- Knowing that God has not changed, that He still requires His people to walk in one-hundred percent obedience to Him…
- Knowing that God compares our acts of obedience to spiritual adultery and whoredom…
- Knowing that the disobedient will not inherit the kingdom of heaven and that God's wrath will be poured out upon us…
- Knowing that God has made it possible for us, through the power of the Holy Spirit, to live in one-hundred percent obedience…

…How much more does God require of us who have entered into a covenant relationship with Him under the New Covenant?

> *For if the message given through angels (that is, the Law spoken by them to Moses) was authentic and proved sure, and every violation and disobedience received an appropriate (just and adequate) penalty, How shall we escape (appropriate retribution) if we neglect and refuse to pay attention to such a great salvation…*
>
> *(Hebrews 2:2-3, TAB)*

Have you entered into the New Covenant with God? If so, how will you escape if you refuse to pay attention to such a great salvation?

Make this declaration: "I will walk in one-hundred percent obedience to God because He has made it possible for me to do so through the power of the Holy Spirit."

Set Free From Sin

Scripture reading: Ezekiel 36:23-28

You may be questioning, "Brother Cerullo, is it really possible for us today to walk in one-hundred percent obedience to God? "You may look at your own weaknesses and failures and think, "I just can't do it. I might as well give up. The harder I try, the more I fail. I just don't see how it is possible for anyone to walk in one-hundred percent obedience to the Word!"

It is impossible for man to walk in one-hundred percent obedience to God and His Word in his own strength. Under the Old Covenant, man was required to walk in obedience to the law God set forth on Mount Sinai. He was unable to do it and he lived under the curses that were pronounced upon those who were disobedient. If you are struggling to obey God and live in accordance with His Word in your own strength, stop! It is impossible in your own strength. You can't do it!

Under the New Covenant, however, God has made it possible. The New Covenant is not based upon your ability to live your life in one-hundred percent obedience to God. God has placed His Spirit within you, giving you the power to obey Him and keep His commandments. God promised, *"I will put my spirit within you, and cause you to walk in my statutes, and ye shall keep my judgments, and do them"* (Ezekiel 36:27).

You have been totally set free from the power of sin! Paul told the Romans, *"For sin shall not (any longer) exert dominion over you, since now you are not under Law (as slaves), but under grace – as subjects of God's favor and mercy"* (Romans 6:14, TAB). There are those who take this verse out of context and use it to encourage permissiveness, self-indulgence, and the satisfaction of carnal desires and lusts of the flesh. They have the attitude that since they are no longer under the law they are no longer obligated to live their lives in one-hundred percent obedience to God and His Word.

In this passage, Paul was not telling the Romans that they were free from the law to live their lives fulfilling the lusts of the flesh. He was emphasizing their freedom from the power of sin. He was teaching them that since they were no longer bound by the power of sin, they were to stop yielding themselves to the cravings of their flesh and yield themselves instead to righteousness. Through the power of the Holy Spirit living within you, you are set free from the power of sin. You have been given power to obey.

There are Christians today who yield themselves to the lusts of their flesh and afterward say, "I just couldn't help myself. I have no control over my actions." Do not be deceived! You do have control through the power of the Holy Spirit. Whatever you face today, Satan cannot make you rebel against God. He cannot make you disobey God and His Word. He cannot make you yield to temptation. He has absolutely no power over you, except what you allow him to have.

Make this declaration: "Sin no longer has dominion over me because I am no longer under the law. I am under grace, an object of God's favor and mercy."

ELIMINATING DISOBEDIENCE

Scripture reading: Deuteronomy 30:1-20

God has made it possible for you to live in one-hundred percent victory over sin and walk in obedience to Him through the sacrificial offering of His own Son. Whenever you rebel and disobey God it is because you choose to do so. No one can make you disobey God. When you rebel and fail to walk in obedience to God's Word, it is not because you do not have power over sin or no control over your actions. It is because you yield your will to disobey.

- You are willfully rejecting God's will and asserting your own will.
- You are willfully rejecting His Word.
- You are willfully opposing God and His authority.
- You are willfully disregarding the blood of Jesus that has made it possible for you to have power over sin and walk in obedience to God.

In the garden of Eden, Adam and Eve had a choice to obey God's command not to eat of the tree of knowledge or to disobey and forfeit their blessings. Satan had absolutely no power over them to make them disobey. They yielded their own wills and rebelled.

Israel, also, was given a choice, to obey God's commandments or to disobey and forfeit the promises and incur the curses. Moses told them:

> *...I have set before you life and death, and blessings and curses. Now choose life, so that you and your children may live and that you may love the LORD your God, listen to his voice, and hold fast to him...*
>
> *(Deuteronomy 30:19-20, NIV)*

The children of Israel yielded their wills, rebelled, and willfully turned their backs on God. As a result, they forfeited God's promises and brought destruction upon themselves.

If there are any areas of disobedience in your life, get rid of them. Ask God to reveal to you even the smallest acts of disobedience. Remember – rebellion also involves a failure or refusal to act upon the Word of God. It is not enough for you to simply hear the Word, you must act upon it or you are in rebellion against God.

Rebellion is opposing or disobeying designated authority. When an individual rebels against God or His Word, he is asserting his will above the will of God. He is choosing to follow his own will instead of obeying God. Rebellion against God basically involves two different areas of disobedience: Whenever a Christian does something that is contrary to the Word of God and when he fails to act upon the Word of God and His promises. You must do both: Obey God's commandments and apply to your life what He reveals through the Word.

Are you choosing to walk in single-minded obedience to God and reaping His blessings or are there areas where you have allowed disobedience to gain a stronghold in your life and you are following after your own will?

Make this declaration: "I will not reject God's will. I will not reject His Word or oppose His authority to rule my life. I accept the sacrifice of the blood of Jesus Christ which enables me to have power over sin and walk in obedience to God."

Yielding To The Holy Spirit

Scripture reading: Galatians 6:7-9

The first strategy in breaking down the wall of disobedience is to yield yourself fully to the Holy Spirit. God has placed His Spirit within you, giving you the power to obey Him and keep His commandments. As you yield yourself to the Holy Spirit and are controlled and guided by Him, you are able to obey God and His Word.

You no longer have to strive in your own strength to obey God. You are set free to obey as you depend upon the power of the Holy Spirit working within you and believe God to do exactly as He promised when He said, *"And I will put my spirit within you, and cause you to walk in my statutes, and ye shall keep my judgments, and do them"* (Ezekiel 36:27).

As you continually yield yourself and are guided and controlled by the Spirit, you will not obey the desires of your flesh and you will no longer be in bondage to the law. Paul told the Galatians, *"But if you are guided (led) by the (Holy) Spirit you are not subject to the Law"* (Galatians 5:18, TAB).

As you continually yield yourself and are guided and controlled by the Spirit, you are able to walk in obedience and no longer live under condemnation and guilt because *"There is therefore now no condemnation to them which are in Christ Jesus"* (Romans 8:1). It is those who are yielded to the Holy Spirit and are living their lives being led and controlled by the Spirit who are no longer living under condemnation, not those who continue to operate in the flesh. Christians who yield themselves to the lusts of the flesh and are being led and controlled by their fleshly desires are condemned. Unless they repent and turn away from their sins, they will reap what they have sown:

> *Do not be deceived: God cannot be mocked. A man reaps what he sows. The one who sows to please his sinful nature, from that nature will reap destruction; the one who sows to please the Spirit, from the Spirit will reap eternal life.*
>
> *(Galatians 6:7-8, NIV)*

God has placed His Spirit within you, giving you power to obey Him. You have been set free from the power of sin. Satan cannot make you disobey.

To walk in obedience to God's Word, you must continually remain in the Word. You must know what God has commanded concerning how you are to conduct your life. Then, you must chose to obey and line your life up with the Word.

Your spiritual inheritance, including all the covenant promises of God, legally belongs to you but having the promises is not the same as taking possession of them. To enter into this new spiritual territory, you must be willing to walk in obedience to God and His Word. You must be willing to go into battle and take dominion over the enemy in your life.

Make this declaration: "I am yielding my life to the power of the Holy Spirit. I am advancing in battle and taking dominion over the enemy in my life."

Rekindling The Fear Of God

Scripture reading: Deuteronomy 5:23-29

The second strategy in breaking down the wall of disobedience is to rekindle the fear of God in your heart. The children of Israel rebelled and disobeyed God because they did not fear Him. Without the fear of God in their hearts motivating them, they could not obey God. They were more concerned about falling into the hands of their enemies than they were about displeasing Almighty God and reaping His wrath.

The nation of Israel that God selected to be His prized possession above all other peoples on the face of the earth – the nation He had promised to bless and multiply until they were strong and mighty – was dispossessed, reaped the wrath of God, and was almost destroyed because they refused to fear God and obey Him. God's dealings with Israel – their disobedience, His warnings, pleading, long-suffering, wrath, and judgments – have been given to us as examples and warnings. We need to take heed.

Israel did not fear God. Their hearts were hardened against Him. God told Moses, *"O that they had such a (mind and) heart in them always, (reverently) to fear Me, and keep all My commandments, that it might go well with them, and with their children for ever!"* (Deuteronomy 5:29, TAB). Without this fear, they did not obey God.

Satan's strategy is to steal the fear of God out of your heart so you will walk in disobedience and be defeated. There are ministers today who are living in disobedience to God because they fear man more than they fear God. They are afraid to preach what God has revealed to them because they are afraid they will lose support of the people. They are afraid to preach against sin and about the wrath and the judgment that is coming upon the earth because they are afraid someone will be offended. They would rather please people than obey God.

There are many Christians who are walking in disobedience to God because the fear of God is missing. They are afraid to live their lives according to what God has revealed to them in the Word because they fear what other people will think of them. They are afraid people will think they are fanatics.

There are Christians to whom God has spoken to step out into some area of ministry, but they have rebelled and are walking in disobedience because they are afraid of what their friends or family will say.

How about you? Are you more concerned about what God says or what others think? In order to break through the wall of disobedience you must rekindle the fear of God in your heart.

Make this declaration: "I have a heart and mind to fear God. It will go well with me today and well with my children forever."

April 10

SINGLE-MINDED OBEDIENCE

Scripture reading: Jeremiah 32:37-41

Whenever God speaks to you – regardless of how insignificant it may seem to you, regardless of how foolish it may sound, regardless of the consequences, regardless of how impossible it may seem – do it in the fear of God!

God's people need to rekindle the fear of God in their hearts. We need to have a fresh revelation of God in all His holiness, as Isaiah did when he fell on his face before God and said, *"Woe is me! for I am undone; because I am a man of unclean lips, and I dwell in the midst of a people of unclean lips…"* (Isaiah 6:5). Once we see Him in the beauty and greatness of His holiness, the fear of God will be rekindled in our hearts and we will be empowered to walk in one-hundred percent obedience to Him all the days of our lives.

God is loving and forgiving. He is merciful and long suffering. We have entered into a covenant relationship with Him where we no longer fear to come into His presence, but can enter boldly into the throne of grace, whereby we cry out, "Abba, Father!" (Romans 8:15).

However, God is also a holy, pure, righteous God of judgment and wrath. He hates sin. He hates rebellion and disobedience. He does not overlook these things, but in His great love and mercy He forgives and pardons, just as He did Israel. He has provided salvation through the blood of Jesus and those who fail to repent and turn away from sin will be judged.

Do not be deceived. Every man must give an account of himself before God (Romans 14:12). Paul told the Romans that God will render to every man according to his deeds (Romans 2:7-8). Under the New Covenant God made with us, He has placed His fear in our hearts and has promised:

> *They will be my people, and I will be their God. I will give them singleness of heart and action, so that they will always fear me for their own good and the good of their children after them. I will make an everlasting covenant with them: I will never stop doing good to them, and I will inspire them to fear me, so that they will never turn away from me.*
>
> *(Jeremiah 32:38-40, NIV)*

God has circumcised our hearts (Romans 2:29). He has taken our hearts that were once hardened and rebellious against Him and given us new hearts, placing His fear in us so that we have the desire to walk in obedience to Him. He has placed His Spirit within us giving us the power to obey. It is this fear of God that must now motivate our lives to walk holy before Him in single-minded obedience.

Make this declaration: "God is giving me singleness of heart so I will always fear Him. He has made an everlasting covenant with me and will never stop doing good to me."

Submit Yourself To Obey

Scripture reading: Psalm 5:1-8

The third strategy in breaking down the wall of disobedience is to submit yourself one-hundred percent to obey God's will. Since the beginning of time, Satan has led God's people into rebellion and disobedience by tempting them to reject God's voice and follow after their own self-will. This strategy is one of the most dangerous because many people are so deceived they cannot determine when they have fallen into the snare of the enemy.

Israel was defeated because they rejected the voice of God and followed their own self-will. Instead of going forward to take possession of the land as God commanded them, they rebelled. They refused to listen. They refused to follow God into the Promised Land, but chose instead to follow their own will.

Are there areas of your life where you are following your own self-will instead of submitting yourself to God and following His direction? God has asked you to do something, but instead you have compromised and are doing your own thing?

If you are faced with decisions right now, make a decision before you go one step further that you will no longer compromise. You will no longer follow your own will. Break through the wall of disobedience and submit yourself to do God's will regardless of the cost.

Here are the three major strategies we have meditated on in the past few days that will enable you to break through the wall of disobedience:

- The first strategy in breaking down the wall of disobedience is to yield yourself fully to the Holy Spirit.
- The second strategy in breaking down the wall of disobedience is to rekindle the fear of God in your heart.
- The third strategy in breaking down the wall of disobedience is to submit yourself one-hundred percent to obey God's will.

Examine your life today in relation to these three strategies. God is speaking to you right now, by His Spirit, to move into a new level of obedience.

The Promised Land – with all of God's promises – is before you. God has given you the land, but you must obey if you are to enter in and claim His promises.

Make this declaration: "I am yielding myself to the Holy Spirit. I am rekindling the fear of God in my heart. I am submitting myself to one-hundred percent obedience."

The Road To Obedience

Scripture reading: Matthew 26:36-44

Jesus is our greatest model of obedience. He was obedient unto death. His obedience was not fifty percent, not twenty-five percent, not ninety-nine percent, but one-hundred percent obedience to do the will of His Father. It was His obedience that made it possible for Him to draw on all of heaven's resources and win the ultimate victory over Satan.

Everything Jesus did was an act of obedience to His Father's will. He did not take the responsibility upon Himself to come to earth and die for the sins of the world. He did not come on His own initiative. He came in obedience to His Father's will. When Jesus came to earth He said, *"I have come to do your will, O God"* (Hebrews 10:7, NIV). Jesus denied Himself of heaven's rights and privileges and took upon Himself human form to fulfill God's will (Philippians 2:7). He was subject to human nature and emotions and was *"in all points tempted as we are, yet without sin"* (Hebrews 4:15).

The obedience that Jesus modeled for us involved a determination to do God's will regardless of the cost. Jesus said:

> *I can of mine own self do nothing: as I hear, I judge: and my judgment is just; because I seek not mine own will, but the will of the Father which hath sent me.*
>
> *(John 5:30)*

Because Jesus was willing to give up His life and live in total submission to God's will, He requires the same of you today. The call Jesus makes is clear: Follow me. Follow my example. Follow Me in My obedience. Follow Me in My love.

To follow Jesus in the road of obedience there are five things you must be willing to do:

1. You must have an ear to hear God's voice and know God's will and when He speaks, you must be quick to obey.
2. You must be determined to do God's will regardless of the cost. The driving force behind all your actions must be to do His will.
3. You must be willing to be obedient unto death. You must be willing to give up all claims to your life.
4. You must develop a wholesome fear of a righteous and holy God Who requires one-hundred percent obedience.
5. You must have a fervent, undying love for Christ that will motivate you to follow Him in one-hundred percent obedience.

Make this declaration: "I am making a new dedication to follow Jesus and live in one-hundred percent obedience to God."

The Tongue

Scripture reading: James 3:1-18

Are you ready to begin our next march to demolish the barriers blocking the fulfillment of God's promises in your life? During the next few days we are going to deal with one of the most powerful forces in the world, one that determines the entire course of your life:

> *Death and life are in the power of the tongue, and they who indulge it shall eat the fruit of it (for death or life).* (Proverbs 18:21, TAB)

Your words determine your destiny! You are either speaking forth...

...Blessing or cursing.

...Power or weakness.

...Faith or unbelief.

...Victory or defeat.

...Abundance or lack.

...Life or death.

If you continually speak negative words of fear, doubt, and discouragement, you are setting the course of your life to live in discouragement and defeat. If you constantly talk about your problems, your sickness, and your lack of finances, you will continue to live with your problems, sickness, and lack of finances. The words coming out of your mouth will hinder you from taking possession of God's promises!

However, if you will take control over your tongue, get rid of the doubt and unbelief in your heart and mind, fill your mouth with the Word of God, and begin to speak it forth, you will be setting the course of your life to live in victory. You will take possession of God's promises!

God is bringing us into a new dimension of authority where our words, spoken with authority, invested in the promises of God, will enable us to confront every stronghold of the enemy. Before you can enter this powerful spiritual realm of authority, however, you must gain control of your tongue.

This battle to gain control of your tongue actually begins in your heart and mind. That is why we spent the entire month of March breaking through the barrier of our carnal, natural minds to take hold of the one-hundred percent victorious mind of Christ. Only as your mind is renewed and you take hold of the one-hundred percent victorious mind of Christ will you be able to bring your tongue under the control of the Holy Spirit.

Make this declaration: "I am going to fight and win the battle for my tongue. It will come under the control of the Holy Spirit."

The Treasures Of Your Heart

Scripture reading: Matthew 12:33-37

What is in a man's heart and mind determines his actions and the words coming out of his mouth. If an individual's heart and mind are filled with negative thoughts – hatred, anger, jealousy, fear, worry, pride, bitterness, and strife – that is exactly what will come out of his mouth. If his heart and mind are filled continually with carnal thoughts and lusts of the flesh, it will be manifested in the words coming out of his mouth.

On the other hand, if an individual's heart and mind are filled with positive thoughts – faith, love, joy, peace – that is what will be coming out of his mouth. If his heart and mind are filled with God's Word, his heart is centered on God and fulfilling His will, then faith-filled words and praise will be coming out of his mouth.

Even though an individual claims to be a Christian and has an outward appearance of righteousness, if his heart and mind are not pure, the words coming out of his mouth will not be pure. Out of his mouth will come both blessing and cursing.

A heart and mind that are corrupted by jealousy, anger, pride, and bitterness cannot produce the fruit of good, wholesome words. The heart and mind must first be cleansed and renewed by the Holy Spirit. The good treasure stored up in a believer's heart and mind is the all powerful, living and written Word of God. Jesus said, *"..the words that I speak unto you, they are spirit, and they are life"* (John 6:63). When our hearts and minds are saturated with the Word and the living and written Word are renewing our hearts and minds, we will speak forth pure, powerful faith-filled words that are spirit and life!

On the other hand, the evil treasure of a man's heart are lust, carnal desires, greed, jealousy, pride and bitterness. As long as this evil treasure remains within him he will bring forth evil things.

You are the one who determines what will come out of your mouth – whether you will bring good out of the good treasure of your heart or whether you will bring evil out of the evil treasure of your heart.

Make this declaration: "My heart is filled with faith, love, joy, peace – and that is what will come out of my mouth. My heart and mind are filled with God's Word, my heart is centered on God and His will – so faith-filled words and praise will come out of my mouth."

Sowing To The Flesh

Scripture reading: Romans 8:1-11

As long as you are living according to the carnal desires of your flesh, you will be unable to have victory in the battle to bring your tongue under control of the Spirit. Paul said, *"So then they that are in the flesh cannot please God"* (Romans 8:8). It is impossible for you to please God while you are still walking after the flesh, motivated by your carnal nature and fulfilling the lusts of your flesh.

You cannot win the battle for your tongue as long as you are still living after the flesh. The words coming out of your mouth will be according to the flesh – bitter words, gossip, abusive language, hateful, and evil words.

The word "flesh" is translated from the Greek word, *sarx*, which refers to man's unregenerate, carnal nature with its unbridled passions, desires, and lusts. There are many Christians today who are living in defeat because they are yielding to the lusts of their flesh. They are trying to have one foot in the world and one foot in the Church. Their hearts are divided. They want to serve God and enjoy His blessings but they also want to enjoy worldly pleasures and satisfy the lusts of their flesh. They are bound by habits, selfish and immoral desires, and impure thoughts.

Instead of denying their flesh, they are yielding to it. Many have deceived themselves into believing they can continue to indulge themselves in fulfilling the lusts of their flesh without reaping any consequences and without it affecting their spiritual life. They believe they can fulfill the lusts of their flesh and still walk in close fellowship with the Lord and take hold of His promises.

Do not be deceived! The truth is, as long as you are walking after the flesh you cannot please God! The Apostle Paul warned that if we sow to the flesh we will reap corruption, but if we walk and live according to the Holy Spirit, we will reap eternal life! He said, *"Be not deceived, God is not mocked: for whatsoever a man soweth, that shall he also reap"* (Galatians 6:7).

Do not allow anyone to deceive you, beloved. If you live your life according to the dictates of your carnal flesh, you are going to reap corruption.

- If you sow sexual immorality, you will reap heartache and the consequences of immorality.
- If you sow gossip, slander, backbiting, and strife, you will reap the same.
- If you sow to the flesh – hatred, anger, pride, jealousy, greed, selfishness – you will reap the consequences.

Think about it...how are you sowing today?

Make this declaration: "I will not sow to the flesh. I will sow to the Spirit and I will reap only good in my life."

Walking In The Spirit

Scripture reading: Romans 6:11-14

The flesh, with its carnal thoughts, attitudes, desires and lusts, wars against the Spirit and blocks you from taking hold of the things of God. Paul told the Galatians:

> *But I say, walk and live habitually in the (Holy) Spirit – responsive to and controlled and guided by the Spirit; then you will certainly not gratify the cravings and desires of the flesh – of human nature without God. For the desires of the flesh are opposed to the (Holy) Spirit, and the (desires of the) Spirit are opposed to the flesh (Godless human nature); for these are antagonistic to each other – continually withstanding and in conflict with each other – so that you are not free but are prevented from doing what you desire to do.*
>
> *(Galatians 5:16-17, TAB)*

As long as you are walking in the Spirit – yielding to the Spirit, being controlled and guided by the Spirit – you will not gratify the desires of your flesh. You will continually resist and deny the flesh.

The flesh with its carnal desires and lusts, struggles and wars against the Spirit. It resists the move and work of the Spirit in your life and everything that is spiritual. The battle rages. The Holy Spirit within you continually opposes and resists the flesh. As a result of this war between the Spirit and the flesh, you are hindered from doing the things you would like to do. You are at a spiritual standstill, unable to take hold of the things of the Spirit. To break through this barrier of the flesh holding you back from entering in to possess all God has provided for you, you must consider yourself dead to sin and your relationship to it broken.

Paul said:

> *Even so consider yourselves also dead to sin and your relation to it broken, but (that you are) alive to God – living in unbroken fellowship with Him – in Christ Jesus. Let not sin therefore rule as king in your mortal (short-lived, perishable) bodies, to make you yield to their cravings and be subject to their lusts and evil passions*
>
> *(Romans 6:11-12, TAB)*

There can be no compromise with the flesh. When you are walking in the Spirit and are continually yielding and being controlled by the Spirit, you will not fulfill the lusts of the flesh. I did not say you wouldn't be tempted, but you will not yield to your fleshly desires and appetites.

Make this declaration: "I will not compromise with the flesh. I will be yielded to, and controlled by the Holy Spirit."

Satan Has No Power Over You

Scripture reading: 1 John 4:1-4

Satan, the tempter, has absolutely no power over you. There is no trial too great, no temptation too strong if you face Satan from a position of knowing. God has given you the ability to walk in one-hundred percent victory during every test, trial, and temptation.

Jesus has destroyed the works of Satan in your life, including the power of sin which once ruled your will and held you in its grasp. You are no longer bound to the desires of your flesh (your carnal nature). You have the power to say "no" to every temptation. You have the power to break every habit that has you bound.

God has placed the Spirit of His Son within you and you are free from all sin! Paul said: *"For the power of the life-giving Spirit – and this power is mine through Christ Jesus – has freed me from the vicious circle of sin and death"* (Romans 8:2, TLB). God has provided total and lasting victory for us. There is not one reason why Christians should ever stumble and fall into Satan's snare through temptation. Not one! Yet, there are many Christians who are still bound by habits they cannot break and who are ensnared by the temptations of the flesh.

Men and women of God who have been faithfully serving Him all their lives, in a weakened moment, are yielding to temptation and being destroyed.

There are Christians who have been deceived by the enemy into thinking they have no control over their will. They yield to the temptations of their flesh and say, "I just couldn't help myself. The temptation was too strong, I didn't have the power to resist." The truth is, every born again child of God, no matter how weak or strong they may be, has the power to resist every temptation. When God has made full provision for victory over every trial and temptation, why are Christians falling into temptation? Why are so many Christians living in such a weakened condition? Eighty to ninety percent of Christians today are setting themselves up for defeat. They are actually giving Satan the ammunition he needs to defeat them.

How? They are giving Satan access through the "eyegate" and "eargate." They are giving Satan access into their minds and are being bombarded by evil thoughts and lusts of the flesh.

Christians are being defeated, not because Satan's power as tempter is so great, but because they do not understand Satan's strategy of temptation and do not know how to resist. As you rip the mask off Satan and expose his strategies against your flesh and your tongue, you are going to know how to position yourself beyond defeat. You will have in your hands the knowledge that will enable you to walk in victory over every trial and temptation in your life!

Make this declaration: "I am free from the vicious cycle of sin and death. The power of the life-giving Spirit is mine."

The Tempter's Trap

Scripture reading: James 1:12-15

There are Christians today who think that Satan has the power to compel men to sin against their free will. He does not! As the tempter, he has the ability to place a temptation before you. He can seduce, lure, and entice you, but he cannot make you sin.

Since he cannot make you sin, Satan's strategy is to gain control of your will. Like a hunter stalking his prey, he sets a trap. He lures you with bait by placing temptation before you to arouse the desires and lusts of your old carnal nature. Then he just sits back and waits until you fall into his trap.

Look at James 1:12-15, in the Amplified Version, which gives a clearer understanding of the original Greek:

> *Blessed, happy to be envied is the man who is patient under trial and stands up under temptation, for when he has stood the test and been approved he will receive (the victor's) crown of life which God has promised to those who love Him. Let no one say when he is tempted, I am tempted from God; for God is incapable of being tempted by (what is) evil and He Himself tempts no one. But every person is tempted when he is drawn away, enticed and baited by his own evil desire (lust, passions). Then the evil desire, when it has conceived gives birth to sin, and sin when it is fully matured brings forth death.*

The battle takes place in your heart and Satan baits you by placing a temptation before you. It could be a person other than your husband or wife; worldly pleasures like houses, lands, or money; or a high-ranking position on your job.

We are tempted when we are drawn away by our own desires. If we allow Satan to fill our minds with thoughts that continually feed those desires, we will eventually yield our will to these desires and sin will be conceived in our hearts and minds even before it manifests itself in our actions.

There are many Christians who have been deceived into thinking that as long as they do not outwardly sin by their actions, they have not sinned. They allow their minds to be filled with all sorts of evil desires, longings, and fantasies. Do not allow Satan to deceive you. Sin is conceived in your heart and mind. Jesus said:

> *...every one who so much as looks at a woman with evil desire for her has already committed adultery with her in his heart. If your right eye serves as a trap to ensnare you or is an occasion for you to stumble and sin, pluck it out and throw it away. It is better that you lose one of your members than that your whole body be cast into hell (Gehenna).*
>
> *(Matthew 5:28-29, TAB)*

When we allow our hearts and minds to be filled with impure thoughts, evil desires, and lusts of the flesh and entertain these thoughts, we have already sinned just as if we had outwardly committed the act.

Make this declaration: "I will not allow my heart and mind to be filled with impure thoughts, evil desires, and lusts of the flesh. I will not entertain these thoughts."

Step-By-Step

Scripture reading: Genesis 3:1-24

Step by step, Satan works through our uncontrolled flesh to lure us through temptation into his trap.

1. He "baits" us by placing some form of temptation that arouses a desire within us.
2. We become tempted when we are drawn away by our own evil desires.
3. Sin is conceived in our hearts and minds when we yield our wills to our desires and entertain those desires in our minds.
4. Sin manifests itself through our actions.
5. Sin brings death.

Satan's major strategy in luring and seducing us into temptation is to gain access into our minds through what we see and hear. Eve gave Satan access to her mind by first listening to him. He gained a foothold in her mind through her thought life. His next strategy was to deceive her. As the arch deceiver, he lied to her about the consequences of eating the forbidden fruit and said, "You surely shall not die!"

God had made it clear what the consequences of Adam and Eve's disobedience would be. To disobey God's command was rebellion against Him. The consequence was death, yet Satan made the temptation appear so satisfying and rewarding to Eve that she ignored the consequences. He told her that when she ate the forbidden fruit she would have the knowledge of good and evil and be like God.

Satan did not have the power to make Eve sin. When God created Adam and Eve, He made them in His own image. He gave them a free will. Eve had power to say "yes" or "no" to the temptation. Satan's strategy was to place a temptation so strong before her that she would use her will to disobey God.

Having gained access into Eve's mind through her thoughts, Satan filled her mind with temptation:

> *When the woman saw that the tree was good for food, and that it was a delight to the eyes, and that the tree was desirable to make one wise, she took from its fruit and ate; and she gave also to her husband with her, and he ate.*
> *(Genesis 3:6, NAS)*

Eve gave Satan access into her mind through the eye gate – what she saw with her eyes.

When she saw the tree with the forbidden fruit, she desired it. What she saw with her eyes triggered a desire and longing within her. Her mind became filled with thoughts of the satisfaction it would bring. The more she looked, the greater the desire filled her mind.

Eve knew the fruit of the tree was forbidden. She knew if she ate it she would be disobeying God, but the desire that filled her heart was so strong that instead of using her will to reject the temptation, she yielded to it. Sin was conceived in her heart and mind and manifested in her actions when she took the fruit and ate it.

Make this declaration: "I will not partake of what God has forbidden. Sin will not be conceived and birthed in my life."

Crucify The Flesh

Scripture reading: Romans 13:11-14

As long as there is an area in your life where you are yielding to the lusts of your flesh, you will experience four major limitations that will prevent you from accessing God's blessings:

1. You will be unable to please God.
2. You will reap corruption.
3. You will be bound, unable to do the things you would do.
4. You will be hindered from taking hold of God's promises.

During the next few days we will meditate on strategies that will demolish the walls of the carnal flesh and the tongue.

Strategy One: Don't make allowances for your flesh – crucify it. This is the first strategy you must follow to break through the barrier of the carnal flesh and bring your tongue under the submission of the Holy Spirit. Paul said:

> *And those who belong to Christ Jesus, the Messiah, have crucified the flesh – the Godless human nature – with its passions and appetites and desires.*
>
> *(Galatians 5:24, TAB)*

The Apostle Paul didn't say we are to pamper our flesh. He didn't say we should make excuses or allowances for the flesh. He said we should crucify it! God doesn't want you to pacify your flesh all by yielding to your carnal, selfish desires and lusts. He expects you to crucify these things and to put them to death.

Peter said, *"...abstain from fleshly lusts, which war against the soul"* (1 Peter 2:11).

Paul told the Romans, *"But clothe yourself with the Lord Jesus Christ, the Messiah, and make no provision for (indulging) the flesh – put a stop to thinking about the evil cravings of your physical nature – to (gratify its) desires (lusts)"* (Romans 13:14 TAB). You must abstain from the lusts of the flesh and make no provisions for it. Nowhere in the Scriptures can we find one place where we are ever told to make allowances for our flesh.

Christ has set you free from the power of sin. He has given you power through the Holy Spirit, Who lives within you, to crucify the lusts of the flesh and walk in the Spirit. You don't ever have to be overcome by temptations of the carnal flesh. Christ paid the supreme sacrifice of laying down His own life so that you can live in one-hundred percent victory over your flesh. You owe a debt, so crucify your flesh!

Make this declaration: "I will make no allowances for my flesh today. The flesh is crucified."

We Owe A Debt

Scripture reading: Romans 8:12-15

In Romans 8:12-15, the Apostle Paul said if we live according to the carnal flesh and are continually yielding to the lusts and desires of the flesh we will die. How would it ever be possible for us to claim God's promises and expect His blessings as long as we are still living after the flesh? On the other hand, if we are crucifying our flesh, living in the Spirit, continually denying and putting to death the flesh, and walking after the things of the Spirit, we will live. We will please God and reap His blessings.

Due to the tremendous sacrifice Christ made on the cross as the sinless Son of God, we owe a debt. We are debtors not to ourselves but to God. We are debtors, not to our carnal flesh, but to the Spirit of God Who lives within us. We have an obligation to no longer live our lives to please our fleshly desires but to please Him.

The major strategy you must use to demolish the barrier of the flesh is to crucify it.

You cannot do this in your own limited strength. You must draw upon the power and anointing of the Holy Spirit to continually put to death every thought, every fleshly desire, and every evil work of the flesh!

Your words are a major obstacle that will block you from taking possession of God's promises. Once you have broken through the barrier of your carnal flesh and have crucified it, you will be able to bring your tongue under the control of the Holy Spirit

As we march forward to demolish the barrier of the tongue and flesh, think about any areas of your life where you have been yielding to your flesh. Think about any carnal, fleshly desires or habits that are hindering you from walking in the fullness of the Spirit and from taking hold of God's promises.

Don't hesitate one minute longer! Get rid of them! Put them to death. Place yourself on the altar before God and ask Him to cleanse you of all works of the flesh. Continually exercise the power and authority Christ has given you, through the Holy Spirit, to get rid of every trace of them.

Make this declaration: "Death and life are in the power of my tongue. I choose life!"

Keep Your Tongue From Evil

Scripture reading: Psalm 39:1-7

The Word of the Lord to you today is that you can change the entire course of your life by bringing your tongue under the control of the Holy Spirit. If you are living in discouragement, failure, or defeat in any area of your life, you can change it by changing how you speak, restraining your tongue, and speaking forth life, blessing, and words of faith. Here is the next strategy:

Strategy Two: Keep your tongue from evil. God requires us to discipline and bring our tongues into submission and under the control of the Holy Spirit. It does not happen automatically. We are commanded: *"Keep your tongue from evil and your lips from speaking deceit"* (Psalm 34:13, TAB).

King Solomon warned: *"Do not allow your mouth to cause your body to sin"* (Ecclesiastes 5:6, TAB).

The Apostle Peter said: *"Whoever would love life and see good days must keep his tongue from evil and his lips from deceitful speech"* (1 Peter 3:10, NIV).

Like King David, we must purpose in our hearts that we will not sin with our mouths. David said, *"I am purposed that my mouth shall not transgress"* (Psalm 17:3). He also said, *"I will take heed to my ways, that I sin not with my tongue: I will keep my mouth with a bridle, while the wicked is before me"* (Psalm 39:1). The word *keep* is translated from a Hebrew word which means to guard or protect. You must guard your tongue and keep close watch over the words you are speaking. You must restrain your tongue and refuse to yield to the flesh by speaking evil, vain, or idle words.

We are to guard our tongues from any form of evil speaking or vain talk. We must set a watch and keep a sharp lookout for any sign of the enemy using our tongues to sin against God. Satan's objective is to gain control of your tongue and use it to destroy you. Every time you yield your tongue to criticize, backbite, gossip, lie, slander, or spread discord among other members of the body, you are yielding your tongue to Satan!

Paul told the Ephesians: *"Let no foul or polluting language, nor evil word, nor unwholesome or worthless talk (ever) come out of your mouth; but only such (speech) as is good and beneficial to the spiritual progress of others, as is fitting to the need and the occasion, that it may be blessing and give grace (God's favor) to those who hear it"* (Ephesians 4:29, 31-32, TAB). Paul said we are to speak blessing instead of cursing!

Have you made a strong determination within your heart to keep your tongue from evil? Are you carefully watching the words coming out of your mouth? If not, this is the day to start!

Make this declaration: "I will not let any foul or polluting language, or evil word, or unwholesome, worthless talk come out of my mouth. I will speak only what is good and beneficial to the spiritual progress of others or what is fitting to the need and occasion."

Limiting God By Words

Scripture reading: Numbers 14:26-35

We are to guard and keep our mouths from speaking evil and to speak only words that edify and bless others. To walk in the power and victory God has planned for you, you must determine in your heart that through the power of the Holy Spirit you will restrain your mouth from sins of the tongue – gossiping, slandering, backbiting, criticizing, lying, deceiving, flattering, and every form of evil speaking.

The children of Israel forfeited God's promises and blessings through sins of their tongue. At Kadesh Barnea, on the border of the Promised Land, they began to speak words of doubt and unbelief and began to murmur and complain against Moses and Aaron. When they heard the negative, evil report of the ten spies concerning the giants in the land and the great walled cities, they believed their report and rejected God's promises.

Look carefully at the words they spoke: *"We are not able to go up against the people (of Canaan); for they are stronger than we are"* (Numbers 13:31, TAB). Regardless of the promises God had given them and the mighty miracles He had performed in delivering them out of Egypt, they said, "We can't do it!" They limited what God had promised by the words coming out of their mouths.

They began to murmur and complain:

> *And all the children of Israel murmured against Moses and against Aaron; and the whole congregation said unto them, Would God that we had died in the land of Egypt! Or would God we had died in this wilderness!*
> *(Numbers 14:2)*

The negative words of fear, doubt, and unbelief that the ten spies spoke brought cursing and death. Their negative words were like a deadly cancer spreading through the people until the entire congregation of approximately two million were contaminated.

God heard the murmuring and complaining of the people and was ready to destroy the entire congregation, but Moses interceded on their behalf and their lives were spared. As punishment for their murmuring and complaining, none of them were allowed to enter and take possession of the land God had promised them. The land and all the promises of God were legally theirs, but they forfeited them because of the words coming out of their mouths. They limited God by their own words.

Beloved, are you limiting what God has promised to do for you by the words coming out of your mouth?

Make this declaration: "I will declare only God's Word and in so doing I will not limit God by the words of my mouth."

Watch Your Mouth

Scripture reading: James 4:11-12

Beloved, many Christians today are forfeiting the promises of God because of the words coming out of their mouths. God had planned to establish them as a mighty people in the land "flowing with milk and honey," but through their words they changed the entire course of their lives.

Just as murmuring and complaining brought curses, death, and prevented the children of Israel from taking possession of their Promised Land, they have allowed Satan to gain a stronghold in the lives of many believers and have prevented them from taking possession of the promises of God. As a result, they have brought curses upon themselves. They have forfeited God's blessings and are living in defeat. Through murmuring and complaining, they are speaking death into their circumstances instead of life.

Change the course of your life by changing your words. Repent of your murmuring and complaining. Set a watch over your mouth and refuse to allow any negative words of fear, doubt, discouragement, or defeat to come out of your mouth.

Gossip is one way Satan has weakened the Church. It is time for us to recognize gossip for what it is – a sin. Those who yield their tongues to gossip are yielding their tongues to Satan and bringing curses upon themselves. Set a watch over your mouth and guard your tongue from gossip. Refuse to listen to gossip. Do not allow yourself to become part of the "grapevine" in your church or on the job. When someone comes to you with gossip, take authority over it and quench it. Stop it in its tracks!

Backbiting is another method through which Satan has gained entrance into the Church and has built a major stronghold. Backbiting is a form of gossip with the specific purpose of saying mean or spiteful things about another individual. Backbiting is also referred to in the Word as "slander" or "false accusation." It is finding fault with others and spreading innuendos and criticism.

When a Christian yields his tongue to backbite and slander someone, he is speaking curses upon others and is bringing curses upon himself. Slander brings curses to three persons: the individual who is speaking, the individual spoken to, and the individual spoken about. Not only must we not speak evil of our brothers and sisters in the Lord, we are commanded not to speak evil of any man (James 4:11).

Make this declaration: "I will not slander or gossip about others. I will not speak evil of my brothers and sisters in the Lord."

How Are You Using Your Tongue?

Scripture reading: Proverbs 6:16-19

One of the deadliest sins of the tongue which we must guard our mouths from is lying. When we allow our tongues to lie, mislead, or deceive others, we are bringing cursing upon ourselves.

Among the seven things listed in the Word that God hates is a lying tongue:

> *There are six things the Lord hates, seven that are detestable to him: haughty eyes, a lying tongue, hands that shed innocent blood, a heart that devises wicked schemes, feet that are quick to rush into evil, a false witness who pours out lies and a man who stirs up dissension among brothers.*
>
> *(Proverbs 6:16-19, NIV)*

When you speak the truth, you speak life!

When you speak lies or deceive, you speak death!

When we make promises to others or to God and do not keep them, we are lying and it gives Satan an opportunity to keep us from receiving the blessings of God. If you realize you have failed to be honest or if you have made a promise to someone or to God and have not fulfilled it, acknowledge it. Repent and ask God to enable you to keep your tongue from speaking lies. Before you make promises, be determined that you will fulfill them.

Satan has entered into the Church and gained a major stronghold where he is bringing confusion, strife, and division through believers who have failed to discipline their tongues and are lying and breaking promises.

We must have a breakthrough! We cannot continue to lie to others. We cannot continue to use our tongues to speak forth both blessing and curses, blessing God and cursing our brothers and sisters in Christ.

Pray about each of the following areas before you leave your quiet place today. Are you using your tongue to...

Murmur and complain?

Lie?

Deceive?

Gossip?

Slander?

Backbite?

Make this declaration: "With God's help, I will no longer complain, lie, deceive, gossip, slander, or backbite. I am taking control over my tongue."

Speak The Word

Scripture reading: John 11:20-44

We have dwelt on two powerful strategies for dealing with the flesh and the tongue: Don't make allowances for your flesh and keep you tongue from evil. Here is the third strategy to break through the barrier of the carnal flesh and bring your tongue under control of the Holy Spirit...

Strategy Three: Speak the Word in a new dimension of authority. Not only must you guard your tongue and restrain from speaking evil, God desires to bring you into a new dimension of authority where you are speaking the Word and His supernatural power is released to fulfill the words He has given you to speak.

Jesus spoke words directed by the Father in the power and authority of God and it was done. There was absolutely no margin for error, no faltering, no wondering. He spoke and it was done. This was the key to releasing the supernatural power of God through Him.

Jesus was sent by God, anointed by the Holy Spirit, and given power and authority to speak and act in His stead, to speak His Words, to do His will: *"...God anointed Jesus of Nazareth with the Holy Ghost and with power: who went about doing good, and healing all that were oppressed of the devil, for God was with him"* (Acts 10:38).

Today, as you speak forth the Rhema Word under the unction and direction of the Holy Spirit, it is Christ the Living Word within you Who will fulfill the Word and manifest His power through you! As you speak forth the Word, He will do the works.

The key for the fulfillment of the Word is in the Word itself – the written and living Word! It is in developing a relationship with Christ where He is in you and where you are continually dwelling in Him and allowing His Word to rule and reign in your life.

"Speaking the Word" is not merely picking verses out of the Bible, repeating them over and over again, hoping they will eventually come to pass. It is not based upon mind over matter. You must first develop a relationship where the living Word has become flesh within you, where you are staying in His presence until you are receiving His direction concerning what you are to say and do. You need to stay in His presence, waiting upon Him, fellowshipping with and worshiping, until He gives you the Rhema Word to speak forth.

Jesus said: *"If you live in Me – abide vitally united to Me – and My words remain in you and continue to live in your hearts, ask whatever you will and it shall be done for you"* (John 15:7, TAB).

Make this declaration: "God's Word abides in me and I am learning to speak the Word in a new dimension of power and authority."

Abiding In Christ

Scripture reading: Psalm 119:33-40

If you are to speak God's Word in a new dimension of authority, then you must spend time abiding in Christ, the Living Word. Wait upon Him for His direction concerning what you are to say and do. Allow the powerful resurrection life that is within Him to flow through your entire being. Draw your life from Him. It is the life within the Living Word that will quicken you – give life to you – and enable you to go beyond your own human limitations.

Stay in the written Word and fill your heart and mind with it until it comes alive within you. Jesus said, if we are vitally united to Him and His words live in our hearts, then we will be able to ask anything and it will be done!

Now, are you ready for the next powerful strategy to use in confronting the flesh and the tongue?

Strategy Four: Speak the Word to demolish Satan's strongholds. Life and death are in the power of your tongue. As you bring your tongue under control of the Holy Spirit and enter a new dimension of authority, you will be able to confront the enemy and, by speaking the Word, destroy his strongholds and conquer him in the power of the Holy Spirit.

By the very words coming out of your mouth, you will be able to demolish every barrier, every obstacle, every work of the enemy that is blocking you from taking possession of God's promises.

God has planned for His Word to dwell in you. As you speak the Rhema Word He directs you to speak, it will be a mighty, indestructible weapon in your mouth driving the enemy out of your life and circumstances.

We are no longer on the defensive, running, afraid, and intimidated by Satan. God has given us the same anointing of the Holy Spirit that was upon Christ. Through the Holy Spirit you have the same power and authority that Jesus had over all the power of the enemy. You have the Living Word within you and God has given you the eternal, all-powerful, living, infallible, impregnable Word through which you are able to tear down Satan's strongholds and fulfill the will of God upon the earth.

The everlasting Word of God has not changed since the day God spoke the worlds into existence. It is timeless! It is not the product of time, but eternity. The Word of God in your mouth today is the same as when Jesus spoke the Word and opened blind eyes, healed the lame, cast out devils and raised the dead. The Word of God today in your mouth...

...Cannot be destroyed!

...Cannot be defeated!

...Cannot be bound!

Make this declaration: "I am speaking the Word of God over my circumstances today. I cannot be destroyed. I can not be defeated. I cannot be bound."

It Will Be Done

Scripture reading: Acts 4:29-33

In this end-time battle where Satan has declared war on the Body of Christ and is pouring out his fury upon God's people in an attempt to bring weariness, confusion, and frustration, God has turned the battle around. When you speak the Word forth in the power and authority that is yours as a son or daughter of the living God, the Word is going to break through every barrier the enemy has built and pull down the strongholds of evil powers and principalities that are over your home, community, city, and nation.

As we take our position of power and authority, God is going to give us spiritual utterance -words of wisdom and knowledge beyond our natural understanding. God has given me four prophecies regarding the new dimension of authority He is bringing us into.

Prophecy #1: As we take our position of power and authority, God is going to give us spiritual utterance – words of wisdom and knowledge beyond our understanding.

Prophecy #2: There will be a new boldness in speaking the Word and seeing blind eyes open; speaking the Word and seeing the crippled walk; speaking the Word and seeing cancer and all types of sickness and disease healed; speaking the Word and seeing demons cast out.

Prophecy #3: There will be a new fearlessness that will enable us to speak the Word, knowing God will protect and deliver us. Regardless of the opposition we face, we will continue to speak the Word! We will not be stopped!

Prophecy #4: The Word of God coming out of our mouths is going to pierce and break through the darkest satanic stronghold. The word in our mouths will be the weapon which will enable us to overcome Satan and all his horde of demons.

When Jesus spoke words in the power and authority of His Father, there was absolutely no margin for error, no faltering, no wondering. He spoke and it was done!

Today, as you speak the Word in the power and authority He has given you, there is absolutely no margin for error. When you know the written, Living Word is in you and you speak the words He directs you to speak in the power and authority of His Name, there will be no struggle, no striving to make it come to pass, no uncertainty. You will speak and it will be done.

Make this declaration: "I have a new position of power and authority, a new boldness, and a new fearlessness. The Word of God coming out of my mouth is piercing the darkest satanic stronghold."

Speak What God Has Said

Scripture reading: Hebrews 4:12-16

The Word of God in your mouth is alive with the power of God. Paul said, *"For the Word that God speaks is alive and full of power – making it active, operative, energizing and effective"* (Hebrews 4:12, TAB). God wants to bring you into a new dimension of authority where your words, spoken with authority and invested in the promises of God, will enable you to confront every stronghold of the enemy.

God's Word – His promises – in your mouth are life! As you speak the promises of God with authority, you are speaking life into your circumstances. You may be facing circumstances in your body where you have a serious physical condition such as heart disease, cancer, diabetes, or some other physical condition that is very painful and hard to cope with. Regardless of how you may feel physically, you must not fall into Satan's trap of confessing and speaking negative words of fear, doubt, and unbelief. You cannot live by how you feel. Instead of continually talking about how bad you feel, all your aches and pains, and the doctor's negative reports, you must begin to speak the Word according to God's promises.

Speak God's promise, *"He was wounded for our transgressions, he was bruised for our iniquities: the chastisement of our peace was upon him; and with his stripes, we are healed"* (Isaiah 53:5).

You may be facing financial problems that in the natural seem impossible. You may be facing a mountain of financial debts without funds to pay them. You may not have enough income to meet the basic needs of your family. It may be that you are discouraged, afraid, and you don't know what to do. You may think you are never going to be able to get out of debt, that there is nothing you can do about your circumstances, and that you might as well give up.

You have a choice! Death and life are in the power of your tongue! Instead of speaking words of fear, discouragement, lack, and doubt concerning your financial problems; instead of speaking about how discouraged, hopeless, and fearful you feel; instead of speaking your discouragement about family relationships – begin to speak forth the eternal, all-powerful, infallible, impregnable, unfailing Word of God. Speak what God has said, according to His promises. Spend time in God's presence, seeking His direction and waiting upon Him to quicken the Word to your spirit. Then speak the Rhema Word, knowing it will be done and that you will have whatever you speak.

In the circumstances you are facing right now, are you speaking words based upon God's promises or upon what you see, feel, or think?

Make this declaration: "The Word of God in my mouth is alive and powerful. I will speak His Word instead of what I see, feel, or think."

DON'T LET SATAN CATCH YOU OFF GUARD

Scripture reading: 2 Samuel 11:1-27

During the past few weeks we have demolished three major strongholds that stand in the way of claiming our spiritual inheritance: Disobedience, the flesh, and the tongue. As we conclude our devotions for this month, I want to warn you that Satan will try to catch you off guard and attack you in each of these areas.

King David, one of the most powerful men of God in the Bible who was called a man after God's own heart (Acts 13:22) fell into Satan's trap through temptation. David's fall came at a time of political and military victory. Have you ever noticed that when everything seems to be caught up, there is money in the bank, and things are going smoothly, how easy it is to slack off in your prayer time and Bible study? Over a period of time, a spirit of complacency sets in and you become an easy target for Satan.

Satan often tempts us after we have experienced a great spiritual victory. This was the case with David. He was tempted at a time of prosperity in the kingdom. He was the king of Jerusalem, lived in a beautiful palace on Mount Zion, had a royal court, and a powerful army. At the time appointed for him to go to battle, he remained in Jerusalem. David had fought and won many battles and apparently he decided it was time to relax.

In our spiritual battle against Satan there is no time for us to relax and let down our guard! The moment we begin to relax – to let down on our prayer time, stop reading the Word, and stop watching for Satan's attacks – the devil will seize the opportunity to attack us. This is exactly what happened to David. Satan set the temptation right in front of David. David took a woman who was not his wife and when she was later found to be pregnant he set her husband up to be killed.

Beloved, don't relax your guard. Don't let down on your prayer and Bible study. Guard what you hear and see. Don't let evil thoughts and desires enter your heart and mind. You must be ever vigilant if you are to remain obedient to God. You must be on guard if you are to conquer the flesh. You must be in control through the power of the Holy Spirit if you are to conquer your tongue.

Make this declaration: "I will remain vigilant. I will maintain control of my tongue through the power of the Holy Spirit."

May

This month you will learn powerful strategies for accessing divine provision from God, drawn from His Word.

May

Accessing Divine Provision

In Joshua chapter 5, God told Israel that the manna He had provided for them in the wilderness would cease. God had provided for His people on one level – the manna level – throughout the forty years of their wilderness journey. Now it was time for Israel to move into a new, supernatural level of provision. They would eat the fruit of their Promised Land:

> *And the manna ceased on the morrow after they had eaten of the old corn of the land; neither had the children of Israel manna any more; but they did eat of the fruit of the land of Canaan that year.*
>
> *(Joshua 5:12)*

Friend, God has provided for us on one level in times past, but as we face the challenges of the new millennium we must learn how to tap into God's supernatural provision. Like Israel, we are "crossing over" into a new era of total provision and continual supply.

You may have been in a financial wilderness for forty years, but God has promised to take you to a land that flows with milk and honey. Hold to the promises God has given to you. Do not be swayed. God is going to bring you spiritually into a land of brooks, water, fountains, vines, fig trees, and pomegranates, a land where you shall eat without want.

During this month you will learn how to access the supernatural provision of God. You will learn about the last great wealth transfer that is going to channel unlimited spiritual resources into the hands of God's people to perpetuate the great end-time, harvest time cycle.

You will learn biblical strategies of sowing and reaping that will affect every area of your life, including your finances.

God does not want you to be constantly struggling financially. He wants you to experience divine sufficiency. He wants you to know how to enter into His divine cycle of supernatural provision – and during this coming month you will learn powerful strategies that will enable you to do this.

When Israel entered their Promised Land, God blessed them with the good of the land. They ate fruit from trees they had not planted. They harvested grain from fields they had not sown. As they battled their way across the land that God had promised them, they didn't have to worry about food for their armies. God supernaturally provided. The wealth of the wicked Canaanites fell into Israel's hands.

As circumstances in the economic world deteriorate, God wants you to know how to tap into His abundant provision. Like Israel, He wants you to eat of the good of the land. No more manna. No more scrambling around each day wondering where you will get your next meal. Abundant provision. Supernatural supply. That is God's promise. It is your inheritance.

Are you ready to claim it?

Can You Take The Blessing?

Scripture reading: Deuteronomy 8:1-10

God is going to have an end-time people. You and I are going to have to know how to go past the natural and into the supernatural, because the natural man is always going to fail.

I challenge you to raise the level of your faith. God reminded Israel: *"Thy raiment waxed not old upon thee, neither did thy foot swell, these forty years"* (Deuteronomy 8:4).

How would you like to have God provide for you a pair shoes so that as your feet grew, the shoes grew? That's what happened in the wilderness! For forty years their clothes and their shoes grew!

If you are to enter into the Promised Land of supernatural provision God has reserved for you, then you must hold on to the Word:

> *Therefore thou shalt keep the commandments of the Lord thy God, to walk in his ways, and to fear him. For the Lord thy God bringeth thee into a good land, a land of brooks of water, of fountains and depths that spring out of valleys and hills.*
> *(Deuteronomy 8:6-7)*

You may be in the wilderness for forty years, but God has promised to take you to a land that flows with milk and honey. Hold to the promise. Do not be swayed. God is going to bring you into a land of brooks, water, fountains, vines, fig trees and pomegranates, a land where you shall eat without want:

> *A land wherein thou shalt eat bread without scarceness, thou shalt not lack any thing in it; a land whose stones are iron, and out of whose hills thou mayest dig brass. When thou hast eaten and art full, then thou shalt bless the Lord thy God for the good land which he hath given thee.*
> *(Deuteronomy 8:9-10)*

As we enter into this month's devotions on supernatural provision, the Holy Spirit is asking, "Can you take the blessing? Are you ready to experience supernatural provision?"

Make this declaration: "I am ready to receive the blessing. I am ready to experience supernatural provision."

Remember God's Promises

Scripture reading: Deuteronomy 8:11-18

God told Israel:

> *And thou shalt remember all the way which the Lord thy God led thee these forty years in the wilderness, to humble thee, and to prove thee, to know what was in thine heart, whether thou wouldest keep his commandments, or no.*
> (Deuteronomy 8:2)

The spirit of remembrance needs to be impregnated into your being. When Israel forgot what God did for them, they stopped worshiping at the altars. When they thought that the blessings they received were at their own hands, they stopped tithing which is a remembrance of what God has given to us. When you stop tithing, you forget that it is God Who has blessed you and that you are not your own source.

Israel was reminded that God led them through the wilderness. He preserved them despite fiery serpents, scorpions, and drought where there was no water. He fed them with manna. When they enter the Promised Land, God warns them:

> *But thou shalt remember the Lord thy God: for it is he that giveth thee power to get wealth, that he may establish his covenant which he sware unto thy fathers, as it is this day.*
> (Deuteronomy 8:18)

Satan's strategy is to try to get you to forget the promises of God, to forget that God always does what He says He is going to do!

During this month, we are going to break the cycle of Satan's financial defeat over your life from now until Jesus comes, and one of the key strategies we will use to defeat the devil is to remember the promises and the actions of God!

You can no longer accept Satan's lie that "You were born to be poor" when you remember that it is God Himself Who gives you the power to get wealth! It is time for you to tell the devil, "Satan, from this day forward you have no right to touch my finances."

The Word of God says that you need to get violent. Inside your spirit, say "I am going to let God raise the level of my faith to where I am going to understand and comprehend once and for all His covenant of blessing for me. By the grace of God, nothing is going to hold me back. I am going to get violent. I am drawing that line and I am saying, "Devil, you stick your ugly head up. I dare you. Raise your head against me. I dare you. Lift your hand against me. I dare you. Lift your hand against my family. I dare you. Lift your hand against my body. Lift your hand against my ministry. Lift your hand against my finances." You have got to get violent!

Make this declaration: "I will be spiritually violent! Satan will not control my finances, those of my family, or those of my ministry."

The Harvest Time Cycle

Scripture reading: Joel 2:21-27

When Israel entered Canaan, God promised them an abundant harvest. The word *harvest* comes from a Greek Word *therismos*. This means to reap and refers to the act of harvesting. In biblical days, the harvest varied according to fluctuating conditions, but usually occurred during the middle of the year. Harvest time was a festive time. There was feasting and rejoicing over God's abundance.

Many people equate the term "harvest time" with the harvest of lost souls throughout the world, but it is much more than that. The end-time, harvest time cycle is a climactic era during which all the teaching God has given will be translated into dynamic action by God's people.

When Israel entered their Promised Land, God set in motion a divine cycle of sowing and reaping. This natural cycle of harvest can be applied spiritually to every area of life – including your finances. Just as God provided for Israel, He wants to provide for you.

God uses the harvest analogy frequently in Scripture because the people of Israel had an agrarian economy. They were farmers, sowers and reapers. They understood the principles of harvest in the natural world, so God used this natural knowledge to draw dynamic parallels which they could understand and apply in the spiritual world.

Strategies are principles, methods, and plans that enable you to reach a certain goal. Strategies for spiritual harvest are revealed in God's Word as natural principles of harvest which parallel spiritual truths.

God wants you to experience abundance in every area of your life, but many people are still struggling to bring in their financial harvest. God wants you to experience complete and total restoration:

> *And I will restore to you the years that the locust hath eaten, the cankerworm, and the caterpillar, and the palmerworm...*
>
> *(Joel 2:25)*

God instituted the basic law of perpetual harvest at the beginning of time. This is the law of sowing and reaping, and as long as the earth remains, this law will be in effect:

> *While the earth remaineth, seedtime and harvest, and cold and heat, and summer and winter, and day and night shall not cease.*
>
> *(Genesis 8:22)*

The seeds for natural, as well as spiritual harvest, must be sown during seed time and harvested during harvest time. The key to the whole process is spiritual timing. Applied in the financial realm, this means that when you understand and get plugged into God's cycle of sowing and reaping, you will experience a perpetual financial harvest.

Make this declaration: "I am getting plugged into God's cycle of sowing and reaping. I will experience a perpetual financial harvest."

Sowing Seed

Scripture reading: Mark 4:26-33

During the next few days we will meditate on principles of sowing and reaping that will enable you to enter into God's divine cycle of sowing and reaping in every area of your life.

Each seed is capable of reproduction: In the natural world, each seed is capable of reproducing itself:

> *And God said, Let the earth bring forth grass, the herb yielding seed, and the fruit tree yielding fruit after his kind, whose seed is in itself, upon the earth: and it was so.*
>
> *(Genesis 1:11)*

Just as each plant in the natural world is capable of reproduction, each financial seed you sow has the ability to reproduce. You have tremendous potential to reproduce when you invest it in God's Kingdom:

> *And he said, So is the kingdom of God, as if a man should cast seed into the ground...and the seed should spring and grow up, he knoweth not how. But when the fruit is brought forth, immediately he putteth in the sickle, because the harvest is come.*
>
> *(Mark 4:26,27,29)*

You must sow in order to reap: In the natural world, you must sow seed in order to reap a crop. In the spiritual world you must also sow in order to reap. God provides the spiritual resources for sowing:

> *And (God,) Who provides seed for the sower and bread for eating will also provide and multiply your (resources for) sowing...*
>
> *(2 Corinthians 9:10, TAB)*

God multiplies your resources so you can invest back into the work of His Kingdom. God does not multiply your finances so you can accumulate great fortunes.

The seed must die: A seed looks dead. There are no green leaves, no tender shoots or branches. In order for a seed to grow, it must be buried. In essence, it must die.

This is a natural parallel of a great spiritual truth. Spiritual life depends on sacrificial death: *"Every time you plant seed you sow something that does not come to life (germinating, springing up and growing) unless it dies first"* (1 Corinthians 15:36, TAB).

In order to bring forth life, it was necessary for Jesus to die on the cross. His ministry seemed in vain because it ended in death, but it was His sowing time. What an abundant harvest resulted from that one kernel of wheat which fell into the ground. Through His death millions have found life!

Sowing your financial seed requires sacrificial death. It requires death to covetousness. It requires death to selfishness. It requires death to holding on to what you have for fear that you won't get more.

Make this declaration: "I am dying to covetousness and selfishness. I am releasing what I have so that God can use it and bless me."

Sowing Beside All Waters

Scripture reading: John 12:23-26

Jim Elliott, martyred for the cause of Christ in an attempt to reach the Auca Indians of Ecuador with the Gospel, wrote in his diary: "He is no fool who gives what he cannot keep to gain what he cannot lose." After Jim Elliott's death, his own wife and many others stepped forward to take his place and win this heathen tribe to the Lord Jesus Christ. Through Jim Elliott's death, salvation came to the Auca Indians.

Jesus said,

> *Verily, verily, I say unto you, Except a corn of wheat fall into the ground and die, it abideth alone: but if it die, it bringeth forth much fruit.*
> *(John 12:24)*

A seed cannot produce anything until it is first sown into the ground and dies. The potential is within each seed to produce a harvest, but it is dormant until it is sown into the soil. Jesus taught the principle of gaining through loss and receiving through giving (Matthew 10:39; Luke 18:29-30).

The harvest principles we are meditating on relate to the economy of God's Kingdom, sowing and reaping and eating the good of the land. Here is another principle to think about:

You must sow regardless of the circumstances: Those who regard circumstances will never enter into the divine cycle of financial harvest. Negative financial circumstances will defeat them, just as a farmer who neglects his field because of the cold, wind, and clouds:

> *He that observeth the wind shall not sow; and he that regardeth the clouds shall not reap.* *(Ecclesiastes 11:4)*

An important strategy of spiritual harvest is that you must sow regardless of your circumstances:

> *Happy and fortunate are you who cast your seed upon all waters (when the river overflows its banks; for the seed will sink into the mud and when of the waters subside will spring up; you will find it after many days in an abundant harvest...)* *(Isaiah 32:20 TAB)*

You must "sow beside all waters" which means in all circumstances. You must sow financially in good times and bad times. Even when the rivers of financial difficulties threaten to overflow the banks of your life, you must continue sowing. God has promised it will yield an abundant harvest.

...Isaac sowed his seed and reaped a harvest in a time of famine.

...Jesus commended the widow lady who did not have much to give because she gave sacrificially.

...The widow with only a little oil and meal used it to minister to God's prophet and received her financial breakthrough.

Make this declaration: "I will sow my finances in good times and bad. I will reap abundantly in good times and bad."

Reaping Abundantly

Scripture reading: Matthew 13:18-23

Today, we continue meditating on natural principles of harvest and applying them spiritually.

You must cultivate the new seed: After a farmer puts seed in the ground, he cultivates it. He waters it, hoes the weeds, and breaks up the ground so it will grow. He guards the seeds and young plants from pests and disease.

Each day during this month, you are receiving tremendous seeds of revelation regarding your finances in this devotional guide. If they are to produce a harvest in your life, however, you must cultivate them. You must guard them from the enemy and refuse to let him snatch them from you. (Take time right now to study more about this in Matthew 13:18-23.)

You reap in proportion to the amount sown: Seed reproduces after its own kind in the natural world. If you plant apple seeds you will harvest apples. The same is true in the spirit world:

> *Be not deceived; God is not mocked: for whatsoever a man soweth, that shall he also reap.*
>
> *(Galatians 6:7)*

If you plant abundantly, you will reap abundantly:

> *(Remember) this: he who sows sparingly and grudgingly will also reap sparingly and grudgingly, and he who sows generously and that blessings may come to someone, will also reap generously and with blessings.*
>
> *(2 Corinthians 9:6, TAB)*

This principle applies in every area of spiritual life. If you give grudgingly of your time, talents, and finances to the work of God, then your harvest will be small. If you sow generously, you will reap abundantly.

There are four levels of productivity in spiritual harvest described in John 15. You can bring forth:

- Fruit: John 15:2
- More fruit: John 15:2
- Much fruit: John 15:5
- Permanent fruit: John 15:16

In the realm of finances, God's desire is that you sow abundantly so you can reap abundantly.

Make this declaration: "I will sow generously and I will reap fruit, more fruit, and much fruit. My fruit will be permanent. Satan will not be able to steal it."

WAITING FOR THE HARVEST

Scripture reading: Hebrews 6:13-15

In the beginning God established the seasonal principle of seed time and harvest in the natural world. This brings us to our next principle of spiritual harvest:

You do not harvest in the same season you sow: Sowing is necessary in order to have a harvest, but the harvest does not come in seed time. Many who try to enter into this cycle of financial harvest become discouraged because they do not understand this principle. They give sacrificially, but don't see immediate results.

After sowing, you must wait for the seed to germinate, just as a natural seed germinates when it is put into the ground. Psalms 1 speaks of this process in the life of believers:

> *And he shall be like a tree planted by the rivers of water, that bringeth forth his fruit in his season...* *(Psalm 1:3)*

If there are fruitful seasons of life, then this means there are also periods which seem unfruitful. These are periods of preparing the ground, sowing, cultivating, and waiting for the crop to grow and the harvest to come in. But these seasons are not wasted. They are important and necessary to bring in the harvest.

The Apostle Paul admonished:

> *Let us not be weary in well doing: for in due season we shall reap, if we faint not.* *(Galatians 6:9)*

You will reap if you are patient: One vital key to your financial harvest – your spiritual breakthrough – is timing. Timing is the hidden, mysterious ingredient to a supernatural harvest. Timing requires patience. Patience is the forgotten, but powerful season, between sowing and reaping. Patient waiting reveals your trust in God:

> *The LORD is good unto them that wait for him, to the soul that seeketh him. It is good that a man should both hope and quietly wait for the salvation of the LORD.* *(Lamentations 3:25-26)*

Sometimes waiting may be painful, but our spiritual walk is a life-long journey – it is one long continual cycle of spiritual harvest, sowing and reaping, seed time and harvest.

Some of you have lost your harvest because of the pain of waiting, but if you will just hold on, the joy of harvest will surely come:

> *...weeping may endure for a night, but joy cometh in the morning.* *(Psalm 30:5)*

Waiting is burdensome and has a tendency to breed frustration, doubt, unbelief, anger, and criticism. All of these can abort your harvest. You may not have the resources at hand immediately, but you must learn to trust God, just like the farmer who waits patiently for the seed to sprout.

Make this declaration: "I am not weary in well doing because I know I will reap if I am patient. I am like a tree planted by the river and whatever I do will prosper."

Winning The War For Your Finances

Scripture reading: Isaiah 55:8-13

As you wait for your financial harvest, you must realize that God has divine purposes in the waiting period – even though you may not understand the delay:

> *For my thoughts are not your thoughts, neither are your ways my ways, saith the Lord. For as the heavens are higher than the earth, so are my ways higher than your ways, and my thoughts than your thoughts. For as the rain cometh down, and the snow from heaven, and returneth not thither, but watereth the earth, and maketh it bring forth and bud, that it may give seed to the sower, and bread to the eater: So shall my word be that goeth forth out of my mouth: it shall not return unto me void, but it shall accomplish that which I please, and it shall prosper in the thing whereto I sent it.*
>
> *(Isaiah 55:8-13)*

While waiting for your financial harvest to mature, don't doubt. Don't criticize. Don't give up. Have faith! God's Word will be fulfilled just as He promised!

While you are waiting, refuse to sin with your mouth. Your enemy reacts to your words. There is tremendous power in words:

> *Death and life are in the power of the tongue: and they that love it shall eat the fruit thereof.*
>
> *(Proverbs 18:21)*

If you feed on doubt and unbelief, you energize Satan. Don't speak words that encourage your enemy because...

> *...by thy words thou shalt be justified, and by thy words thou shalt be condemned.*
>
> *(Matthew 12:37)*

Instead of grumbling and complaining while you wait, praise God for the harvest you are going to reap:

> *In every thing give thanks: for this is the will of God in Christ Jesus concerning you.*
>
> *(1 Thessalonians 5:18)*

If you are faithful, God has promised you will reap your harvest.

Winning the war for your finances – as in any area of life – involves more than one battle. You may have ups and downs during this battle, but remember...

> *The Lord upholdeth all that fall, and raiseth up all those that be bowed down.*
>
> *(Psalm 145:14)*

Never give up. Keep getting up again, and again, and again.

You can make no plans for retreat. You can only advance. Your miracle is on its way!

Make this declaration: "I have no plans for retreat. I can only advance. I will never give up."

Reaping A Mega Harvest

Scripture reading: Psalm 37:16-19

Here are some more important principles that will assure abundant spiritual harvest in every area of your life – including your finances:

You must reap during harvest time: You must sow in seedtime and you wait patiently during the time of growth and development – but you must also recognize when it is harvest time and act promptly before the crop perishes.

The Bible reveals that there is…

> *A time to plant, and a time to pluck up that which is planted.*
> *(Ecclesiastes 3:2)*

This is true in both the natural and spiritual worlds. The harvest must be reaped when it is ripe or it will rot in the fields. You can lose your harvest if you don't act in time!

Now is your time. It is harvest time! Don't miss the supernatural cycle of this spiritual season. This is the end-time harvest and God is going to have a people. You are part of that people, that chosen generation!

God has a work for you to do, but in order for you to accomplish it, you must have a breakthrough in your finances. You must be set free from debt. That is why God wants to release an end-time, harvest time blessing upon you. It is time for the fulfillment of the words of the prophet Amos:

> *Behold, the days come, saith the LORD, that the plowman shall overtake the reaper, and the treader of grapes him that soweth seed; and the mountains shall drop sweet wine, and all the hills shall melt.*
> *(Amos 9:13)*

The financial seeds we are sowing now into God's Kingdom are so supernatural that the harvest will begin as soon as it is sown – as soon as you release the seed. God is going to supernaturally multiply what we sow into His Kingdom.

Take some time now to meditate on the principles of sowing and reaping which we have learned during the past few days:

- Each seed is capable of reproduction.
- You must sow in order to reap.
- The seed must die.
- You must sow regardless of the circumstances.
- You must cultivate the new seed.
- You reap in proportion to the amount sown.
- You do not harvest in the same season you sow.
- You will reap if your are patient.
- You must reap during harvest time.

Make this declaration: "I am receiving God's principles of sowing and reaping. They will become a reality in my spiritual life."

God's Supernatural Cycle

Scripture reading: Philippians 4:11-19

During each Sabbath Year and the Jubilee Year, Israel did not sow or reap. They depended solely upon the supernatural provision of God to meet their needs. As spiritual Israel, there are two spiritual cycles that have the potential to affect your finances, and you will be plugged in to one or the other of these two supernatural cycles. You will either be caught up in the cycle of financial defeat or God's cycle of supernatural provision.

Satan's cycle of financial defeat: Whenever Satan attacks your finances and you struggle with debt, then fear, worry, and doubt begin to enter in. If you keep your eyes on your negative financial circumstances, you will stop giving to God. This blocks the flow of His provision in your life and leads to increased debt. Unbroken, the cycle will continue to repeat and you will incur even more debt.

God's cycle of supernatural provision: The second cycle that has the potential to affect your finances is the cycle of God's supernatural provision. Beloved, God wants you to break out of Satan's cycle of financial defeat and enter into this divine cycle!

To break out of the cycle of financial defeat you must cast out fear, worry, and doubt; get your eyes off your circumstances; and continue to give to God, expecting, believing, and trusting Him to fulfill His promises.

You must declare war on the devil's war against your finances and face your financial problems from a new position of strength! To do this, you must get plugged in to God's divine cycle of giving and receiving. In the midst of financial crises – even poverty – you must continue to obey God and give your tithes and offerings. When you do this, you can be at rest knowing God will make total provision for your needs through His covenant with you.

Here's how God's supernatural cycle of provision works: When you give to God, you receive from Him supernatural provision and increase. When you give to God from this increase, you receive additional supernatural provision and increase – and the divine cycle continues.

When you get plugged into this cycle, you can declare, as Paul did, *"I have all, and abound"* (Philippians 4:18). You can know that God will *"supply all your need according to His riches in glory by Christ Jesus"* (Philippians 4:19).

Make this declaration: "I am breaking out of Satan's cycle of financial defeat. I am entering into God's cycle of supernatural provision."

Divine Sufficiency

Scripture reading: Acts 2:44-45; 4:34-35

You cannot face your financial problems with "head knowledge" because it doesn't work! You must learn to take the very thing Satan wants to defeat you with – your finances – and use them to strengthen your faith!

Most people facing financial need in their lives – a mountain of overdue bills, mounting debts, a lack of sufficient finances – are looking for more money to help them out of their financial problem. They think borrowing, changing jobs, or manipulating circumstances will solve their problems.

Yet more money is not the answer. You may get more money, but that, too, will soon be gone unless you learn how to enter into God's Cycle of Supernatural Provision. God wants you to enter into a new realm of the Spirit: Divine provision. It is a supernatural experience where you can rest from your financial struggles, even as Israel rested in their Year of Jubilee.

The Early Church experienced God's Cycle of Supernatural Provision and had divine sufficiency. Following Pentecost, Christians sold their houses, lands, and possessions and placed the proceeds in a common fund. They did not selfishly cling to their worldly possessions, but gave them willingly.

From this fund, distribution was made according to need: *"And all that believed were together, and had all things common; And sold their possessions and goods, and parted them to all men, as every man had need"* (Acts 2:44-45). As a result, there was not any among them who lacked (Acts 4:34).

There was supernatural provision and divine sufficiency because the Church was functioning in unselfish unity as God intended. Believers were walking together in covenant relationship with one another. They had all things in common. They did not consider their possessions solely their own, but as gifts from God for the good of all. If a Christian brother or sister needed clothing, they provided it. If someone needed shelter, they opened their homes. If food was needed, they shared what they had.

These believers did not allow Satan to use their financial needs to keep them from fulfilling the commission Christ gave them. They continued to preach and teach the Word of God from house to house until they turned the world upside down.

Later, the believers in Jerusalem and Judea were facing a great financial crisis through war, famine, and persecution. Many of them had lost their means of support – their jobs and businesses – because of their refusal to deny their faith and their commitment to Jesus Christ. Yet even in the midst of their great financial need, there was divine sufficiency!

God wants you to experience this divine sufficiency also, but you will only experience it as you enter into God's supernatural cycle of provision.

Make this declaration: "God wants me to experience divine sufficiency as I enter into His supernatural cycle of provision."

The Key To The Cycle

Scripture Reading: 2 Corinthians 8:2-5

Be sure to read the scripture reading for today before you read this devotion. In this chapter, Paul was writing to the believers in Corinth concerning a contribution they began to gather for the Christians in Jerusalem but had not yet finished collecting. To encourage the Corinthians to complete this offering, Paul used the churches in Macedonia – Philippi, Thessalonica, Berea and others – as examples.

Verses 2-5 in the Amplified Version give us a better understanding of the original Greek:

> *For in the midst of an ordeal of severe tribulation, their abundance of joy and their depth of poverty (together) have overflowed in wealth of lavish generosity on their part. For, as I can bear witness, (they gave) according to their ability, yes, and beyond their ability; and (they did it) voluntarily, Begging us most insistently for the favor and the fellowship of contributing in this ministration for (the relief and support of) the saints (in Jerusalem). Nor (was this gift of theirs merely the contribution) that we expected, but first they gave themselves to the Lord and to us (as His agents) by the will of God – that is, entirely disregarding their personal interests, they gave as much as they possibly could, having put themselves at our disposal to be directed by the will of God.*
>
> *(2 Corinthians 8:2-5, TAB)*

These believers gave liberally to meet the needs of the Church at Jerusalem, knowing that God would supernaturally provide for them.

The Macedonian congregations who gave to the Jerusalem Church were themselves facing a time of great trial and poverty. Macedonia had been overrun by armies, the country had been devastated, and the inhabitants stripped of their wealth, but their focus was not on their own financial situation. They were not moaning, groaning, or complaining because of Satan's attacks on their finances. They were rejoicing! They knew God would supply their need!

Their eyes were not on their own meager resources, but on God's supernatural provision. They gave liberally out of their deep poverty, trusting God to provide supernaturally for them.

One of Satan's major strategies to keep Christians in financial bondage is to convince them they cannot afford to give. He tries to keep you focused on your own financial needs so you stop giving or give sparingly because your eyes are fixed on your limited finances instead of God's promises. You can't afford not to give because it is the key to this divine Cycle of God's Supernatural Provision.

Make this declaration: "I can't afford not to give because it is the key to the divine Cycle of God's Supernatural Provision."

Give Your Best To God

Scripture reading: Romans 12:1-3

There are several principles that the New Testament church at Macedonia followed in their giving that we should emulate in our own giving. We will be meditating on these truths during the next few days.

They gave liberally: The New Testament church at Macedonia would never have given anything had they limited their giving by their financial circumstances. Although they were in deep poverty, they did not give a small token offering. They gave sacrificially, even beyond their ability.

This is what you must do if you are to receive your financial breakthrough. You must declare war on the devil's war against your finances, get your eyes off your own financial needs and limited resources, and give as God directs.

They gave willingly: Not only did the believers in the Macedonian churches give liberally beyond their ability, they also gave willingly. Many who are experiencing financial problems resent being asked to give. The believers in Macedonia did not consider giving as something to be resented or avoided; rather a privilege to be sought after.

In the Church today, our whole system of giving is based on the strength of emotional appeals. Christians wait to be urged, begged, and convinced to meet the needs of their brothers and sisters in other parts of the world who are crying out for the tools and training to win their nations for God. Yet God wants us to give liberally and willingly, not grudgingly or based on how moving the emotional appeal might be.

What was it that enabled the Early Church to give liberally and willingly in the midst of their own great need? They gave their best to God. They gave themselves to the Lord. God does not want your money first. He wants you!

They gave sacrificially: The Macedonians gave willingly and unreservedly because they had totally surrendered themselves to God. This is one of the most vital keys to walking in financial victory. To enter into God's Cycle of Supernatural Provision you must first offer yourself as a living sacrifice upon the altar of God. Paul urged the Romans:

> *I beseech you therefore, brethren, by the mercies of God, that ye present your bodies a living sacrifice, holy, acceptable unto God, which is your reasonable service.*
>
> *(Romans 12:1)*

There are Christians today who think they can buy their way into God's favor. They give large offerings, but have not fully surrendered themselves to God. They are withholding areas of their lives, thinking their gifts will make them acceptable to Him. Don't be one of them. Give your best to God!

Make this declaration: "I will give liberally, willingly, and sacrificially. I will give my best to God."

According To Your Ability

Scripture reading: 2 Corinthians 9:6-11

Just as God required the Israelites to offer unblemished sacrifices, He requires us today to give Him offerings that are unblemished. We must not withhold anything, but must give Him our very best. God would not accept from the Israelites animals which were blemished as sacrifices. Neither would He accept offerings which were not given in proportion to the individual's ability to give. He required their very best. If they had the ability to offer a bullock or a lamb, He would not accept a pigeon.

Today we continue meditating on principles that the New Testament church at Macedonia followed in their giving which we should emulate in our own giving.

They gave cheerfully: God loves a cheerful giver (2 Corinthians 9:7) – one who gives willingly and ungrudgingly, without compulsion. He approves of those who, instead of waiting to be asked, look for opportunities to give to the work of God.

They followed Christ's example: After citing the churches of Macedonia as an example, Paul set forth Christ as the supreme example for our giving:

> *For ye know the grace of our Lord Jesus Christ, that, though he was rich, yet for your sakes he became poor, that ye through his poverty might be rich.*
>
> *(2 Corinthians 8:9)*

There can be no greater example for us to follow in our giving than that set by the Lord Jesus. He gave up His supreme position and unmeasurable riches in Heaven, came to earth as a man, and lived in poverty (according to the world's standards) so that we might be made rich. He not only took our sins and sicknesses upon Himself, He also took our poverty.

They gave proportionately: Following these examples of giving, Paul encouraged the Corinthian believers to finish the collection they started a year before for the relief of the saints in Jerusalem. Although they had been first to respond to the need of their brethren in Jerusalem and were anxious to give, a year had passed and little had been done. Paul told them:

> *So now finish doing it, that your (enthusiastic) readiness in desiring it may be equaled by your completion of it according to your ability and means. For if the (eager) readiness to give is there, then it is acceptable and welcomed in proportion to what a person has, not according to what he does not have.*
>
> *(2 Corinthians 8:11-12, TAB)*

Paul urged the Corinthians to give according to their means, in proportion to what they had. This is God's plan of supernatural provision. He does not intend giving to be a hardship on His people. He does not want to increase the financial burdens we carry, but rather to relieve them. In His plan of supernatural provision, He calls for every member of the Body of Christ to give according to their ability.

Make this declaration: "I will follow Christ's example. I will give cheerfully and proportionately. I will give according to my ability."

Supernatural Multiplication

Scripture reading: Luke 5:4-11

This is God's plan for us today. It is how we enter into the cycle of His supernatural provision. As each member of the Body of Christ gives…

…sacrificially
…liberally
…willingly
…the best he has
…cheerfully
…following Christ's example
…proportionately

…then there will not only be supernatural provision to meet your needs, but also to finish the work God has given us to do in evangelizing the world.

When you enter into this divine cycle, God will literally multiply your finances. The Greek word for multiplication is "*plethuno*" which signifies "to increase, to cause to multiply." Within God's Cycle Of Supernatural Provision, a supernatural multiplication occurs.

After the Day of Pentecost when the Holy Spirit was outpoured, there was supernatural multiplication which took place as God added to the Church daily:

> *And the word of God increased; and the number of the disciples multiplied in Jerusalem greatly… (Acts 6:7) And believers were the more added to the Lord, multitudes both of men and women.* *(Acts 5:14)*

This multiplication was not the result of man's efforts but was a supernatural move of God. God multiplied the number of disciples, *"And the Lord added to the church daily such as should be saved"* (Acts 2:47).

This is the type of increase occurring during this end-time, harvest time cycle. We are witnessing the greatest multiplication of souls we have ever seen. It is not the result of man's efforts, but it is supernatural multiplication by God's Spirit.

The same supernatural principle of multiplication occurs when you give to God. Not only does God provide the seed for you to sow, He multiplies back to you what you give until there is divine sufficiency in your life:

> *And (God) Who provides seed for the sower and bread for eating will also provide and multiply your (resources for) sowing…*
> *(2 Corinthians 9:10, TAB)*

The multiplication that occurs is not according to man's measurement, but according to God's unlimited supply. God multiplies what you give until there is abundant provision. Not only are your needs met, but you have more to sow back into the Kingdom of God.

Make this declaration: "God, who provides seed for the sower and bread for eating, will also provide and multiply my resources for sowing."

Act Upon The Word

Scripture reading: John 6:1-13

In John 6:1-13, we find a hungry multitude and a meager food supply of five loaves of bread and two fish. Let these loaves and fish represent your own limited financial resources. You may look at your finances, and think how insufficient they are to meet the needs you face right now. You may look at your limited income and your small offerings for the work of God, and wonder, "What can my small gift do to meet such a great need as evangelizing the world?"

But the little boy with the bread and fish gave his limited resources to Jesus. The Lord took them, praised God for His supernatural provision, and multiplied the food until there was an abundance. Not only was there enough for more than five thousand people to eat and be satisfied, there were twelve baskets of food left! How would you like God to multiply what you give until all your needs are supplied with twelve baskets left over?

Paul declared:

> *... Now he that ministereth seed to the sower both minister bread for your food, and multiply your seed sown, and increase the fruits of your righteousness;) Being enriched in every thing to all bountifulness, which causeth through us thanksgiving to God.* *(2 Corinthians 9:10-11)*

When you give from your limited resources, believing God for His supernatural provision in your life, He will multiply back into your life His abundance.

God wants to you to bring your five loaves and two small fish – your limited finances – to Him. As you give what you have, He will multiply it and you will literally be catapulted into the divine Cycle of God's Supernatural Provision.

The Church has sat back passively for too long. We have let Satan run roughshod over us. We need to learn to say what we want from God, and then we need to manifest it. No longer can we hide. It is time to come out from the closet and let the world know who we are – the children of the living God!

God hath not given to us the spirit of fear. We are not afraid of what the devil can do, or what he can say, or how he can act. The God you and I serve has given us His power!

Fear is a demon spirit that controls, dominates, possesses, oppresses, vexes, and torments eight out of ten people in the world today – including born-again Christians! Fear binds you. Nothing I know of in this world can bind us any more than fear. Fear keeps us from walking out and taking that step of faith. Fear short-circuits the Word of God. If you are to claim the promises of God then, like the little lad with his loaves and fishes, you must step out in faith.

You must get your eyes off your circumstances – off of the bigness of your debt and onto the supernatural provision of Almighty God! The time for you to act upon God's Word is now!

Make this declaration: "I am taking my eyes off my financial circumstances and focusing them onto the supernatural provision of Almighty God."

Financial Jubilee

Scripture reading: Leviticus 25:8-19

One of the greatest keys to spiritual harvest is timing. God always works in times and seasons. He moves in cycles with specific beginning and ending points. Within these divine cycles, various supernatural things occur which lead us to conclude that these are God's cycles, not man-made cycles.

Israel was God's chosen people and He established for them a Jubilee – a supernatural cycle – during which specific, liberating events were to occur. As a born-again, Spirit-filled believer, you are spiritual Israel – chosen and ordained of God – not because your flesh is circumcised – but because your heart is circumcised by the blood of Jesus:

> *In whom also ye are circumcised with the circumcision made without hands, in putting off the body of the sins of the flesh by the circumcision of Christ.*
>
> *(Colossians 2:11)*

When Moses went up to Mount Sinai to meet with God, he was given instructions regarding specific days, times, and seasons which were to be consecrated and set apart as holy days. Each of these appointed days was very important to God and had a specific purpose. The observance of these days were foreshadows of God's plan of redemption through Christ and revealed His desire to pour out the fullness of His blessings upon His people.

Among these special observances was the Year of Jubilee. It was not just a man-made tradition or ceremony, but it was God-ordained! God planned for Israel to observe it throughout the generations as a testimony of God's faithfulness and supernatural provision.

The word "jubilee" is taken from the Hebrew word "yovel," and over the years, it became known as a year of liberty, freedom, and victory! It means *"to cause to bring back."*

In Leviticus 25, God established five essential features which characterized Jubilee in ancient Israel:

1. **Release:** Liberty was proclaimed to all who were in bondage – those enslaved were set free.
2. **Restoration:** Lost possessions were returned – property lost through debt or poverty reverted to its original owner.
3. **Debt cancellation:** All debts were canceled.
4. **Protection:** The people would dwell in the land in safety because the enemy – who came to rob, steal, and destroy – would be defeated.
5. **Increase:** There would be divine sufficiency, a supernatural cycle of God's provision.

As New Testament Israel, God wants you to experience financial jubilee – continual provision, total supply – as did His people, Israel.

Make this declaration: "I am entering into a financial jubilee of continual provision and total supply."

Spiritual Israel

Scripture Reading: Deuteronomy 15:1-23

During Israel's Jubilee, there was to be a release of property, debt, slavery – all that had been lost was to be restored. This release was not just something God suggested that Israel do, but it was *"the Lord's release!"* God was declaring complete and total liberty for His people!

God wants you to experience a Spiritual Jubilee. It is time for you to reclaim everything the devil has stolen from you!

...Reclaim your health.

...Reclaim your family.

...Reclaim your ministry.

...Reclaim it all!

God wants you to have liberty in every area of your life...physically, mentally, spiritually, and financially. He wants you to experience total freedom from every bondage of the enemy – sinful habits, wrong attitudes and emotions, and your past!

This is the time for taking back all Satan has stolen from you! It is time to take back property, finances, lost loved ones, ministries, dreams – anything Satan has robbed you of in the past. It is time for the devil to give it all back!

God wants you to experience a new, fresh release of the Spirit. He wants you to enter into a new dimension of rest from your financial struggles and get plugged in to God's Cycle of Supernatural Provision.

Too many times we are confused about God's will. We question...

...Is it God's will for my family to face these financial difficulties?

...Is it God's will for my children to live in sin, rebellious, and away from God?

...Is it God's will for me to have this physical infirmity in my body?

Often, we are mistakenly taught that these things are the will of God. Yet God never intended you to live in defeat. He does not want you to be in poverty, barely getting by in this life. He never intended you to be held in the cruel bondage of sin and sickness. Jesus died for your sin on the cross of Calvary. He bore your sickness. He bought your freedom! Total. Complete. No exceptions!

How can you know that this Spiritual Jubilee is God's will for you? God established Jubilee for His children – Israel – and as a New Testament believer, you are spiritual Israel! The same promises God gave to Abraham are yours because you are a spiritual Israelite (Galatians 3:29). You have an even greater covenant established on new and better promises through the Lord Jesus Christ (Hebrews 8:6).

Make this declaration: "I am reclaiming my health, my family, and my ministry. He bought my freedom, so I can experience jubilee in every area of my life."

Don't Lose Hope

Scripture reading: John 5:1-9

Eighty percent of the letters I receive in my office concern financial problems. This is how I know that many of you are still struggling in this area of your lives. Did you know that God wants to prosper you in every area – physically, mentally, spiritually, and financially?

You may say, "Brother Cerullo, I've struggled with finances – I've battled for so long. How can I really know that God wants me to prosper?"

Listen to the Greek translation of 3 John 2: *"Beloved, I pray that you may prosper in every way. And that your body may keep well even as I know your soul keeps well and prospers."*

In this passage God reveals His desire for you to prosper, be in good health, and for your mind, body, soul, and spirit – including your finances – to be kept healthy, even as He has provided for the salvation of your soul!

It is time for spiritual harvest in every area of your life – and that includes your finances. This is God's will for you! Regardless of the financial pressures you face, before you can experience the breakthrough you need you must know it is God's will that you prosper. Whatever crises you face regarding your finances you must take your eyes off your circumstances, set your focus on God as your divine Source of provision, and act upon God's promises to you in His Word.

You must face your financial problems with a solid scriptural foundation concerning God's will for your finances. If you are not convinced that it is God's will to supply your needs and prosper you, then when Satan attacks your finances your faith will waver and you will begin to worry. You must settle in your heart and mind – to know beyond a shadow of doubt – that God wants you to prosper.

There are specific reasons why people do not move into their end-time financial harvest…

…Some don't believe God wants them to prosper financially.

…Some continue to rely on their own ability instead of the supernatural power of God.

…Some refuse to sow during times of crisis. They hoard their limited resources instead of giving sacrificially to God.

…Others wait too long and the timing of the anointing passes them by. The supernatural cycle concludes, and they have missed it!

You may have struggled with financial problems so long that you have given up hope of anything ever changing and tried to just adjust to your situation. Remember the crippled man at Bethesda? He had given up all hope of getting well, but then one day – in the perfect timing of God – he got his breakthrough. He was totally healed, took up his bed, and walked! God wants to touch your finances supernaturally in the same way. Don't lose hope.

Make this declaration: "I will not lose hope. God will touch my finances supernaturally. I declare it upon the authority of His Word."

God's Commanded Blessing

Scripture reading: Deuteronomy 7:12-23

From the time Israel became a nation, God expressed His desire to bless His people. He declared:

> *Wherefore it shall come to pass, if ye hearken to these judgments, and keep, and do them, that the LORD thy God shall keep unto thee the covenant and the mercy which he sware unto thy fathers: And he will love thee, and bless thee, and multiply thee: he will also bless the fruit of thy womb, and the fruit of thy land, thy corn, and thy wine, and thine oil, the increase of thy kine, and the flocks of thy sheep, in the land which he sware unto thy fathers to give thee.*
>
> *(Deuteronomy 7:12-13)*

Does this sound like a God who wants His people to live in poverty? No! It is God's plan for you to have sufficiency in every area of your life – and that includes finances:

> *And God is able to make all grace abound toward you; that ye, always having all sufficiency in all things, may abound to every good work.*
>
> *(2 Corinthians 9:8)*

God said that His people will…*"not be ashamed in the evil time: and in the days of famine they shall be satisfied"* (Psalm 37:19).

As a born-again, Spirit-filled believer, you are no longer under Satan's control. He is not in control of your life or your finances.

You are an heir of God's promises. In the natural world, when someone is named as an heir, certain conditions are usually set forth which must be met in order for the inheritance to be claimed. The same is true in the Spirit world. Your finances won't just automatically be blessed. God has certain conditions that must be met in order to receive the blessing:

> *And it shall come to pass, if thou shalt hearken diligently unto the voice of the LORD thy God, to observe and to do all his commandments which I command thee this day, that the LORD thy God will set thee on high above all nations of the earth: The LORD shall command the blessing upon thee in thy storehouses, and in all that thou settest thine hand unto; and he shall bless thee in the land which the LORD thy God giveth thee. And the LORD shall make thee plenteous in goods, in the fruit of thy body, and in the fruit of thy cattle, and in the fruit of thy ground, in the land which the LORD sware unto thy fathers to give thee.*
>
> *(Deuteronomy 28:1,8,11)*

You cannot walk in disobedience and expect God to bless you. God's blessings are commanded upon you when you are obedient to His Word – all of His Word – including biblical commands concerning finances.

Make this declaration: "I will walk in obedience to God and He will bless me financially and in every area of my life."

Get It Right

Scripture reading: Malachi 3:8-10

During the restoration of Jerusalem after seventy years of Babylonian captivity, God used Nehemiah, Haggai, and Malachi to reveal why Israel was not prospering. There were many varied reasons: They were offering polluted sacrifices, failing to keep the Sabbath, committing immorality, and men were divorcing faithful wives. But one major area that God emphasized was their disobedience in the area of finances:

> *Will a man rob God? Yet ye have robbed me. But ye say, Wherein have we robbed thee? In tithes and offerings. Ye are cursed with a curse: for ye have robbed me, even this whole nation. Bring ye all the tithes into the storehouse, that there may be meat in mine house, and prove me now herewith, saith the LORD of hosts, if I will not open you the windows of heaven, and pour you out a blessing, that there shall not be room enough to receive it.*
>
> *(Malachi 3:8-10)*

God declared that if His people would come into obedience in every area of their lives – and specifically in their giving of tithes and offerings – He would open the windows of Heaven and pour out a blessing upon them.

God's Word is His bond and He will not go against His Word to meet your need. No matter what dire situation you are in, He will not disregard His Word for your sake. He does not make exceptions. He commands obedience in tithes and offerings – but He isn't looking to take from you, He is looking to give to you. He wants to open His windows of financial blessing upon you!

But if you are to experience this great financial Jubilee, then you must be willing to obey God and surrender every area of your life – including your finances. You must exercise your will and choose God's way in order to receive a supernatural breakthrough in your finances:

> *See, I have set before thee this day life and good, and death and evil; In that I command thee this day to love the LORD thy God, to walk in his ways, and to keep his commandments and his statutes and his judgments, that thou mayest live and multiply: and the LORD thy God shall bless thee in the land whither thou goest to possess it.*
>
> *(Deuteronomy 30:15-16)*

This verse does not present us with options or suggestions, but it commands...

1. Love the Lord your God.
2. Walk in His ways.
3. Keep His commandments.

These are the criteria. God's people had a choice. They had to exercise their will. God told them, "You have robbed Me. You failed to keep My commandments. If you get it right this time, I will open the windows of Heaven and send you a blessing that there is no room to receive!"

Make this declaration: "I will love the Lord, walk in His ways, and keep His commandments. I will not rob God. I will be blessed under the open windows of Heaven."

Today Is The Day

Scripture reading: Deuteronomy 4:22-29

You have the power of choice regarding your finances. You must get your eyes off your financial problems and obey the commandments of God's Word which declare that if you are faithful in tithes and offerings, you will be blessed. God has set before you financial death or life – which way will you choose?

If you are obedient to God's commandments – including those pertaining to tithes and offerings – you will possess the spiritual territory for which you are battling: *"God shall bless thee in the land whither thou goest to possess it"*... *(Deuteronomy 30:16)*

...That lost loved one.

...The physical healing you need.

...That struggling business.

...Your ailing finances.

God said, *"See, I have set before thee this day life and good, and death and evil." (Deuteronomy 30:15)* This day – today! Again – God is a God of specific timing. Today is the day to make your choice.

From Genesis to Revelation, God clearly reveals that financial blessing is His will. The prosperity He has planned for you to enjoy is total provision and well-being – abundant blessings, both temporal and spiritual. If you will receive this prophetic revelation into your spirit, you will prosper – the Word guarantees it:

> *Believe in the LORD your God, so shall ye be established; believe his prophets, so shall ye prosper.*
> *(2 Chronicles 20:20)*

Satan wants to hinder you from walking in the financial freedom and victory Christ has provided for you by making you so overly burdened and worried about providing for your needs that you begin to doubt God's provision, doubt God's promises, and doubt His love for you.

Instead of trusting and believing God to provide your needs, Satan wants you to struggle, worry, and depend on your own limited abilities to provide for yourself. The devil's objective is to keep your mind in a constant state of turmoil so that you fail to release your faith and receive from God what you need.

The moment you begin to worry about your finances, set your eyes on your circumstances, or allow Satan to fill your mind with worry and doubt, you will be caught in a cycle of defeat. To break out of this cycle, cast worry and doubt out of your life and get your eyes off your circumstances. Present your needs to the Father in faith today, knowing He will do exactly as He has promised.

Make this declaration: "Today, I am breaking out of the cycle of financial defeat. I am casting out worry and doubt and focusing on my Heavenly Father in faith."

Satan Can't Stop It

Scripture reading: Genesis 13:1-18

God has promised that the wealth of the wicked (the world) will be transferred into the hands of the righteous: *"... and the wealth of the sinner is laid up for the just"* (Proverbs 13:22). The Patriarch Abraham is an excellent illustration of this.

From the beginning of God's dealings with Abraham when He entered into covenant relationship with him, God prospered him. The transfer of the wealth of the wicked began when Abraham obeyed God, left his own country, and went to Canaan. God told Abraham that He was going to give him the wealth of Canaan (Genesis 13:15). The Promised Land, with all its wealth, did not belong to the Canaanites. The wealth of the wicked was just being stored for the righteous!

Later, when there was a great famine in Canaan, Abraham went down to Egypt. You know the familiar story of how he told the Egyptians Sarah was his sister because she was beautiful and he was afraid they would kill him and take her for themselves. Pharaoh lavished gifts upon Abraham because of Sarah. He gave him sheep, oxen, donkeys, camels, and servants (Genesis 12:16). When the Lord sent plagues upon Pharaoh because of Sarah, Pharaoh sent Abraham and his family out of Egypt. When Abraham left Egypt, he took with him great herds of cattle and other livestock, servants, gold and silver (Genesis 13:2).

God so blessed Abraham while he was in Egypt that the land where he and his nephew, Lot, settled was not big enough for them and they had to go separate ways (Genesis 13:6). Later, when Lot was taken captive along with all his possessions, Abraham pursued the enemy, rescued Lot, and took back all his belongings plus the spoils of Sodom and Gomorrah. The wealth of the wicked was transferred into Abraham's hands!

Generation after generation, we see this transfer continue. God transferred the wealth of Abimelech, king of the Philistines, to Isaac. While Isaac was living in the land of the Philistines, God poured His blessings upon him. Even during a famine, Isaac sowed seed and reaped one hundredfold in one year (Genesis 26:12-14). God continued to prosper Isaac until King Abimelech told him, *"...Move away from us; you have become too powerful for us"* (Genesis 26:16 NIV).

God transferred Laban's wealth to Jacob (Genesis 31:6-9). Jacob served Laban fourteen years for Laban's daughters, Rachel and Leah. He served an additional six years for his cattle. Laban changed his wages ten times, but God poured out His blessings and prospered Jacob.

Beloved, God has declared that you will prosper. There is nothing Satan can do to stop it.

Make this declaration: "I will prosper! God has declared it and there is nothing Satan can do to stop it!"

Building Character

Scripture reading: Genesis 41:37-40

The story of Joseph is another excellent biblical example of the transfer of wealth from the wicked to the righteous. This young man was content to be the favorite in his father's house, wearing better clothing and being preferred above his brethren, but God had greater plans for him. You, too, may be content with your present circumstances. Many of us settle for a lesser role than God intends for our lives. Sometimes He has to let us experience conflict, as Joseph did, in order for us to fulfill our potential in God – even in the area of wealth!

Joseph stood to inherit a great deal from his father and it was very likely that Jacob would have given Joseph more of an inheritance than his brothers because he was his favorite. But God had more in mind for Joseph and He has more in mind for you because you are a key part of this end-time, harvest time cycle.

In order for us to fulfill God's plan, there must be one last great wealth transfer. To enable us to receive that transfer, God must get us out of our comfort zone just as he got Joseph out of his comfort zone. Sold into slavery in Egypt and wrongly accused and incarcerated, Joseph did not forget the dreams God gave him when he was a young man. In the face of seeming defeat, he continued to believe and maintain his integrity and character.

When he was a slave, Joseph had a heart to be the best he could. When he was in prison, he was an exemplary inmate (Genesis 39:21-23).

In order to participate in this last great wealth transfer, we must be people of integrity like Joseph. Without character, great wealth will destroy us: *"... the prosperity of fools shall destroy them" (Proverbs 1:32).*

God knows that many people in the Church have so little character when it comes to finances that a great wealth transfer into their lives would cause destruction. God is not interested in giving us something that would destroy us, but He is interested in tempering us and building character to get us to the place where we can handle anything He would grant us.

How is His character building process coming along in you?

Make this declaration: "God is building character in me so I will be properly equipped to can handle anything."

God Will Preserve You

Scripture reading: Genesis 41:41-46

Trials like those experienced by Joseph build character. If you want to be blessed financially beyond your wildest imagination, you must go through a character-building process that tempers you to the point that you will be able to properly handle the resources God wants to put into your hands. After you develop the proper character in your life, there will be a natural flow of the power of God that will result in a transfer of wealth.

God will not transfer great amounts of wealth for ministry into the hands of those who lack character and refuse to support the kingdom of God with their finances! God has a purpose for your wealth, even as He had a purpose for Joseph's control of the wealth of the greatest nation on the earth!

Joseph rose from being a slave and prisoner to being the ruler of all Egypt, second only to Pharaoh. He had authority over all the wealth of what was then the world's greatest kingdom. This position eventually gave him the ability to spare God's people from death during a terrible famine.

God transferred control of Egypt's wealth into Joseph's hands for a powerful purpose: To preserve God's people in the midst of the famine. God will once again make a powerful, tremendous transfer of wealth – the last great wealth transfer. The wealth of the wicked will be transferred into the hands of the righteous, and again, it will be for a specific purpose: To support the greatest evangelistic thrust the world has ever known.

God never intended the wealth of the world to be in the hands of the ungodly. Never! The wealth of the world belongs to God and He wants to release it into the hands of His people to accomplish His purposes.

Remember – we wrestle not against flesh and blood, but we wrestle against principalities and powers and spiritual wickedness that are in the high places of the earth. If you are going to break the cycle of financial defeat in your life and receive the wealth of the world then you must learn how to ward off fear, doubt, and unbelief.

Joseph supplied all of the world with the resources of Egypt's storehouse. God's storehouse is even bigger. His storehouse is inexhaustible. Regardless of the financial needs you have, there is more than enough to supply all of those needs with a vast surplus. God's supply is never diminished. His supernatural supply cannot be measured.

God will preserve you regardless of the circumstances around you. He will prosper you and raise you up to be mightily used of God in the greatest end-time evangelistic thrust that this world has ever witnessed.

Make this declaration: "God will preserve me regardless of the circumstances around me. He is raising me up to be mightily used of God in the great end-time harvest."

Focus On The Kingdom

Scripture reading: Exodus 12:29-42

After the great wealth transfers from the wicked to Israel in the days of the patriarchs, Abraham and Joseph, God prepared Israel to receive another great wealth transfer. Following the end of Joseph's administration in Egypt, the Hebrew people were enslaved by the Egyptians and became impoverished under the strict rule of Pharaoh. In time, God heard their cry and revealed a plan to Moses that would not only set them free, but would also provide a great transfer of wealth (Exodus 3:1-2). (A great wealth transfer is always preceded by a revelation from God.)

Did you ever wonder why God hardened Pharaoh's heart so many times when Moses requested that Israel be allowed to leave? Have you ever wondered about the lengthy process and why all the plagues were necessary? It was part of God's plan from the beginning. He fully intended to bring ten horrible plagues upon the Egyptians so they would be so glad to see Israel go that they would help them on their way by giving them all the gold, silver, and treasure of Egypt. In the divine plan and providence of God, the transfer of wealth from the wicked to the righteous occurred once again.

Later on in the biblical record, when King Solomon received his visitation from God, he was asked what he wanted. He could ask for anything and God promised to give it to him.

Solomon's mind was not on riches, fortune, or fame. His concern was with God's people. Solomon had his priorities right, and so he illustrated an eternal principle of God that Jesus would reiterate hundreds of years later:

> *But seek ye first the kingdom of God, and his righteousness; and all these things shall be added unto you.*
> *(Matthew 6:33)*

When God asked Solomon what he wanted, he didn't use the opportunity to satisfy his own desires. He asked for an understanding heart that would give him the ability to lead God's people, Israel. He demonstrated character and integrity, and as we learned earlier these qualities are necessary if we are to receive the transfer of wealth God has prepared for us.

You may be sitting amidst a mountain of bills right now wondering how you are ever going to finance the end-time harvest of God when you can't even pay your own debts. You may pray daily for God to give you finances to pay your bills and you may be wondering why God has not done anything for you.

The answer is this: We shouldn't expect to receive a great wealth transfer as long as our hearts are only on dealing with our own personal needs. We must change the focus of our hearts and set our affections on God and the work of His Kingdom. This enables us to receive wealth by God's sovereign decision.

Make this declaration: "My heart is set on God and the work of His Kingdom. I am ready to receive the wealth transfer He has planned to channel through me for the extension of the Gospel."

True Prosperity

Scripture reading: 1 Kings 3:3-13

When you seek first God's Kingdom, you will become a candidate for the last great wealth transfer!

- When you can't see light at the end of your financial tunnel, seek first the Kingdom of God!
- When you can't give to God the way you desire, seek first His Kingdom!
- When King Solomon asked for wisdom instead of riches, he received what he asked for, but God had even greater things in mind for him:

> *And the speech pleased the Lord, that Solomon had asked this thing. And God said unto him, Because thou hast asked this thing, and hast not asked for thyself long life; neither hast asked riches for thyself, nor hast asked the life of thine enemies. . .I have also given thee that which thou hast not asked, both riches, and honour: so that there shall not be any among the kings like unto thee all thy days.*
>
> *(1 Kings 3:10-13)*

God was so pleased that Solomon put the kingdom before any personal desires that He gave him all the things that most men would have asked for instead of wisdom. In fact, God gave Solomon so much wealth that no king before or since has ever been able to match the splendor of his kingdom.

How would you like to receive a blessing from God unlike any blessing you have ever received? You would not struggle to get it. God will command the blessing. It will literally overtake you supernaturally.

God told me in prophecy that His blessings are going to literally come upon His people, run after them, and overtake them. They are not going to have to struggle to reach up and get anything that God has promised because it will run after them!

Yet promises are always dependant upon obedience. God is looking for people who will be obedient to Him, walk in covenant relationship, and put His program and His kingdom ahead of their own selfish desires.

Prosperity is not the accumulation of wealth or material things of this world, such as expensive cars, luxury homes, jewels, and designer clothes. That is the world's concept of prosperity. True prosperity is the abundance of God's blessing, both spiritual and temporal. It is total provision, continual supply, and you can only enter in to receive it as you walk in covenant relationship with God.

Make this declaration: "I reject the worldly view of prosperity. I receive true prosperity from God – His blessing upon my life in both spiritual and temporal areas."

Examine Your Motives

Scripture reading: Matthew 6:1-4

We have traced the riches of the wicked being transferred to the children of Israel generation after generation. We have talked about the wealth of kings and kingdoms being transferred to Solomon. Now we come to the really exciting part...the last great transfer of wealth – the end-time, harvest time, transfer of wealth from the wicked to God's people.

Beloved, you have received dreams and visions from God and it doesn't matter what your present circumstances are, or what may seem to hinder you. God has given you a dream, and like Joseph, you will hold onto that dream in the face of adversity and persecution.

But remember...if you are to fulfill your destiny and embrace your dreams, you must keep your priorities focused on God's Kingdom. Now is the time for this final great wealth transfer. I believe God is preparing to unleash the greatest flow of wealth from one place to another in the history of mankind! Now is the time for the wealth of the wicked to be transferred to the righteous!

Don't get confused. God is not just going to drop a bag of money in your lap. He is preparing to unleash a flow of creative power into His saints that will release the last great transfer of wealth. He is going to give His people power to get wealth:

> *But thou shalt remember the LORD thy God: for it is he that giveth thee power to get wealth, that he may establish his covenant which he sware unto thy fathers, as it is this day.*
>
> *(Deuteronomy 8:18)*

Many of God's people have not been blessed financially because God knows that they would love and trust in money instead of Him. God wants to bless His people, but to some people great financial blessing would be a snare because they have not developed the character of God in their lives. Let that sink deep into your spirit. Examine your own motives.

- Is your heart for God?
- Do you earnestly desire to see the Word of God preached all over the world?
- What is your motive for financial gain?

A great financial harvest is coming. God is draining the world of wealth and flooding the Church with finances so we can rise up in power for the final end-time harvest!

- Now is the time to prepare by rekindling your lost dreams and visions.
- Now is the time to prepare by concentrating on the Kingdom of God.
- Now is the time to prepare yourself by letting God develop integrity in you.

We must be as Joseph, Moses, and Solomon – secure in our dreams and callings, people of integrity, with our priorities focused on God's Kingdom, unafraid to pursue the goals God has set before us.

Make this declaration: "I am preparing to receive all God has for me. I am rekindling my lost dreams and visions. I am concentrating on God's Kingdom. I am letting God develop integrity in me."

ENTERING THE CYCLE

Scripture reading: Matthew 6:31-34

Many misunderstand Matthew 6:31-34 and say that Jesus does not want us to plan ahead. Nothing could be further from the truth! Otherwise, why would He talk about counting the cost before building a tower?

> *For which of you, intending to build a tower, sitteth not down first, and counteth the cost, whether he have sufficient to finish it? Lest haply, after he hath laid the foundation, and is not able to finish it, all that behold it begin to mock him, Saying, This man began to build, and was not able to finish.*
>
> *(Luke 14:28-30)*

Jesus is telling you not to worry about tomorrow or be anxious about it, but He does want you to plan for your economic future. It is time to make sure that your destiny and your resources are controlled by the Spirit of God and not the spirit of this age.

Here are five steps to take to ensure that you will enter into God's supernatural cycle of divine provision:

1. Develop a godly attitude to receive the money God is going to pour into your heart. If you believe money is somehow evil, then you have misunderstood God's Word. Only the love of money above the things of God is wrong. If you faithfully tithe and give offerings, you are obeying God and doing what He tells you to do, loving Him above all else. *(1 Timothy 6:10, NAS)*

2. Recognize that God uses money to direct our lives: He promises to bless our seed-planting into the Gospel by returning a financial harvest. *(Galatians 6:9, NAS)*

3. Recognize that God uses money to demonstrate His power: Offer to God a sacrifice of thanksgiving and, pay your vows to the Most High. *(Psalm 50:14-15, NAS)*

4. Recognize that God uses money to satisfy the needs of others: How can you give to the poor and to the Gospel if you have not been liberally blessed by God? He fully intends to bless you so you can freely give to the poor and to His work! *(Deuteronomy 15:11, NAS)*

5. Understand that God wants you to plan financially for the end-time hour: Prayerfully seek His will regarding your finances. (Proverbs 16:9, NAS)

Make this declaration: "God will use finances to demonstrate His power in my life, direct my life, and work through me to meet the needs of others."

Set Your House In Order

Scripture reading: 2 Kings 4:1-7

God wants you to be prepared to receive His promises for prosperity. He wants you to take practical steps to prepare yourself to be a good steward of what He will let flow into your life.

For years, I have been urging Christians that now is the time to...

- Reduce debt, build up cash and liquid assets, and a healthy cash reserve.
- Refinance high interest business debt and home mortgages at lower rates.
- If you own a home, accelerate the pay-off time.

As you hear God and heed His voice, you will be prepared to rise above the world in your finances!

God wants you to plan, but also know this: He is able to bless His people regardless of circumstances. He only expects you to plant whatever seed you can into His Kingdom. He will multiply your sacrifice back to you far beyond any worldly investment!

> *Give, and it shall be given unto you; good measure, pressed down, and shaken together, and running over, shall men give into your bosom. For with the same measure that ye mete withal it shall be measured to you again.*
>
> *(Luke 6:38)*

When you plant into the Gospel, God promises a miracle harvest – far exceeding what you plant:

> *But other fell into good ground, and brought forth fruit, some an hundredfold, some sixtyfold, some thirtyfold.*
>
> *(Matthew 13:8)*

In this end-time hour, God is going to explode His financial blessings beyond anything He has ever done before.

As you follow the biblical command to honor the Lord with your tithes and offerings, your house will literally overflow with grain and your vats will be fat with wine and oil despite what is happening in the world!

> *The first of the firstfruits of thy land thou shalt bring into the house of the Lord thy God.*
>
> *(Exodus 23:19)*

If you do your part, God will do His part. Be prepared. Take practical steps regarding your finances. Follow the scriptural guidelines for giving and receiving, sowing and reaping. When you step out in faith and begin to act, God will fulfill His promises.

Is your financial house is in order? Are you adhering to the Bible-based standard of living which stresses faith in God and individual self-reliance? Are you exhibiting self-control rather than riotous excess and debt? Are you living within your means rather than stealing from others?

If you are not following God's commands, it is time to put your house in order.

Make this declaration: "I am putting my financial house in order today. I will adhere to the Bible based standard of faith in God and self-reliance, self-control, and living within my means."

Eating The Good Of The Land

Scripture reading: Isaiah 16:19

The recent economic shockwaves in the economies of major nations around the world only mirror what will soon happen in every major nation in the world. To prepare properly for God's prosperity plan, you need to understand the coming financial events of the end-time hour so you can prosper as His child – even in the midst of worldwide economic chaos!

God does not do anything without first warning His people through His prophets. *"Surely the Lord God will do nothing, but he revealeth his secret unto his servants the prophets"* (Amos 3:7). The purpose of prophecy is not to scare us, but to prepare us. Do not be afraid of the economic conditions around you. Do not wring your hands and fret about what you are going to do. Do not try to figure out your finances in the natural realm with head knowledge.

Hear me on this – God wants you to control your financial destiny and your resources through the Holy Spirit and not the spirit of this age. God wants you to prosper despite the economic chaos that will affect this world in these final days of time.

God is going to release an end-time financial anointing upon His people – unlike anything ever seen before in spiritual history – so they can rise above the world's financial problems and enter into God's divine business cycle. He will bless you regardless of current credit or stock market circumstances! God will prosper you even in the midst of end-time economic chaos! Even the heathen will see you are blessed and will actually tremble in your presence!

> *And it shall be to me a name of joy, a praise and an honour before all the nations of the earth, which shall hear all the good that I do unto them: and they shall fear and tremble for all the goodness and for all the prosperity that I procure unto it.*
>
> *(Jeremiah 33:9)*

When Israel entered their Promised Land, God blessed them with the good of the land. They ate fruit from trees they had not planted. They harvested grain from fields they had not sown. As they battled their way across the land that God had promised them, they didn't have to worry about food for their armies. God supernaturally provided. The wealth of the wicked Canaanites fell into Israel's hands.

That is exactly what God wants to do for you during the days, weeks, months, and years ahead. As circumstances in the economic world deteriorate, God wants you to tap into His abundant provision.

Like Israel, He wants you to eat of the good of the land. No more manna. No more scrambling around each day wondering where you will get your next meal. Abundant provision. Supernatural supply. That is God's promise. It is your inheritance – so claim it!

Make this declaration: "I will eat the good of the land. No more manna. Abundant provision, supernatural supply. That is God's promise to me."

June

These deserted wheelchairs, crutches, and braces left behind after our crusade in Ecuador in 1998 are mute but powerful testimony to the proof producing power of God. This month I will show you – step-by-step exactly how to become a Proof Producer.

JUNE
PRODUCING THE PROOF

At the city of Jericho, the nation of Israel produced the proof that God's power was greater than that of the enemy. God's people marched around the massive walls of the city for seven days, and on the final day the walls fell flat, the Israeli army invaded, and Jericho was conquered. This supernatural victory was executed for a divine purpose:

> *That all the people of the earth might know the hand of the LORD, that it is mighty: that ye might fear the LORD your God for ever.*
> *(Joshua 4:24)*

It was the miracles that convinced the people, their enemies, and entire nations of the reality and power of Almighty God. Even the prostitute, Rahab, declared:

> *For we have heard how the LORD dried up the water of the Red sea for you, when ye came out of Egypt; and what ye did unto the two kings of the Amorites, that were on the other side Jordan...And as soon as we had heard these things, our hearts did melt, neither did there remain any more courage in any man, because of you: for the LORD your God, he is God in heaven above, and in earth beneath.* *(Joshua 2:10-11)*

In the new millennium we are entering, the future success of the Kingdom of God will lie in the hands of those who find the answer to this question: *"What shall we do, that we might work the works of God?"* (John 6:28). Jesus did not say, "Ye shall receive a blessing after the Holy Spirit comes upon you." What Jesus said was this: "Ye shall receive power..."

For too long, we have stopped at the point of blessing – at the tongues, at the beginning of the joy of a relationship in the Holy Spirit. We stop there. Very seldom does someone press past this point. Very few, indeed, go on to experience the miracle-working power of God. Very few learn to produce the proof.

During the coming month our devotions are drawn from the *Proof Producer* teaching, one of the most profound revelations God ever gave me. This message has been shared around the globe, impacting hundreds of thousands of people, and raising up a virtual army of proof producing believers who know how to work the works of God.

In the devotional readings this month you will learn how to respond to the negative forces of unbelief. You will learn how to have faith without struggling and how to go beyond the point of blessing into the realm of power. You will learn to produce the proof that God is still in the soul-saving, miracle-working business.

Get ready – you are about to become a proof producer.

How Far Do You Want To Go?

Scripture reading: John 6:22-29

The Church of Jesus Christ has prayed for years for a deeper, experience with God.

It has had revivals, fasts, rallies, retreats, and promotions. People have been blessed – clapping their hands, enjoying the presence of the Lord, dancing in the Spirit, experiencing times of great joy and blessing. The trouble is that the Church, year after year after year, has stopped at the point of blessing and has not entered into the depth of relationship that God intends it to have. We have not gone far enough. The big question is how far do you want to go?

We have beautiful edifices, lovely choir robes, artistic crosses, and elaborate stained glass windows. We have great choirs singing tremendous anthems. We have fellowships and spaghetti dinners. We have men in the pulpit who have been to a theological seminary or a three-or four-year Bible course and know the meaning of homiletics. They can deliver a beautiful sermon point-by-point. They baptize children, marry them when they are grown, listen to their problems, wipe their tears, bury their dead…but as needful as the day-to-day religious chores are, something is missing. It is not enough!

What I am sharing with you about the general condition of the Church today is what may be called "negative input." No one likes to deal with negatives. However, in order to learn how to work the works of God, we must deal with things as they are and not as we would like them to be. We must "take the mask off." We must face the truth even if that truth hurts or is completely contrary to what we have believed all our lives.

Here is a truth that, if you learn, absorb, and put it to use, will be one of the most valuable assets you will ever have in dealing with people or things. It is this: All success is based on dealing with people, places and things exactly as they are and not as you would like them to be, or as you are.

If you are dealing with a crook, you must deal with him as a crook. You may be the most honest person in the world, but you cannot deal with the crook successfully if you deal with him as if he were honest, after your standards of morality or ethics. He will steal you blind. In order to be successful you must deal with him as a crook. You must make him sign his name on the dotted line, have his signature notarized, secure collateral from him, and lock it in a safe deposit box for which you have the only key. You deal with a crook as a crook if you want to be successful.

Now, all truth is parallel. We must also deal with situations as they are. The Bible says: *"And ye shall know the truth and the truth shall set you free"* (John 8:32). The truth about the situation of the Church today is that as churches, ministers, and lay people, we are in in desperate trouble. How will we ever reach this world for God if we continue to do business as usual inside the walls of our churches?

Make this declaration: "I don't want to do business as usual in the spiritual realm. I want to go all the way with God!"

THE POWER OF THE HOLY SPIRIT

Scripture reading: Acts 2:1-4

Two thousand years after Jesus rose from the grave, there are billions who have not yet even heard the Name of Jesus! We will never change this condition by ignoring it or pretending it is not there. It is time to take the mask off and stop going down the same ruts of harvest we have been in for the past hundreds of years of Christianity. We must face the problems and zero in on a new revelation, a new breakthrough, that will carry us into realms far beyond the present status.

One of the main problems is that the Church has missed one of the greatest truths Jesus ever taught. They know the Scripture. They can quote it by heart. But they have overlooked the real truth of it. It has to do with the Holy Spirit and it is a spiritual "bombshell." The Baptism of the Holy Spirit has to be something much more than we are experiencing under what we refer to as the "charismatic outpouring." It has to be much, much more than what is being experienced in the traditional Pentecostal churches.

It is clear from the Scriptures that God intended the power of the Holy Spirit to be our greatest source of reaching the world, but the Church has missed this by and large. Here is the bombshell regarding this truth: When Jesus Christ promised us that we would receive the blessed Third Person of the Trinity, the Holy Spirit, He never promised us that we would just speak in other tongues.

There is more to Pentecost than just doctrine. There is an experience of power! This power has to be experienced. You may have memorized doctrine, but doctrine will not give you what it takes to face the negative circumstances of this life and rise above mediocrity.

All these other things may fail you, but when you have experienced the burning bush, it will be no problem to face Pharaoh and declare, "Thus saith the Lord God of Israel, let My people go!"

When you have had an experience with God, you can face your circumstances. You can face disease! You can face the devil! You can face anything and come out victorious because you have had an experience! This experience goes beyond blessings and the manifestation of even the gift of speaking in another tongue. I believe fully in the manifestation of the prayer language of the Holy Spirit. I exercise it daily. Paul said: *"I thank my God, I speak with tongues more than ye all"* (1 Corinthians 14:18).

I am not trying to tell you that the gift of other tongues is not an evidence of what we call the Baptism of the Holy Spirit. There is great scriptural evidence that this manifestation flowed through their lives, as they received the Pentecostal experience. But the real power of the Holy Spirit is more than just blessings. It is more than "goose bumps." It is power!

Make this declaration: "I am going to move beyond goose bumps into the realm of power. I will have an experience where I can face my circumstances, disease, the devil – and emerge victorious."

At The Point Of Blessing

Scripture reading: Acts 1:1-9

Jesus didn't tell His disciples to tarry in Jerusalem until they had one goose bump on top of another! He didn't say, "Wait until you speak in tongues!" The disciples didn't even know the manifestation of the gift of tongues would come. They had never heard of it before!

Jesus said: *"...ye shall receive power, after that the Holy Ghost is come upon you..."* (Acts 1:8). They were waiting for the power of the Holy Spirit!

Today, people rarely keep going on in their spiritual experience of tarrying before the Lord. They stop at the point of blessing. They are camped there. They are camped at speaking in tongues. They are camped at certain spiritual gifts of the Spirit. They have failed to break past the point of blessing to the point of Jesus' promise.

What is the point of Jesus' promise? Power! He does not say: "You shall receive power, after that the Holy Ghost is come upon you and that power will enable you to just speak in other tongues or the prayer language of the Holy Spirit." He says, "That power will enable you to be witnesses unto Me."

I can show you millions of people who are blessed. I can show you millions of people who can go out and talk about Jesus Christ. I can show you millions of people who can speak in tongues or who experience great joy as part of their worship. Yet I have not found many people to whom I can point and say, "I can see in that person the ability called the power, the enduement of the Holy Ghost, to give witness and evidence of the resurrection of Jesus Christ" – knowing that to give evidence of Jesus Christ is to produce the proof that He is the resurrected Lord. Perhaps that is why half the world today has never yet heard about Jesus Christ.

I do not like to be negative, but you cannot produce this ingredient called power that is missing from our churches and experience today without this negative input. You must have negative input before you can produce power. Electricians tell me that in order to produce electrical power to turn on a light to dispel darkness there must be two wires. One is a positive wire and the other is a negative wire. You must take the positive wire and the negative wire, strip down the insulation from them and hook them up on each side of a switch or receptacle. You must strip the insulation from the wires or it will not work.

This stripping down process is not easy, but it is necessary to produce power. If you hook up only the positive wire and go to the switch, you do not get any power. You get no electrical current. There is no light to dispel the darkness. You must put the negative and the positive together. When you do that and then flip the switch, the light comes on.

Make this declaration: "I am going through the stripping process that is necessary to produce power. The light is dispelling spiritual darkness."

Breakthrough To Power

Scripture reading: Mark 16:15-20

Light dispels darkness, but in order to achieve it you must have negative input or the light does not come on. In the spiritual realm you must deal with the negative input in order to produce the light we seek. You must be willing to face things as they are, not as you might like them to be.

We are in desperate trouble as soul winners for God. The Christian Church is losing ground as the heathen are multiplying faster than the Church. Despite the tremendous moving of the Holy Spirit in the nations of the world, we are losing the race. Much of the population of the entire world has never yet heard the Name of Jesus.

I did not say that half the world has rejected Christ. I did not say that much of the world had turned away from Him. I said that billions have never yet, one time even heard the Name, Jesus. That is a devastating fact, a terrible reality,

The reason this situation exists is because we have stopped at the point of blessing and we have failed to press on into the place of power God means us to have in order to do His work.

The future success of the Kingdom of God will lie in the hands of those who find the answer to this question:

> *Then said they unto him, What shall we do, that we might work the works of God?*
>
> *(John 6:28)*

Spirit-filled people, Pentecostal people, get many emotional blessings, but Jesus did not say, "Ye shall receive a blessing..." What Jesus said was this: "Ye shall receive power..."

We stop at the point of blessing, at the tongues, at the beginning of the joy of a relationship in the Holy Spirit. We stop. Very seldom does someone press past this point. Very few, indeed, ever enter into the power area at all.

Many receive powerful blessings, but they do not have the element needed to cast out devils. They do not have the ingredient needed to heal the sick. Yet that is the work that Jesus commanded us to do:.

> *...In my name shall they cast out devils; they shall speak with new tongues; they shall take up serpents; and if they drink any deadly thing, it shall not hurt them; they shall lay hands on the sick, and they shall recover.*
>
> *(Mark 16:17-18)*

What is the answer? What must we do to work the works of God? To work the works of God we must be willing to go beyond where we are to where we need to be. We must not stop when we get to a point of blessing in our lives. We must break through into the spiritual realm of God's power.

Make this declaration: "I will break through into the spiritual realm of God's power. I will learn to work the works of God."

The Lord Working With Them

Scripture reading: Matthew 28:16-20

Since the task given to the early disciples and to us, as the disciples of Jesus today, is basically the same – producing the proof of the resurrection power of Jesus Christ in a needy world – we need to give careful consideration to the highly successful methods the early disciples used to affect this task.

The methods of the apostles led to miracle ministries. The same methods in our lives will bring us into miracle ministries. When we talk about miracles, we must be careful to understand we are not talking about just physical miracles. All of God's wonderful workings in the life of humanity is His miraculous working in our souls, our families, our loved ones, our finances – at the point of every natural or spiritual need we may have.

The reason for the success of the early disciples is very simple: It was the Lord "working with them," and with the Lord there is no room for failure. He will confirm His Word with unmistakable miracles.

The apostles would not have been successful if they had not been able to produce evidence of the resurrection. When they preached, things happened just as Jesus said they would when He gave them the great commission. Jesus said:

> *And he said unto them, Go into all the world, and preach the gospel to every creature, He that believeth and is baptized shall be saved; but he that believeth not shall be damned, And these signs shall follow them that believe, In my name shall they cast out devils; they shall speak with new tongues; They shall take up serpents: and if they drink any deadly thing, it shall not hurt them: they shall lay hands on the sick. and they shall recover. And they went forth, and preached every where, the Lord working with them, and confirming the word with signs following. Amen.*
>
> *(Mark 16:15-18,20)*

They had signs and wonders following them because the Lord was working with them.

Some seem to read that Scripture this way: "They, working with the Lord..." That is the way many try to minister, but the method that proved so successful for Peter and Paul was "the Lord working with them."

How did Jesus work the works of God in meeting human needs? In the ministry of Jesus Christ, as far as we can ascertain from the Scriptures, Jesus never prayed for one sick person as we know and understand praying or interceding. It certainly is not wrong for us to pray for the sick. In fact, there are many ways to pray for the sick – scriptural, biblical methods – which God has given us such as the laying on of hands (Mark 16:18) and anointing with oil (James 5:14-15).

However, what we are looking for and what we are believing God to do for us as we pursue these truths, is a breakthrough in the spirit world that will completely revolutionize our ministries – the Lord working with us.

Make this declaration: "Lord, I am open to new truth that will revolutionize my life and ministry. Work with me, in me, and through me."

The Will And Purpose Of Christ

Scripture reading: Matthew 8:5-13

The Roman army captain whose story is related in the eighth chapter of the Book of Matthew came to Jesus Christ and said, "Lord, my servant lieth at home sick of the palsy, grievously tormented." Without asking one question, Jesus said, "I will come and heal him."

I can tell you this unequivocally, that Jesus Christ died for the sins of the whole world.

It is the will of God that every sinner be saved. That does not mean they will all get saved, but it is God's will to save them. I also tell you this just as unequivocally: Sickness has no place in the Body of Christ. Sickness is part of the curse. It came through the disobedience and judgment of Adam in the Garden of Eden.

When God created man there were three things He never intended man to possess: sin, sickness and death. God never intended man to sin. He never intended man to get sick. He never intended man to die. He intended man to live forever in righteousness and health. Sin, sickness, and death came as a result of man's disobedience. Deliverance from sin, healing, and eternal life came through the life, death, and resurrection of Jesus Christ.

Where the first Adam was disobedient, the Second Adam (Jesus) was obedient even unto the death on the cross. He bought us forgiveness of sin, healing and resurrection life (Romans 5:15 and 19; 1 Corinthians 15:22).

It is God's will for all who will accept the blessings of Christ Jesus to be well and whole (3 John 2). That is why to the centurion's request for healing for his servant, Jesus said, "I will."

There is no theological seminary, no doctrine, no tradition that can change the will of God. A will in the natural world is a legal, official document that someone executes in his lifetime to determine the legacy he will leave to his loved ones after his death. The human will can be, and often is, changed many times. However, once the death of the will-maker occurs, there can be no further changing of that will. The inheritance has been sealed by that person's death. The entire power of the state's judicial system is put behind enforcement of that will.

God has never changed His will. He sent Jesus Christ here to do His will (John 4:34).

The will of God and the sum, total purpose of Christ coming to this earth and undergoing the things that transpired during the time He was here in the flesh is this:

> *...For this purpose the Son of God was manifested, that he might destroy the works of the devil.*
>
> *(1 John 3:8)*

Make this declaration: "Jesus destroyed the works of the devil. Therefore, Satan has no authority in my life."

An Unlimited God

Scripture reading: Matthew 21:17-27

The will of God is a tremendous thing. You can't change it. It is already written. It is attested to with the blood of our Savior. Through His death, He has left us the legacy of the complete destruction of the works of Satan in our lives. Heaven's highest court will see that this will is enforced as we step out to claim its provisions.

Before we can demonstrate the power of God in our lives, we must have our faith tied to a known factor. Those who pray, "If it be thy will..." do not have their faith tied to a known fact. The Bible declares:

> *But let him ask in faith, nothing wavering, For he that wavereth is like a wave of the sea driven with the wind and tossed. For let not that man think that he shall receive any thing of the Lord.* (James 1:6-7)

We must know how to pray in faith believing:

> *And all things, whatsoever ye shall ask in prayer, believing, ye shall receive.* (Matthew 21:22)

It is not enough, Hebrews tells us, just to believe God exists. We must also believe that He rewards those who diligently seek Him:

> *But without faith it is impossible to please him: for he that cometh to God must believe that he is, and that he is a rewarder of them that diligently seek him.* (Hebrews 11:6)

We can pray with authority only when we know that we know that we know that we know that we know! We must know that – no matter what some preachers may say – God does not put sickness and evil upon us. He permits them as in Job's case, but He does not send them upon His children!

The centurion who asked Jesus to heal his servant told the Lord, "It is totally unnecessary for You to come to my house." He knew that if it was God's will for his servant to be healed, there was only one thing needed to make that healing a reality. This soldier had grasped one of the greatest truths that God has given my ministry. It is this: The God that we serve is a God who knows no limit!

One of the greatest reasons why we limit an unlimited God, is because we do not see Him as He is. The centurion saw Jesus as He was. He said to Jesus, "You don't have to come to my house. Stand right here and speak the word only and my servant shall be healed."

Jesus was moved to such a great extent by the centurion's declaration of faith that He said, "There never has been any greater faith in all the world than that which this man has demonstrated."

Make this declaration: "I serve an unlimited God. His unlimited power is resident within me."

You Have The Authority

Scripture reading: Luke 7:1-10

Jesus knew what He was talking about, because He is the great I Am. He said, "I Am. I know what I'm talking about because I Am." He said, "Many will come from the East and the West. They will sit down with Abraham and Isaac and Jacob in the Kingdom of Heaven, but there will never be any greater faith than that demonstrated through this centurion."

The centurion whose story is told in Matthew 8 perceived a truth in Jesus Christ that was parallel to his own situation. He said, "Lord, look at me. Do you see this uniform of war? Do you see these Roman army captain's bars? When I say to a man, 'Go,' he goes. When I say to a man, 'Come,' he comes. When I say to someone, 'Do this,' he does it. Do you know why? Because I am a man under authority, and when I speak, my soldiers obey my voice."

He said, "It is not necessary for You to come to my house. When I look at You, Jesus, I see that all truth is parallel." This man transferred his allegiance from a man to the Son of God right on that street comer. The only Lord that he had acknowledged before this time was Caesar. However, on the street corner before the multitude, he bowed before Jesus and called Him "Lord."

"You don't have to come to my house," he told Jesus, "because You are just like I am. I am a man of authority. I speak, and when I speak there is action. When I give a command, it is obeyed. Behind me is all the influence, power, and authority of the Roman Empire including Caesar himself and the military forces that I am a part of. When I speak, it is not just my word, but those who are under my command know and recognize where my authority comes from. They know and they recognize that I have been given this authority."

The centurion continued, "Lord, You don't have to come to my house. If You will stand right here and speak the word, my sick servant who is stricken with paralytic palsy, unable to move, will be healed. You do not have to come. You do not have to lay hands on him. If You will just stand right here and speak the word, my servant will get up."

A man given authority, a man under authority, the centurion recognized that in Jesus Christ there was a special anointing. Jesus had power. He had the authority of God Almighty. The omnipotent God, the only real God, the mighty Creator of heaven and earth was acting through Jesus Christ. He delegated the authority to His Son. Jesus said,

> *... All power is given unto me in heaven and in earth.*
>
> *(Matthew 28:18)*

Whatever your need is today, remember that Jesus Christ has authority over it. He has delegated that authority to you, which means you also have authority over every situation, every problem, and every difficulty you encounter.

Jesus spoke the word, and the centurion's servant was healed from that very hour!

Make this declaration: "I have the authority to speak the Word and it will be done. I will speak the Word to every situation I encounter during this day."

UNDER DIVINE AUTHORITY

Scripture reading: Mark 3:1-5

There was a uniqueness in the ministry of Jesus. He never really prayed for a sick person. He used the authority His Father gave Him.

One day Jesus met a man in the synagogue who had a withered hand. Jesus did not pray for him to be healed. He looked at the man with the withered hand and said, "Sir, stretch forth thy hand." Jesus spoke the word. In obedience to His spoken word the withered hand came out and became perfectly whole (Mark 3:1-5). "Speak the word."

A leper came to Jesus and bowed before Him on the street comer. He said, "If Thou wilt, Thou canst make me clean." Jesus did not pray for him. He looked at him and said, "I will; be thou clean." When He spoke the word, the leprosy disappeared. The man's flesh instantly became perfectly whole like that of a new baby under the authority of the spoken word of Jesus Christ (Matthew 8:2-3).

- He said to the man at the pool, "Rise." (John 5:8)
- He said to the blind, "See." (Luke 7:21)
- He said to the deaf, "Hear." (Mark 7:32-35)
- He said to the demons, "Come out" (Matthew 9:32-33) and the religious leaders of His day said:... *What thing is this? what new doctrine is this? for with authority commandeth he even the unclean spirits, and they do obey him.* (Mark 1:27)
- He said to the winds and the waves, "Peace, be still." (Mark 4:39)

When Jesus stood outside the tomb of Lazarus as the stone was rolled away, He looked up at His Father and said, "Father, the power is so strong, the anointing is so strong, the authority You have given to Me is so strong, I could just look in that tomb and Lazarus would come out. But because of all the people who are standing by here, I am going to say it."

. . .Jesus did not go into the tomb and massage Lazarus.

. . .He did not give him spiritual mouth-to-mouth resuscitation.

. . .He did not even pray for him.

Jesus looked in that grave and said, "Lazarus, come out of there. Come forth!" He spoke the word and Lazarus came out still wrapped with grave clothes. Jesus said, *"Cut him loose. Let him go free"* (John 11:44).

Jesus was a Man under divine authority. That same divine authority is available to you today.

Make this declaration: "Jesus was a Man under divine authority. That same authority is available to me today as I come under God's power and authority."

Authority And Responsibility

Scripture reading: Mark 5:22-24; 35-43

Jairus, the ruler of the synagogue, came to Jesus and said, "My little daughter is sick." Jesus said, "I will come to your house and I will heal her."

In Mark 5:22-24, we find Jesus on His way to Jairus's house, but He was delayed in His journey because of the multitude that thronged Him and the miracle He did for the woman with the issue of blood. Jesus was met with this message: *"While he yet spake, there came from the ruler of synagogue's house certain which said, Thy daughter is dead: why troublest thou the Master any further"* (verse 35).

By the time Jesus got to Jairus's house, the girl was dead and Jesus and Jairus were met by mockers who laughed them to scorn. Jesus put them all out of the room, then He and the child's parents went into the room where her body lay. Jesus took the little girl by the hand and said, *"Damsel, I say unto thee, arise!"* He spoke the Word with divine authority and whenever He did, miracles occurred.

You cannot give someone responsibility without giving Him authority to carry out that responsibility. Jesus has given us the responsibility of bearing witness to Himself throughout the length and breadth of this world. He also has given us the authority to carry out this task. We may speak the word and have the same miracles that Jesus had, for whatever He says, we may boldly say.

Some people think they have a free-for-all license in their relationship of responsibility and authority. Such a relationship does not exist. No one is a law unto himself. If you think you are a law unto yourself, you are still in a spirit of rebellion. You need to get on your face before God and let Him break that rebellion until you learn how to be submissive to the Holy Spirit and the plan of God for the Body of Christ.

Some people walk around saying, "I can do what I want to do. I am free in the Spirit of God." You are not free in the Spirit of God. You are free with divine authority to do what God tells you to do.

Make this declaration: "I have both responsibility and authority. I am empowered with divine authority to do everything God tells me to do."

All Authority Is Of God

Scripture reading: John 20:19-23

God sent Jesus here for a purpose. He gave Him a responsibility and He gave Him the authority to fulfill that responsibility. Jesus came to earth for the sublime, supreme purpose of dying on the cross, but He also came to engage Satan in spiritual battle.

> *...For this purpose the Son of God was manifested, that he might destroy the works of the devil.*
> *(1 John 3:8)*

Jesus came as the representative of God to work God's works against the kingdom of Satan and to take from the devil the keys of the kingdom that he stole from Adam and Eve.

He came with God's authority. He spoke the Word. He worked the works of God.

After the death and resurrection of Jesus Christ, He took that authority vested in Him by God Almighty. He did not take that authority away from the earth. He did not take that authority back to heaven and leave us here, powerless, without it.

Jesus passed that authority along to us as co-workers of the works of God. He gave His authority to the disciples who followed Him while He was here on earth and He gives it to us as His disciples now:

- *...as my Father hath sent me, even so send I you.* (John 20:21)
- *Behold, I give unto you power to tread on serpents and scorpions, and over all the power of the enemy: and nothing shall by any means hurt you.* (Luke 10:19)
- *...All power is given unto me in heaven and in earth.* (Matthew 28:18)

The disciples followed the method and example of Jesus and used the authority He transmitted to them as they went out. They literally turned the world upside down for God.

> *...These that have turned the world upside down are come hither also.*
> *(Acts 17:6)*

These are the same men who once denied Him and doubted His resurrection, but now:

> *And they went forth, and preached every where, the Lord working with them, and confirming the word with signs following. Amen.*
> *(Mark 16:20)*

We have always thought of our working with God, not stopping to realize the depths of the spiritual reality that the Lord longs to work with us. God began to work with them when they were changed from fearful, unbelieving, doubting men, to men of authority.

When you are changed to a man or woman of authority, you have the basis for doing the works of God. How do you become a person of authority? By whose authority do you do the works of God? All authority begins and ends in Jehovah God!

Make this declaration: "I have power to tread on serpents and scorpions and over all the power of the enemy. Nothing will hurt me."

It Shall Be Done

Scripture reading: Matthew 18:18-20

When Jesus came to this earth, He came as the Son of God, with power and authority. Where did He get it? He received it from His Father, God. Do you think that God would send His Son into this world to face the devil without the ability to achieve all that He had sent Him to accomplish?

God sent Jesus here for a divine purpose to destroy the works of the devil (1 John 3:8). What would you think of a God Who knew the strength of the enemy and did not give His Son the necessary weapons to defeat the enemy's power?

Jesus said: *"All power is given unto me in heaven and in earth"* (Matthew 28:18). He recognized that the Father had given Him the authority. Jesus gave His own testimony in the synagogue when He opened the Book of the prophet Esaias and read:

> *The Spirit of the Lord is upon me, because he hath anointed me to preach the gospel to the poor; he hath sent me to heal the brokenhearted, to preach deliverance to the captives, and recovering of sight to the blind, to set at liberty them that are bruised.*
>
> *(Luke 4:18)*

Now, do you think that same Jesus, the Son of God, would send you, would send me, to face this same devil without power and without authority? No! When Jesus sent out His disciples, He told them as His Father had sent Him, with all authority, even now He was sending them:

> *And when he had called unto him his twelve disciples, he gave them power against unclean spirits, to cast them out, and to heal all manner of sickness and all manner of disease.*
>
> *(Matthew 10:1)*

He not only commissioned them to preach that the Kingdom of Heaven was at hand, but also to...

> *Heal the sick, cleanse the lepers, raise the dead, cast out devils: freely ye have received, freely give.*
>
> *(Matthew 10:8)*

In Matthew 18:18-19 He put powerful weapons into the hands of believers with these words:

> *...Whatsoever ye shall bind on earth shall be bound in heaven: and whatsoever ye shall loose on earth shall be loosed in heaven. Again I say unto you, That if two of you shall agree on earth as touching any thing that they shall ask, it shall be done for them of my Father which is in heaven.*

You have powerful spiritual weapons in your hands to face the battles of life today. Be sure to use them!

Make this declaration: "I have powerful spiritual weapons in my hands to face the battles of life today. I will use them with God-given authority."

Working The Works Of God

Scripture reading: Acts 3:1-10

Jesus demonstrated the power He was giving His disciples in the miracle of the withered fig tree (Matthew 21:19-20) and then gave them this great promise:

> *...Verily I say unto you, If ye have faith, and doubt not, ye shall not only do this which is done to the fig tree, but also if ye shall say unto this mountain, Be thou removed, and be thou cast into the sea; it shall be done. And all things, whatsoever ye shall ask in prayer, believing, ye shall receive.*
> *(Matthew 21:21-22)*

Jesus delegated His authority. We have the tools. We can speak to mountains and they shall be removed – mountains of oppression, sickness, depression, and pain.

When confronted with the sick, brokenhearted, and captive, could it be that we have used the excuse, "I don't have any power to heal," because in reality we are not in touch spiritually with the divine flow?

Look at Peter in the third chapter of Acts. He said, "Such as I have!" No matter how you interpret it, it comes out the same: "Such as I have!"Peter had the authority and he used it just as Jesus did: He spoke the word! He said, *"In the name of Jesus Christ of Nazareth rise up and walk"* (Acts 3:6). Jesus spoke the Word and it was accomplished. Peter and the other disciples spoke the Word and it was done. We may speak the Word with that same authority, with the same invincible forces behind us.

The word "authority" in the Greek is *exousia* which literally means "to be out." *"Ex"* means "out of" and *ousia* means "being." One writer has defined it as "the ability to go beyond oneself." It is used of that authority which a person has which is delegated to him from someone else. The Greek word also denotes complete mastery, superhuman power, complete jurisdiction.

The person delivering the authority is in a sense out of himself and acting in and through the person to whom he has delegated that authority. Do you see how totally powerful this is?

Jesus has delegated His authority to us. He lives in us, acting in us and through us. In this sense, we are out of ourselves, we go beyond ourselves and we are in Him, the Lord working with us (and through us) with signs following. That is why the methods of the disciples were so powerful, decisive, and so dynamically effective.

Look at your hands. Let the Lord work through your hands, as the disciples did.

They grasped the keys. They spoke the Word. They acted in the power and authority of God Himself. Such a method can never fail. It works the works of God.

Make this declaration: "I have faith. I do not doubt. Whatever I ask in prayer, believing, I will receive."

June 14

Our Badge Of Authority

Scripture reading: Mark 1:21-28

Let me use a story to illustrate the power of delegated authority. In the United States we still have many small towns we call "whistle stops," the kind where if you come down one side of a steep hill and go up the other side, you have passed through the town. At the bottom of the hill in this story there was one main store which housed groceries, a post office, and the city gas pumps.

A big thirty-two-foot truck once came down the hill at about sixty miles per hour, violating the speed limit of twenty-five miles per hour. A man sitting in front of the all-purpose store, dressed in a blue uniform, saw this truck speeding down the hill. Without a moment's hesitation he stepped out into the middle of the street, held up his hand, and brought the truck to a screeching halt.

This man was an elderly gentleman, the town's only police officer, a retired coal miner. About seventy years of age, he was about five feet six inches in stature and weighed about 150 pounds.

That elderly man slowly walked over to the cab of that truck and spoke with a voice that echoed down the street. "Get out of that truck, mister!" When the truck cab opened, a burly man of about six feet and two-hundred and forty pounds crawled out. The short, elderly policeman shook his old bony finger at this giant of a truck driver and shouted, "Where do you think you're going? What are you trying to do, kill somebody?"

"No, sir!" the truck driver replied. "I'm sorry, sir. I didn't mean to violate the speed limit."

"Follow me," the policeman ordered. He took him into the general store which was also the police station, wrote him a ticket and fined him right on the spot. The driver meekly paid the fine and left silently.

My question is this: Do you think the truck driver, six feet tall, two-hundred and forty pounds, middle aged and strong, was afraid of the old retired coal miner? No. It is absurd to think so. It was his police badge of authority which said, "This man represents the state of Pennsylvania and behind him is all the power of that state and the United States."

You stand in authority over all the power of the enemy. Jesus said, "As My Father sent Me, even so send I you." You are a man of authority. God's badge of authority gives you power which the devil cannot violate. Satan and evil must obey this authority for it comes from God Who is all powerful.

No, the devil is not afraid of us, but he is afraid of Jesus. He is afraid of the badge of authority that we wear because we do not stand alone. Behind us stands Jesus and behind Him stands God the Father. When we catch sight of these facts, we have the keys to power in our hands.

Make this declaration: "I wear the badge of authority of Jesus Christ. Behind Him stands the power of God Almighty."

When Satan Challenges Your Authority

Scripture reading: Luke 8:41-42; 49-56

We must realize that as long as we are here on earth Satan will challenge our authority at every turn through every means he can. This means he will use other people, often even our loved ones, to bombard us with negative forces of unbelief and render us powerless to do the works of God.

When Jesus went to the home of Jairus to raise his little girl from the dead, He encountered a situation which is very similar to situations and circumstances in which we find ourselves today. When Jairus and Jesus stepped into Jairus's house, they ran headlong into the same condition you will sometimes encounter as you work the works of God: They were met by "professional mourners" (Mark 5:38). Professional mourners are "Job's comforters." Whenever you begin to believe God for something, they are always around to tell you why it cannot happen, why it should not happen, and why it is not God's will for it to happen.

When Jairus walked in, he had to face these people. The first thing they said to him was, "Aren't you ashamed of yourself? You trusted in this Healer called Jesus. You went running after the Healer, and now look what has happened! This is what you get for trusting the Healer. Your daughter is dead."

I can hear them now... "You are a leader. What is everyone going to say? What are the Jews going to do? Do you know what disgrace you have brought on the synagogue? You trusted in this Man – this charlatan, this crook, this psychology worker. It serves you right! Your daughter is dead."

Thank God that Jesus was there on the scene with a very positive statement of victory. He declared boldly to them all: *"... the damsel is not dead, but sleepeth"* (Mark 5:39). His declaration elicited a very negative reaction from these professional mourners, for the Bible says that ... *"they laughed him to scorn..."* (Mark 5:40).

We are not dealing only with circumstances that occurred 2,000 years ago. We are dealing with issues right down where we live, relating them to crises we meet in our world of reality today. What are you going to do when you face these kinds of circumstances and problems? How are you going to deal with the forces of unbelief when Satan challenges your authority?

Make this declaration: "When Satan challenges my authority I will stand in the same strength and power demonstrated by the Lord Jesus Christ."

PUT OUT THE MOURNERS

Scripture reading: John 14:1-14

To become a Proof Producer, you must learn to face the negative forces of unbelief. You must have something to carry you through when someone tries to pour cold water all over your experience!

You will never be able to meet these challenges with logic. You cannot meet them with intellect. The devil does not listen to logical arguments, nor does he respond to theological discourses.

Jesus Himself set the example for us in how to solve this problem effectively. How much better off every one of us would be if we always followed this example with diligence. When Jesus confronted unbelief in Jairus' home He put all the scoffers completely out of the way:

> *...But when he had put them all out, he taketh the father and the mother of the damsel, and them that were with him, and entereth in where the damsel was lying.*
>
> *(Mark 5:40)*

When the negative forces of unbelief come – the professional mourners who say miracles are not for today, voices which speak contrary to what God has told us in His Word and in our spirits – never give way to them, not for one minute! Turn your back on them. Look at the bigness of God, not the doubts or scoffing of others.

> *For what if some did not believe? shall their unbelief make the faith of God without effect? God forbid: yea, let God be true, but every man a liar...*
>
> *(Romans 3:3-4)*

When the negative forces of unbelief come against you, you must remain true to your task, acting upon the authority of your heavenly instructions rather than the negativism of man. You must stay true to heaven's task no matter who or what method the enemy uses against you. You must turn your back on the negative forces of unbelief and continue your work for God.

Look at your hands...in them lies the destiny of your family, your nation, even of this entire world. You have the power to produce the proof that will shatter the negative forces of unbelief and raise the dead spiritually, mentally, and physically through the delegated authority of God which rests upon you.

It's time to put the professional mourners out of your life and get to work.

Make this declaration: "I have the power to produce the proof that will shatter the negative forces of unbelief and raise the dead spiritually, mentally, and physically through the delegated authority of God which rests upon me."

RISE UP IN POWER

Scripture reading: Leviticus 20:24-26

The children of Israel would have been forty years better off if they had listened to the positive statements and promises of God instead of the negative report of the spies who went in to spy out the land of Canaan. God had told them:

> *But I have said unto you, Ye shall inherit their land, and I will give it unto you to possess it, a land that floweth with milk and honey: I am the LORD your God, which have separated you from other people.*
>
> *(Leviticus 20:24)*

He had led them out of the bondage of Egypt and right up to the border of this promised land, demonstrating many mighty miracles all along the way. He had fed them, protected them, given them water. Their clothes and shoes did not wear out. There was not an infirm one among them. God Himself led them in a pillar of fire by night and a cloud by day. Every time they turned around, every time they had a need, God met them and He filled their needs.

Yet when they came to the borders of that land which God had promised them, they sent a committee in to look the land over and vote on whether they should possess it and the vote came out ten to two with the negatives leading. These men saw the devil as a giant and themselves as grasshoppers:

> *...We be not able to go up against the people; for they are stronger than we. And they brought up an evil report of the land which they had searched unto the children of Israel, saying, The land, through which we have gone to search it, is a land that eateth up the inhabitants thereof, and all the people that we saw in it are men of a great stature. And there we saw the giants, the sons of Anak, which come of the giants: and we were in our own sight as grasshoppers, and so we were in their sight.*
>
> *(Numbers 13:31-33)*

There were two men who did not believe the negatives, Joshua and Caleb. Caleb delivered the minority report to Moses and the people. He said:

> *...Let us go up at once, and possess it (the land); for we are well able to overcome it.*
>
> *(Numbers 13:30)*

Caleb was overruled, however, and ten negative votes caused the entire host of the children of Israel to miss out on their land of opportunity and to endure countless hardships in the wilderness, wandering for forty years.

It is time for us to stop wandering in the wilderness of confusion, uncertainty, and negativity which is in the Church today and go in to possess everything that God has for us and that He has for others through us!

Make this declaration: "I will no longer wander in the wilderness of confusion, uncertainty, and negativity which is in the Church today. I will go in to possess everything that God has for me and for others through me!"

June 18

The Spirit Of Negativity

Scripture reading: Nehemiah 1:1-11

The strategy of coming against God's people with negativity is a strategy the devil has employed since the beginning. Satan inquired of Eve, *"... Yea, hath God said...?"* (Genesis 3: 1). He implied that God was withholding something good from Eve by preventing her from eating of the Tree of the Knowledge of Good and Evil. The real situation was that God was protecting Eve and all of mankind from the negatives of the curse that would come upon them once she and Adam partook of the forbidden fruit. Eve believed the negatives. She also believed Satan's other negative, *"Ye shall not surely die..."* (Genesis 3:4).

We need to know Satan's devices and how we can combat them. Paul declared: *"... we are not ignorant of his devices"* (2 Corinthians 2:11).

In the time of Nehemiah when the Jews began to rebuild the walls of Jerusalem, Satan threw a whole book of negatives at the Jews. Through enemy leaders named Sanballat and Tobiah, the builders were subjected to all sorts of ploys and ridicule.

> *And he (Sanballat) spake before his brethren and the army of Samaria, and said, What do these feeble Jews? will they fortify themselves? will they sacrifice? will they make an end in a day? will they revive the stones out of the heaps of the rubbish which are burned? Now Tobiah the Ammonite was by him, and he said, Even that which they build, if a fox go up, he shall even break down their stone wall.*
>
> *(Nehemiah 4:2-3)*

This did not move the Jews at all. They just called on God and kept working. Nehemiah reports: *"So built we the wall..."*(Nehemiah 4:6).

Next the enemy tried another tactic:

> *And conspired all of them together to come and to fight against Jerusalem, and to hinder it.*
>
> *(Nehemiah 4:8)*

The spirit of negativism is very contagious if we allow it to get started. Soon it had infected even some of the Jews who began to worry.

> *And Judah said, The strength of the bearers of burdens is decayed, and there is much rubbish; so that we are not able to build the wall. And our adversaries said, They shall not know, neither see, till we come in the midst among them, and slay them, and cause the work to cease.*
>
> *(Nehemiah 8:10-11)*

Thank God that He was still on the scene and again the enemy was defeated and the work for God continued (Nehemiah 8:15).

Have you been infected by the spirit of negativity? If so, be sure to make today's declaration!

Make this declaration: "I rebuke the spirit of negativity from operating in my mind, heart, or spirit."

Stay True To The Task

Scripture reading: Nehemiah 6:1-19

Another tactic used by the enemy against Nehemiah and the building project in Jerusalem was sowing division among the Jews themselves – brethren cheating brethren, brethren accusing brethren. What a great negative that is, even today. That is a tactic of Satan, a negative to stop the work of God. It did not work on the Jews building the wall. They solved their problem and the work went on.

However, the devil is not one to give up easily and he had other tricks up his sleeve. You will find that when he is defeated in one area, he will emerge in another area. That is why we must ever be watchful, prayed up, and full of the Holy Spirit. Sanballat and the enemies had another plan ready when the others failed. This time they invited Nehemiah down for a talk. They wanted to "reason" with him.

There is no reasoning with the devil, and Nehemiah knew that. His message back to the enemy was very positive:

> *I am doing a great work, so that I cannot come down: why should the work cease, whilst I leave it, and come down to you?* *(Nehemiah 6:3)*

The enemy came with these negative suggestions not once but five separate times. Nevertheless, Nehemiah continued to work the works of God and continued to speak with positiveness and authority.

> *Then I sent unto him, saying, There are no such things done as thou sayest, but thou feignest them out of thine own heart.* *(Nehemiah 6:8)*

You would think the devil would quit after all these negatives failed, but he appeared again – this time in the person of a so-called prophet of God who warned Nehemiah to stop building and hide for his life:

> *...Let us meet together in the house of God, within the temple, and let us shut the doors of the temple: for they will come to slay thee; yea, in the night will they come to slay thee.* *(Nehemiah 6:10)*

Satan often will use even well-meaning religious friends and even leaders to try to discourage us. Nehemiah still held to God's positives:

> *And I said, Should such a man as I flee? and who is there, that, being as I am, would go into the temple to save his life? I will not go in. And, lo, I perceived that God had not sent him; but that he pronounced this prophecy against me: for Tobiah and Sanballat had hired him.* *(Nehemiah 6:11-12)*

Not every prophet comes from God. Some have been hired by the enemy. We certainly need to know the difference! In God's power, on God's Word, by God's revelation, Nehemiah stayed true to his task until he had the victory. The enemy became discouraged. The work got done and the wall was finished (Nehemiah 6:15-16).

Make this declaration: "I will stay true to my task until I win the victory. The work will get done. The wall will get finished."

Disciples or Converts?

Scripture reading: Matthew 7:21-23

The "go ye" that Jesus gave to the disciples who followed Him while He was here on earth is the same "go ye," the same task, that we have as His disciples today: To convince a needy world that the crucified Jesus is the resurrected Christ Who lives and meets the needs of mankind today.

Our task is to preach the Gospel by producing the proof of that resurrection power.

There are many, many people who claim to be "preaching the Gospel." We have more preachers than there have ever been, more ministries, more outreaches, more religious television programs, more churches. Yet the world is not being won for Jesus Christ as He meant it to be. Faced as we are with the exploding population of the unsaved and the rampant evil of our present society, there is only one hope we have of reaching this world for Christ. It certainly is not through the ministry as we know it today in its present form.

I believe beyond the shadow of a doubt that we are going to reach this world, this entire world, for Christ before Jesus comes again. That is the purpose, the core, the heart's throb of our ministry. It is so exciting to be a part of what God is doing in these last days, but it will take an army of God's people to complete it. It will take an army of God's people from every walk of life working together in the power of God.

The biggest need in our churches today is for people to become disciples of Jesus Christ. The reason why we have not taken this world is because we have been making converts, not disciples – sons, not soldiers. A convert, according to the dictionary, is a person who is brought over from one belief or view to another. Converts are people who change their minds concerning their religious concepts, then sit in the church and become lazy "spiritual bumps on a log." They become good-for-nothings who spend more time watching television and reading secular literature than they do in prayer, the Word of God, or out winning lost souls.

Converts can use their time to for entertainment and sports and, in the meantime, let whole cities or nations go to hell. Yet they think that one day the trumpet is going to sound and the Lord is going to rapture them. They will have a rude awakening. The Church Jesus is going to rapture is made up of disciples, not just converts. But the Bible says:

> *Not every one that saith unto me, Lord, Lord, shall enter into the kingdom of heaven; but he that doeth the will of my Father which is in heaven.*
> *(Matthew 7:21)*

Make this declaration: "I will not just say 'Lord, Lord' – but I will do the will of God the Father. I will reach out in power and authority to a lost and dying world."

SIGNS, WONDERS, AND MIRACLES

Scripture reading: Acts 3:11-26

When Peter healed the crippled man in Acts 3, the religious leaders demanded of him, "How can faith in the Name of Jesus Christ raise anybody from a crippled, hopeless condition? This man has been a cripple from his mother's womb. Don't you know that Jesus is dead? We not only killed Him but we buried Him in a grave."

Peter said, *"You killed the Prince of life, whom God hath raised from the dead; whereof we are witnesses..."* (Acts 3:15).

Peter and the other disciples became what I call proof producers. When they went into a city, it was not all talk. It is easy to teach. Anybody can instruct somebody else. To face the power of the enemy and produce the proof that Jesus Christ is alive is another matter.

That was the task the Early Church had to perform. They had to convince a world that had seen Jesus in the flesh. They had to convince a world that although they had seen Jesus act like a man – hungry, tired, weary, thirsty, sleepy – that inside that house of muscle and bone dwelt the Son of the living God.

- They produced the proof.
- They turned cities upside down, but they did not do it by just preaching.
- They turned nations upside down, but they did not do it by just preaching.

They did this by the signs and wonders that took place in their ministries when they first went into a community. Each time a notable miracle would take place that shook the entire region.

The first such recorded incident was the healing of the lame man at the Gate of the Temple as is written in the third chapter of the Book of Acts. The man had been lame from the time of his birth, yet:

> *...Peter said, Silver and gold have I none; but such as I have give I thee; In the name of Jesus Christ of Nazareth rise up and walk. And he took him by the right hand, and lifted him up: and immediately his feet and ankle bones received strength. And he leaping up stood, and walked, and entered with them into the temple, walking, and leaping, and praising God.*
>
> *(Act 3:6-9)*

After this tremendous message a living demonstration of Christ took place!

Make this declaration: "I have the same power resident within me as the Apostle Peter. God will work through me in the same way. I will produce the proof."

The Greatest Sermon

Scripture reading: Matthew 10:5-28

The secret to the success of the Early Church was that they made disciples. A disciple is not simply someone who changes his viewpoint. It is someone who is galvanized into action to further those viewpoints.

The dictionary defines a disciple as "one who accepts and assists in spreading the doctrines of another, a convinced adherent of a school or individual." A convinced adherent to Jesus Christ is going to do something more than warm a pew in a church. A disciple is someone who, once he has an experience with Jesus Christ, does not rest night or day. He tells everyone on the street, everyone in his work, about Jesus. He witnesses until he spreads the Gospel of Jesus Christ throughout the whole area of his personal "Jerusalem," "Samaria," and his "uttermost parts of the earth." A disciple does not sit down.

It is going to take an army of such disciples, dedicated men and women, to perform the task Christ has given to the Church. You are part of that army. God is raising you up.

You may think you are very insignificant, but do not underestimate your strength. It takes a lot of privates to make up an army. What kind of an army would we have if we did not have privates? There is no such thing in the army of God as an unimportant private. The very least person can have a part in preaching the Gospel of Jesus Christ to the whole world and making disciples for Him.

The disciples became Proof Producers, as we must too. God is raising up a special breed of disciples in this day…

But ye shall receive power, after that the Holy Ghost is come upon you: and ye shall be witnesses unto me both in Jerusalem, and in all Judea, and in Samaria, and unto the uttermost part of the earth. (Acts 1:8)

Take this verse in the context of what we have been talking about, not in the context in which we use it in our denominational meetings, when we speak of giving witness to Jesus Christ. Jesus was not talking about walking up to someone and saying, "Are you saved?" He was talking about the ability of His followers to show to the world that He is God.

In Acts 3, when Peter was brought before the religious leaders in the synagogue to be questioned about the crippled man who was healed, he said:

> *…ye men of Israel, why marvel ye at this? or why look ye so earnestly on us, as though by our own power or holiness we had made this man to walk? The God of Abraham, and of Isaac, and of Jacob, the God of our fathers, hath glorified his Son Jesus; whom ye delivered up, and denied him in the presence of Pilate…*
> *(Acts 3:12-13)*

What a sermon! Not just silver-tongued oratory, not just verbosity, big words and a lot of spiritual cliches – but a miracle! The greatest sermon that can be preached in the world is that which the power of the Holy Spirit can demonstrate.

Make this declaration: "When I open my mouth to share God's Word, the Holy Spirit will demonstrate God's power through signs and wonders."

THE ACTS OF GOD

Scripture reading: Acts 9:34-43

When Peter came to Lydda, he found a man named Aeneas who had been bedfast for eight years with palsy, he said to him, *"Aeneas, Jesus Christ maketh thee whole: arise, and make thy bed. And he arose immediately. And all that dwelt at Lydda and Saron saw him, and turned to the Lord"* (Acts 9:34-35).

In Joppa, when a godly woman named Dorcas died...

> *...Peter put them all forth, and kneeled down, and prayed; and turning him to the body said, Tabitha, arise. And she opened her eyes: and when she saw Peter, she sat up. And he gave her his hand, and lifted her up, and when he had called the saints and widows, presented her alive. And it was known throughout Joppa, and many believed in the Lord.*
>
> *(Acts 9:40-42)*

Every time the disciples were used of God in this manner, then they could tell the people that these things happened by faith in the Name of Jesus Christ. After the demonstration of the miracles, they could share all about Jesus. They could tell them, "He is the Son of the living God. He died to cleanse you and to forgive you. You can see these great demonstrations of His power happen in your own life. You can be changed and your sins can be forgiven. You can be revolutionized and be made new by that same power of the Holy Ghost which you see demonstrated."

Many say that miracles such as the disciples performed were for 2,000 years ago. They say the day of miracles is over, but there is no such thing as a "day of miracles." There are no such things as Bible days. There is only a God of the Bible. There is only a God of miracles.

You have the same potential today as anyone in the Bible to write another chapter in God's history book by being able to be moved upon by the Holy Spirit's power to work the works of God.

We talk about the "Acts of the Apostles," but the Book of Acts is not about the acts of the apostles. It is only a brief record of the acts of two men, Peter and Paul. We are not even told anything of the other apostles and their acts. However, none of the acts were really the acts of Peter, John, or any of the apostles. They were the acts of the Holy Spirit.

Peter and Paul are gone, but the Holy Spirit is not gone. Men change, but God does not change. Denominational theories change but God never does. We serve the same God as did Abraham, Isaac, Jacob, Moses, Elijah, Daniel, Shadrach, Meshach and Abednego, Peter, Paul, James and John. The acts of God go on long after man has passed from the scene.

Make this declaration: "God is the same. He does not change. The acts of the Holy Spirit flow through my life and ministry just as they did in Bible days."

Wade Into The Waters

Scripture reading: Joshua 3:1-8

One day, God told a man named Joshua, "I want you to take My people Israel across the River Jordan into the Promised Land."

Joshua said, "Wait a minute, God. Don't You know who this is You are talking to?"

God said, "Yes, your name is Joshua."

Joshua said, "Yes, that is right, but You have the wrong man."

God said, "What are you talking about, Joshua?"

Joshua replied, "Well, God, I am not the miracle worker that Moses was. Moses is dead. I can't take the people across this river into the Promised Land."

God told him, "Joshua, I want to tell you something. Moses never led the children of Israel out of the land of Egypt. I did. I was in the cloud by day and the pillar of fire by night. When Moses did not have a road map or a compass, I was his road map and compass. I led him by day and in the night or he never would have made it. I did it. Moses didn't part the Red Sea. I did. Moses didn't feed those people in the wilderness for forty years. I did."

Joshua said, "I never realized that."

God said, "That's all right, Joshua. You just get the Ark of the Covenant, put it on the shoulders of the priests and have them march out into that river."

God is talking to you in that same manner. He is asking you to tread water that is strange to you. He is asking you to cut away some shorelines, some traditions, some historical aspects, and follow Him only.

Joshua, said, "God, I can't put that Ark of the Covenant on the shoulders of the priests and march out into the River Jordan. Moses would not have done it like that."

Never be afraid of God's firsts. Never be afraid to break with tradition. Do not be so tied to your liturgical, denominational apron strings that you cannot cut them loose. God said to Joshua, "I want you to cut away your shorelines no matter if Moses did not do it this way. You do it this way because you are not Moses and I am teaching you."

God said, "Walk out…"

Joshua said, "I can't do it."

God said, "Now, why not?"

"Because I will make a fool of myself."

Imagine yourself in Joshua's place, taking an Ark of the Lord which is the presence of God – the holiest thing in the kingdom of the Lord – and marching it like an idiot on the shoulders of the priests out into a river! Perhaps God has asked you to do something that appears equally as foolish. If so, wade on in the waters!

Make this declaration: "Whatever God asks me to do today – even if it appears foolish – I will do it in faith."

God Is The Proof Producer

Scripture reading: Joshua 3:9-17

The Bible is filled with miracles…

- They were not Moses' miracles; they were God's miracles.
- They were not Peter's miracles; they were God's miracles.
- They were not Paul's miracles; they were God's miracles.

You and I serve the God of miracles. The men and women of Bible times did just what God told them to do, and that is all you and I are supposed to do. Our only task is to obey God:

> *They (those who believe) shall take up serpents; and if they drink any deadly thing, it shall not hurt them; they shall lay hands on the sick, and they shall recover.*
>
> *(Mark 16:18)*

Just do what God tells you to do. God told us to anoint with oil and pray for the sick. Pray the prayer of faith with a heart that dares to believe that what God said will come to pass.

> *Is any sick among you? Let him call for the elders of the church; and let them pray over him, anointing him with oil in the name of the Lord: And the prayer of faith shall save the sick, and the Lord shall raise him up; and if he have committed sins, they shall be forgiven him.*
>
> *(James 5:14-15)*

Just obey God, that is all. Obey His Word and leave the results in His hands because it is up to Him, not you. You are not the miracle worker. I am not the miracle worker. God is the miracle worker.

When the children of Israel journeyed from Egypt to the promised land of Canaan, Joshua was not the miracle worker. Moses was not the miracle worker. God was the miracle worker Who led them, provided for them, and kept them through the perilous journey.

God told Joshua, "You get that Ark of the Covenant, you march it out into the River Jordan, and when the water gets up over the ankle bones of those priests, just watch Me. I will take My great big hand from heaven and I will roll back the River Jordan. I will hold it up on the other side until two million of you journey across and get into the Promised Land. Then I will let the flood tides loose."

> *And as they that bare the ark were come unto Jordan, and the feet of the priests that bare the ark were dipped in the brim of the water, (for Jordan overfloweth all his banks all the time of harvest,) That the waters which came down from above stood and rose up upon an heap.. And the priests that bare the ark of the covenant of the LORD stood firm on dry ground in the midst of Jordan, and all the Israelites passed over on dry ground, until all the people were passed clean over Jordan.*
>
> *(Joshua 3:15-17)*

When we are obedient to God and have acted on His Word, then it is up to Him. He is the real proof producer.

Make this declaration: "God is the real proof producer and He works through me to produce the proof."

YOUR FAITH IS IMPREGNABLE

Scripture reading: Mark 11:12-14;20-24

In the aftermath of the miracle of the cursed fig tree which withered and died, Jesus told His disciples to have God's faith: *"And Jesus answering saith unto them, have faith in God"* (Mark 11:22).

In Matthew's account of this miracle, Jesus gave added instruction on the power of such faith:

> *Verily I say unto you, If ye have faith, and doubt not, ye shall not only do this which is done to the fig tree, but also if ye shall say unto this mountain, Be thou removed, and be thou cast into the sea; it shall be done. And all things, whatsoever ye shall ask in prayer, believing, ye shall receive.*
>
> *(Matthew 21:21-22)*

Everyone would like to have the kind of faith that removes mountains and does exploits for God. We pray for it, we seek it, we struggle in our spiritual lives for such faith. We have thought that God is depending so much on what we possess, what we have, how we arrive.

In seeking to "obtain faith, step out in faith, trying to have faith" etc., men have in effect, tried to put God in a vulnerable position. In doing this they have caused much confusion and frustration. The struggle to have faith has caused untold stress and perplexity in the lives of people who have not received answers to their prayers.

God never intended His people to struggle for faith. Faith is not the product of human struggling! Can we produce faith? No. Faith, like love, comes only from God. Faith is not a natural life force we can struggle to produce.

Faith never fails. There is no such thing as failing faith. There is no such thing as imperfect faith. There is no unreal faith. There is only real faith. In fact, you cannot have faith without having infallibility! The word "infallible" means no margin for error.

How can we, as humans receive something that requires infallibility? Because our faith comes from God. Going even deeper, we understand that the faith God gives is impregnable! The devil cannot penetrate our faith. He cannot destroy it because he cannot get inside it. It is impregnable.

Make this declaration: "Satan cannot penetrate my faith. He cannot destroy it because he cannot get inside it. It is impregnable."

The Gift Of Faith

Scripture reading: Ephesians 2:1-10

Faith is a spiritual force that comes from God. It is given to the believer as a "sixth sense," one that goes beyond the five senses (taste, touch, hearing, seeing and smelling) with which natural man is endowed. It is not a sense that the unbeliever has inherent in himself.

Even the gift of salvation does not come from any natural faith an unsaved person has. He cannot exercise faith to appropriate God in his life because he does not naturally have this life force flowing within him.

Now someone will ask, "What about the Scripture in Ephesians 2:8 which says 'For by grace are ye saved through faith?" Don't stop there. Read the entire Scripture:

> *For by grace are ye saved through faith; and that not of yourselves: it is the gift of God: Not of works, lest any man should boast.*
> *(Ephesians 2:8-9)*

There is no human being in his natural state, with only the five natural senses God has given him, who has the ability to demonstrate the faith necessary to experience the transforming power of the miracle of salvation.

I want to draw you a verbal picture which will help you to stop struggling for faith from this day forward. Your faith life no longer will be like a roller coaster, thinking one day you have mustered enough faith to believe and the next that you cannot have faith to believe for anything.

Picture an unregenerate, unsaved unbeliever. You can testify to him on the street corner, he can come into a crusade service, or in some way he is witnessed to about the cross of Jesus Christ. What characteristic does that man have? He has a will and power over that will.

When you testify to this man or stand on the platform and invite people to come to Jesus Christ, the Spirit of God goes to that man and begins to call the spirit of that person. Now, this man has a will, he has power over that will, and by determination he can set his spirit to believe or to disbelieve. The Spirit of God comes and deals with that man. The man has the ability through the power over his will to surrender his life to Jesus Christ. At that moment when he surrenders his will, the faith that is not of man but the faith of God – God's gift – is imparted into the life of that sinner. At that moment, the transforming work of God is done!

Make this declaration: "God's transforming power is at work in me every time I speak His Word in power and authority."

Stop Struggling

Scripture reading: Romans 5:1-5

Faith is imparted. It is a gift of faith – God's faith. If this great truth of God's faith were taught to every little child of God, to all our dear people, how many struggles it would save them, how many heartaches, how many problems. If only they knew that what they have inside them never has to fluctuate because it is not theirs, it is God's.

It meets every circumstance. It meets every trial. It meets every test. Because it is not our faith, it does not fail when the trial comes. Because it is not our faith, it can meet the sickness that tries to enter into our bodies. Because it is not our faith, it can encounter every trial, every circumstance, every problem, and can make us more than conquerors over every situation. Faith is not something you work up. It is not "psyching yourself out" or self-hypnosis. It is knowing what God knows.

I heard someone telling people in a prayer line, "Now, before you get to that man who is going to lay his hands on you, just keep saying over and over 'I am healed, I am healed, I am healed.' Then, when he lays his hands on you, you are going to be healed."

We are not Christian Scientists. We are not trying to demonstrate mind over matter. Someone said, "If you go to the Word of God and memorize all the Scriptures of faith, that is how you will become a faith person." We do not take anything away from reading the Bible or memorizing the Scriptures. You should hide God's Word in your heart. This is of utmost importance – but the devil does not mind your memorizing all of the Scriptures in the world as long as the net result is that nothing happens.

You cannot get faith by struggling for it any more than you can get love by struggling for it. A person cannot say, "All right, I am going to go to the Scriptures to find all of the verses on love. I will say, 'I love, I love, I love,' then I will be loving." This won't work, because love is a gift of God shed abroad in our hearts by the Holy Ghost which is given unto us (Romans 5:5).

Let that same thing apply to joy. Saying over and over, "I have joy, I have joy, I have joy" will not give you joy. You don't struggle to create joy. It is God's gift to you. The same thing is true of peace. Peace is a gift from God:

> *Peace I leave with you, my peace I give unto you: not as the world giveth, give I unto you. Let not your heart be troubled, neither let it be afraid.*
>
> *(John 14:27)*

You don't struggle to create peace. You don't struggle to create love, joy, or peace. Now, faith is also a gift of God:

> *And Jesus answering saith unto them, Have faith in God.*
>
> *(Mark 11:22)*

Stop struggling!

Make this declaration: "I will no longer struggle. Faith is God's gift to me."

A Consistent Faith

Scripture reading: John 15:1-5

You can watch the lives of many Christians and see that they are up one day and down the next. That has been the history of the Body of Christ. Do you know why this situation exists? It is because we go in our own strength. We try to produce love. We try to produce joy. We try to produce peace. We think we have to produce faith. But man has nothing in himself. You cannot produce love. You cannot produce joy. You cannot produce faith.

How do you change the structure of the roller coaster up-and-down experience into an experience that never falters but always remains steady?

- Here comes a problem. It hits you, but you keep going on.
- Here comes someone who curses you, but you keep loving them. You never falter.
- Here comes some sorrow in your life, but the joy stays effervescent and real.
- Here comes an experience where you think you are going to lose everything and you do not know where to turn, but the peace remains ever constant.

Nothing causes you to go up and down. Your experience stays steady, steadfast, immovable.

We need to realize that man has nothing in himself but what he receives from above. It is not your joy but it is Christ's joy which is manifesting itself in you. It is not your peace but it is Christ's peace which is manifesting itself in you. It is not your love; it is the love of God.

I want to ask an important question which gets right to the heart of our struggle. If we cannot produce love, if we cannot produce joy, if we cannot produce peace, will you tell me how you think that you or I can produce faith which is one of the greatest fruits of the Spirit?

The Bible says that the Lord Jesus is the Vine and you and I are the branches.

> *I am the vine, ye are the branches: He that abideth in me, and I in him, the same bringeth forth much fruit: for without me ye can do nothing.*
> *(John 15:5)*

The branch cannot bear fruit in itself except it abide in the vine. The life is in the vine. The Father is the Husbandman. He tends it. He gave faith to Jesus Christ. Jesus Christ gives it to you and me. You and I are just branches. That is all. You are just a branch; you bear the fruit. You do not produce it; you bear it.

When it is God's faith, it never changes. It never fluctuates. It is always there. The old roller coaster faith ride is gone. Your faith life is consistent because it is God's faith extended in you and through you from the Vine (Jesus), through the branches (you), to the fruit (love, joy, peace, faith, etc.).

Make this declaration: "God's faith is at work within me. It never changes. It never fluctuates. It is always there. My faith life is consistent because it is God's faith extended in me and through me from the Vine (Jesus)."

None Of These Things Move Me

Scripture reading: Acts 20:17-24

The Apostle Paul declared, "None of these things move me...whether I was naked or whether I was in peril, or whether I was shipwrecked or whether I was being beaten."..."*none of these things move me, neither count I my life dear unto myself*"(Acts 20:24).

Paul had God's faith and nothing moved him. He said:

> *Who shall separate us from the love of Christ? shall tribulation, or distress, or persecution, or famine, or nakedness, or peril, or sword?*
> *(Romans 8:35)*

Paul also declared:

> *For our light affliction, which is but for a moment, worketh for us afar more exceeding and eternal weight of glory.* *(2 Corinthians 4:17)*

Paul could make these declarations because the faith within him was not his faith. It was not affected by the circumstances of life. It was the faith of God.

If God depended upon what you call your faith, He would be in terrible trouble. He is not depending on who you are. He is not depending on what you possess. He is only looking at what He can make of you.

What gives us our confidence is the fact that He is not depending on our faith at all. He is depending on His faith flowing through us, His love flowing through us, His joy flowing through us, His peace flowing through us.

It is not what we are, but what God can make of us by His grace. The faith to accomplish the task He has given us is not anything that can be worked up or struggled for – it must be His faith flowing through us as we undertake the task God has given us to do.

As your life becomes a channel of obedience to Him, God will produce the proof through you. God is still the same today; He never changes. *"For I am the Lord, I change not..."* (Malachi 3:6). There is no shadow of turning with God (James 1:17).

Look at your hands. The task that was placed in the hands of the apostles was also placed in your hands – to share the Gospel. We do it on the authority and the commission of the same miracle-working God Who worked with the disciples "with signs following" to authenticate their teaching.

To get this very vital task done we must follow the methods of the disciples. We must produce the proof that Jesus Christ is the resurrected Son of the Living God, that He is the living Word of God, that He is alive and moving in the affairs of men today. We must stay true to the task and be His proof producers.

Look at your hands. God is not depending on the physical strength of those hands. He is depending on His supernatural anointing flowing through your hands and your entire being to accomplish His works. That is why God has confidence to send us out and that is how the walls of spiritual Jerichos in our lives will crumble.

Make this declaration: "God is not depending on the physical strength of my hands. He is depending on His supernatural anointing flowing through my hands and my entire being to accomplish His works."

July

God's servant displays the trophies of God's miracle power at a crusade in Brazil. There is no defeat – no problem, no sickness, no sin – so great that it cannot be shattered by the miracle-working power of God.

July
Overcoming Defeat

Before the October 1973 war, Israel was confident that nothing could penetrate her security and bring death and destruction to her people. A few days before the October war, the ruling Labor party ran large election advertisements in the newspapers declaring, "There is peace on the Suez, our borders are secure, there is prosperity in our land..."

But the Bible warns:

> *For when they shall say, Peace and safety; then sudden destruction cometh upon them ...* *(1 Thessalonians 5:3)*

In the fateful hours before the war began, the story is told that as the leaders reviewed the intelligence reports telling of the coming attack they said among themselves, "It just won't happen, it can't happen, and if it does the enemy will be crushed just like they were in the Six-Day War." The tragic element overlooked was that the enemy was no longer fighting with the same weapons used in 1967.

As the highly superior Israeli pilots swept in for the initial kill, they were picked off like a flight of geese. The enemy knew their skill and knew they could not meet them on an equal basis. They met them instead with newly-developed Russian missiles that could not be disarmed and destroyed like those in the past. The School of Experience has a very high tuition. For Israel, it was the lives of twenty-five hundred of their best fighting men, officers and pilots.

For months after the war, the nation was torn asunder over placing the blame on someone's shoulders for the disaster. They won the war, but it still was a disaster that will be written in the pages of Israeli history. All of the facts have not been evaluated concerning the war of 1973, but all indications are that Israel did not feel they needed any divine help to keep their enemies out of Israeli territory.

This modern-day account is similar to the story of Israel's defeat at Ai in biblical times.

Joshua chapter 7 recounts this dark picture in Israel's history. It was the one and only defeat suffered by God's people as they battled to claim their inheritance in Canaan. It occurred because of the sin of a man named Achan, who disobeyed God by taking the spoils of the Battle of Jericho. God had commanded that the people not touch the spoils. They were to be dedicated to the Lord.

When Achan's sin was exposed, Joshua dealt with it as God instructed and then Israel returned to battle Ai once again. This time, they were victorious:

> *For Joshua drew not his hand back, wherewith he stretched out the spear, until he had utterly destroyed all the inhabitants of Ai.*
> *(Joshua 8:26)*

In the past, like the nation of Israel, you may have suffered resounding defeats at the hands of the enemy. You may have lost spiritual battles for your children, your ministry, your home, or your finances. Now all of that is about to change. During our devotional readings this month, you will learn how to overcome defeat and take back territory Satan has stolen from you.

Miscalculating the enemy's power in either the natural or the spiritual world can mean disaster. In the natural world it can mean physical death, but in the spiritual world it results in eternal death. This month we are going to continue our journey to claim our spiritual inheritance as we learn how to overcome defeat. We will learn seven steps to victory that will help us avoid the "Ais" of life. We will learn what to do when we face seemingly innumerable odds as did a man named King Jehoshaphat. In the process, we will learn that God has planned no defeats for us!

Victory In Every Circumstance

Scripture reading: Psalm 20:1-9

There is absolutely nothing Satan can bring into your life that can destroy you – nothing!

There is no weapon, no strategy that he can devise against you that is powerful enough to defeat you – not one!

Satan's objective as the destroyer is to destroy everything you have – your home, your family, your health, your finances, your career, your ministry – everything! His strategy is to use your circumstances to defeat you. He will create havoc in the circumstances in your life, bringing affliction and disease upon your body, attacking your children and loved ones, causing strife in your marriage relationship, binding your finances, creating problems on your job.

Satan will grab hold of your mind in a vice-like grip of fear, worry, and doubt and try to keep your mind in a constant state of turmoil. Using your circumstances, Satan will try to wear you down physically, mentally, and spiritually until you are so discouraged and depressed that you are ready to throw your hands up in defeat.

Satan will relentlessly attack you through your circumstances until you become too tired to fight and resist his attacks, until you are caught in a vicious cycle that paralyzes and hinders you from releasing your faith and walking in the victory God has provided for you.

Make no mistake about it, Satan is out to destroy you. But, there is no circumstance, no trial, no persecution, regardless of how terrible it may be that Satan can bring into your life that can defeat you. As the destroyer…

…Satan may wrack our bodies with pain, sickness, and disease.

…Satan may bind our finances and even bring circumstances into our lives to cause us to lose everything we own – houses, lands, or life's savings.

…Satan may do his best to turn our mates against us and even succeed in destroying our marriages.

…Satan may try to destroy our children and loved ones through alcohol or drugs.

…He may take our lives and the lives of our loved ones through death.

…But none of these circumstances – pain, sickness, financial problems, problems in our marriages, family problems, persecution, or death – can threaten us! God has promised us victory in every circumstance we face.

Make this declaration: "God has promised me victory in every circumstance. He has planned no defeats for me."

I Am Ready

Scripture reading: 2 Corinthians 1:8-10

God has planned and provided for us to have complete and lasting victory during our trials, even in the face of death. He hasn't promised that we would never know pain and sickness, never face persecution or suffer hardships. On the contrary, we know that *"...all that will live godly in Christ Jesus shall suffer persecution"* (2 Timothy 3:12). Jesus said, *"In the world ye shall have tribulation..."* (John 16:33). We know that our faith is going to be tested and tried, but He has promised us victory in every circumstance we face.

Paul told the Corinthians: *"We are afflicted in every way, but not crushed; perplexed but not despairing; persecuted, but not forsaken; struck down, but not destroyed"* (2 Corinthians 4:8-9, NAS).

The Apostle Paul faced circumstances that were beyond his strength to endure, even to the point of despairing for his own life. He told the believers in Corinth:

> *We do not want you to be uninformed, brothers, about the hardships we suffered in the province of Asia. We were under great pressure, far beyond our ability to endure, so that we despaired even of life. Indeed, in our hearts we felt the sentence of death. But this happened that we might not rely on ourselves but on God, who raises the dead. He has delivered us from such a deadly peril, and he will deliver us. On him we have set our hope that he will continue to deliver us.*
>
> *(2 Corinthians 1:8-10, NIV)*

Paul did not moan, groan, and complain about the hardships he suffered. He did not question why God allowed these things to happen to him. He knew God was in control of his circumstances.

Paul was not shaken by the hardships and sufferings he endured. He was not trembling and afraid of what Satan might use to try to defeat him. He didn't say, "Satan just about got me that time. I just barely made it. I'm so weak and tired. I just don't know how I'm going to make it."

No, he didn't say that. Something had been forged deep within his spirit. He knew that God had provided victory for him in the present. He said, "He will deliver us." He knew that regardless of what he might face, God had provided victory for him in the future. He said, "On Him we have set our hope that He will continue to deliver us."

It was knowing that God had provided victory for him in the past, present, and future, that enabled Paul to face death without fear in complete victory. As he was preparing to die, he said, *"For I am now ready to be offered, and the time of my departure is at hand"* (2 Timothy 4:6). He wasn't afraid of any of the circumstances he would face. He said, "I am ready."

Make this declaration: "God has provided victory for me in the past. He gives me victory in the present and future. I am ready for any and everything through His power."

What Will You Lose?

Scripture reading: 2 Chronicles 20:1-4

Second Chronicles 20 could be called "The Great Power Play," for the enemies of God put on a display of unified might designed to bring fear into the hearts of the Israelites. This same spectacle is being repeated before our very eyes every day. Satan is the master of the power play.

He accomplishes this on a collective basis against the Church. Sometimes when I look at the carnal programs of evangelism conducted by our churches, I feel that the world is dictating the path the Church should follow far more than the Holy Spirit. I believe in utilizing every means that our technological society provides for the spreading of the Gospel to win men and women to Jesus Christ, but I also believe that the power of the Holy Spirit has to remain the foundation of every effort, and must be the common denominator between us, or we are only a sounding brass or a tinkling cymbal.

What makes a person or church body resort to carnal weapons in spiritual warfare?

It is fear – fear of failure, fear of being the object of ridicule, fear of being a despised minority. Fear has been stated to be man's worst enemy, and this has been proven over and over again to be true. President Franklin Delano Roosevelt said, "The only thing that we have to fear is fear itself."

Where does fear come from? The Bible says, *"Fear hath torment"* (1 John 4:18). We know that this type of fear does not come from God. Paul said to Timothy, *"For God hath not given us a spirit of fear; but of power, and of love, and of a sound mind"* (2 Timothy 1:7). If this kind of fear does not come from God, then it must come from Satan.

In our Scripture reading for today, we see that three nations, Ammon, Moab and Mount Seir, formed a conspiracy to come against Judah and to utterly wipe it out. Intelligence information of the massive army planning to attack Judah reached Jehoshaphat, King of Judah. The first thing that happened was that fear struck his heart, for he knew that in the natural his army was no match for the great numbers that were formed against him.

At this point fear could have completely paralyzed the nation of Judah and they could have been utterly wiped out as a nation – you know the ramifications of that, for the Messiah had to come from the tribe of Judah. If there were no Judah, then there could be no Jesus, there would be no one who could die for the salvation of the world, for the Messiah was to come from the tribe of Judah (Matthew 2:6).

Think about this today: What may be forever lost to you if you are paralyzed by fear? What opportunities will you miss? What victories will be lost?

Make this declaration: "I will take advantage of every opportunity God has for me today. I will not be paralyzed by fear."

THE POSITIVE SIDE OF FEAR

Scripture reading: Psalm 23:1-6

The immediate results of fear determine life or death, paralysis or victory for a nation and an individual. Fear can be our greatest enemy, but it can also have positive results if we react to it in the right way.

In every truth there is a negative side and a positive side. Did you know there is a positive side of fear? Most of us think of fear as our enemy; but fear can also be a blessing in disguise. Can you accept that?

If fear drives us to our knees to begin to travail and pray with a fervency that begins to move the hand of God that moves the world, then that fear has been channeled to be a positive influence in our life. It has forced us into closer relationship and communion with God and has worked for our good.

If fear drives us to the Word of God to seek answers from God, then it has accomplished a great work in our life. Sad to say, in many cases, fear is the only thing that will accomplish that objective.

Some people never become interested in the Baptism of the Holy Spirit until the fear of being overcome by their problems drives them to seek a deeper meaning to their spiritual experience.

Here is a tragic, but true commentary: Some people will never seek God in a meaningful way until tragedy strikes them, their children, or their loved ones. When this happens, all of a sudden we are searching the Word of God and books on more effectual prayer.

We may live for ourselves all our lives with little or no attention to the things of God, but when tragedy strikes and there is no human answer, then our thoughts and hearts are turned in fear toward God. We start recalling the preacher's messages, and the admonition of our godly parents who, up until this time, we may have considered fanatic and out of touch with this generation. Into the house of God we rush with our petitions, "Lord, please help me."

What about you? Are you reacting to fear in a positive or negative way? Is fear paralyzing you and inhibiting your spiritual growth and advancement? Does it loom in your pathway like an impenetrable fortress or is it positively working on your behalf?

Make this declaration: "Fear, get out of my way. You are no longer an impenetrable fortress in my life. You will not block the pathway God has assigned for me."

A Spiritual Formula

Scripture reading: Isaiah 58:6-14

Jehoshaphat was so filled with fear that he called a fast. What do you think would happen to our nation right now if every pastor and denominational leader in America would be so filled with fear concerning the future that they would stand before their congregations next Sunday and tell the people, "Whether you realize it or not we are fighting right now in a life and death struggle and there is only one way that we are going to survive, remain a free nation, and avoid a holocaust and sudden destruction and that is if we all immediately begin to fast and pray that God shall deliver us." Can you imagine what would happen?

I don't think that you can. I know I can't. I can't begin to fathom the repentance and the sweeping move of the Holy Spirit that would begin to cleanse, purge, and purify our government and religious institutions.

The Scripture declares, *"Jehoshaphat feared"* (2 Chronicles 20:3), and when he feared he set himself to seek the face of God. Fear acted as a spur to stir him to seek the Lord. Many people in the world would never be living for God today if there had not been a circumstance which came into their lives which caused fear and drove them to the place of prayer. This fear was the spark to setting their spiritual life in order.

Most people reading this devotional guide will fall into one of two general classifications:

One: You are currently having a crisis in your life and you know the fear that I am talking about.

Two: You do not have a current crisis in your life, but you are dissatisfied. You know that you lack victory and power to live for God. Deep down in your heart you also know that these deficiencies are due to compromise in various areas of your life which you know are not pleasing to God.

Whichever category you are in, during the next few days we are going to meditate on seven steps to personal victory – a spiritual formula that will be a weapon in your hands to bring you deliverance for body, soul, and spirit.

Make this declaration: "I will prepare my mind, in the Name of Jesus, to receive a breakthrough to enable my spirit to comprehend this revelation of truth from Your Word."

The Greatness Of God

Scripture reading: 2 Chronicles 20:5-7

During the next few days we are going to mediate on seven steps to personal victory taught in 2 Chronicles 20 where Jehoshaphat faced an overwhelming challenge as a massive enemy army advanced towards him.

The first step Jehoshaphat took toward the solution of his problem was that he fully recognized Who he was calling upon. Jehoshaphat established that the God that he was calling upon is the God of Heaven, the God of all creation.

To let the magnitude of that statement really grip you, just step outside on a clear night, look up into the starry sky, and locate the family of the Big Dipper. In the bowl of that Dipper, the astronomers now tell us that they can locate more than one million galaxies of stars. They estimate that in the entire universe there are more than one billion galaxies.

Jehoshaphat remembered the words of King David, who said, *"The heavens declare the glory of God; and the firmament sheweth his handiwork"* (Psalm 19:1). If you or I had been in Jehoshaphat's shoes at this moment with his nation facing annihilation, would we simply be recounting the glories of God and what He had done in the past? Or would we be doing something else?

Many belong to what I call the "Gimme Club." That is the vernacular for the "Give me, give me, give me" prayers that are so common in the Christian realm. If doom were hanging over our heads, the first thing we would do is cry out to God to immediately destroy our enemy or to deliver us completely from the enemy's presence.

But what did Jehoshaphat do? It is strange that he never mentions his problems at the beginning of his prayer. He approaches God as the God of creation and the God of his spiritual forefathers, Abraham, Isaac, and Jacob. At the very outset he is recounting the power and glory of his Heavenly Father. He actually is starting out by acknowledging God above all.

Do you think he is saying all of this to build God's confidence in His own ability to deliver Israel? Do you think that he really intends to tell God something which God doesn't already know? Of course not! What Jehoshaphat is saying is for the purpose of bolstering his own faith in order to remind himself of Who he is really speaking to and the potential power of God that can be unleashed on behalf of His people.

If you read between the lines here, you will hear Jehoshaphat say, "Here comes Ammon, Moab and Mount Seir, but God, I just want You to know something: I'm going to ask You to do something that I know You are fully capable of doing, have done before, and will do again."

Instead of coming to God under the heavy burden of those attacking nations, he is acknowledging God's greatness.

Make this declaration: "O Lord God, are you not still God in heaven? Don't you rule over all the kingdoms of the heathen? And in Your hand is there not power and might, so that none is able to withstand You?"

An Unlimited God

Scripture reading: 2 Chronicles 20:8-12

Today let's look at the next portion of Jehoshaphat's prayer. He said, *"Rulest not thou over all the kingdoms of the heathen? and in thine hand is there not power and might, so that none is able to withstand thee?"* (verse 6).

Jehoshaphat exclaims, "You are a God of Power," but when he speaks of power in this context, he is not referring to something ancient, but to an active, moving, living, available, vital power. Not a latent power that just has potential, but a power that is moving and has a proven record of performance.

The God we serve is an unlimited God! Go back and repeat that phrase. Say it over and over again until its mighty truth fills your being. Our God knows no limit.

Jehoshaphat knew that God was able to open the Red Sea to let the children of Israel pass through. The same God fed the entire nation of Israel in the wilderness for forty years. He is the God that could send down fire from heaven. He is the God that could cause an axe head to swim and a donkey to talk. The false prophets could not withstand His power. God's fire consumed the offering and the altar of Elijah. The lion could not withstand the power of God and Samson conquered it. The Philistines could not withstand His power and Jonathan and his armor bearer put them to flight. It was His power that shielded Shadrach, Meshach, and Abednego in the fiery furnace that was heated seven times hotter than normal. It was His power that gave the lions lockjaw when the Prophet Daniel spent the night in their den.

Sickness could not withstand His power as Jesus of Nazareth went about doing good and healing all who were oppressed of the devil. He said to the blind, "See," and the dark scales fell from their eyes. He said to the deaf, "Hear," and the sensation of sound began to register in their brains. He declared to the mute, "Speak," and those paralyzed tongues were released by the power of God and began to function.

The natural elements could not oppose the power of Jesus. When He spoke to them to be calm, they stopped their tempest without putting up a further argument. The devil could not withstand His power as He spoke to the demon possessed and said, "Come out," and immediately they were set free. Jehoshaphat is saying, "God, I know You can do it. God I know You will do it. You will deliver Your people!"

You can make a similar declaration: "God, You will heal me from this cancer. You will deliver me from this rheumatoid arthritis. You will deliver my son from the bondage of drugs. You will bring my wayward, backslidden daughter back home again. You will save my mate and unite our family once again."

Make this declaration: Personalize your declaration today as you complete this phrase: "Lord, I know you will…"

July 8

THE FOUNDATION OF ANSWERED PRAYER

Scripture reading: 2 Chronicles 6:28-31

When faced with a powerful enemy, King Jehoshaphat declared, "God, I know that it is not by might nor by power that we can ever defeat the enemy, no matter how large our armies are, nor how small the enemy's armies. Lord, You have shown Israel over and over again that when they had the stronger army by numbers and trained warriors, they were defeated if You were not fighting for them. We know that our answers are not in numbers, but our answer is in Your power to deliver."

Think of that for a moment. Begin to apply that thinking to your crises, to all the problems that are facing you today. It is not more money that will solve your financial problem – it is the blessing and power of God active and alive upon what you already possess, no matter how much it is. The answer to your sickness is not in greater or more expensive medical treatments or drugs, it is in the delivering power of Jesus Christ.

Jehoshaphat clearly established in his own mind exactly Who he was asking for deliverance and what the potential of that power had for him. In his hour of trial, Jehoshaphat, kept calm – or like our generation would say, "He kept his cool." Fear drove him to prayer, but it was not a panic prayer. It was a prayer that came from a sincere heart and mind that knew exactly where his petitions should be directed. As he heard himself recite the declaration of the power and glory of God, genuine faith began to build in his heart to approach God with his request. When he reached the point of asking, his faith had risen far above wishful thinking into the realm of positive knowledge of the delivering power of Jehovah God.

There is a dynamic spiritual secret of knowing how to come into God's court, recognizing you are not coming to a man, a vessel, or instrument that is being used by God as a channel for blessing. You are coming to your very own Heavenly Father – the God Who created the heavens and the earth. You are actually speaking directly to Him.

You are not saying the Apostles Creed or reciting the Lord's Prayer. You are communicating with the greatest power in heaven and earth. Think of it! God Almighty Himself.

Take this first step now and begin to put the foundation of answered prayer into place by making today's declaration:

Make this declaration: "You are God. Everything in heaven and earth is under Your power. You are above all."

Remembering What God Has Done

Scripture reading: Isaiah 43:1-7

Not only did Jehoshaphat recognize he was calling upon God Almighty, he also remembered what God did for him and the nation of Israel in times past. That is the second step to personal victory: Remember what God has done for you in times past.

> *Art not thou our God, who didst drive out the inhabitants of this land before thy people Israel, and gavest it to the seed of Abraham thy friend for ever?*
> *(2 Chronicles 20:7)*

It is important that you know how God delivered in the past. It strengthens your faith to know God has always delivered His people from every physical and spiritual affliction into which they have fallen.

Of equal importance is the knowledge of your personal relationship with God. Jehoshaphat recited the historical evidence that God was with Israel in the past; but now, he declares that God's power and presence are not only past tense. God not only was, but always is the God of the present, the now, the immediate. God is *"a very present help in trouble"* (Psalm 46:1).

One of the greatest truths ever written in the Bible is in the 23rd Psalm where King David began by declaring, *"The Lord is my Shepherd."* Here we have the personal application that God is not only God of the present moment, but He is "my shepherd" – not just any shepherd, but a personal God.

Jehoshaphat found the secret of knowing God in a personal way. The historical connection was not good enough. Your mother, father, grandfather, grandmother, sister, or brother can know the Lord, but that is not good enough for you. You could be the great-great grandson of Martin Luther or John Wesley, but that doesn't really do you any good in your hour of need unless you can address God as your very own.

Repeatedly in Scripture, God declares His personal relationship with us. In Isaiah, we hear this great promise of the Lord to his people. *"When thou passest through the waters, I will be with thee; and through the rivers, they shall not overflow thee; when thou walkest through the fire, thou shalt not be burned; neither shall the flame kindle upon thee. For I am the Lord thy God, the Holy One of Israel, thy Saviour"* (Isaiah 43:2-3).

Note the personal application here: *"I will be with thee"* – not a messenger; not someone appointed, either heavenly or earthly; but God personally will be right there in the midst of your problem to deliver you. God will deliver you in the times of trouble. The fires may rage, but if you keep your trust in God, you will never be burned. The waters may rise all around you, but God will be right there to see that they never overcome you. You will not drown.

Make this declaration: "When I pass through the waters, He will be with me; and through the rivers, they shall not overflow me; when I walk through the fire, I will not be burned; neither shall the flame kindle upon me. For God is my God, the Holy One of Israel, my Savior."

MOVING FROM STATIC TO ACTIVE

Scripture reading: Ephesians 2:11-22

God is not only big enough to take care of the entire universe, but He is also interested in the smallest details of your life. He is concerned about you personally.

Spiritual progress is impossible without a personal relationship with God. In the Old Testament, the relationship was through the symbols of the Temple and the priesthood, but today our personal relationship with God is directly through Jesus Christ, His Son. Before Christ, the relationship had to be through symbols. As the sacrifice for sin was imperfect, there always was a wall or partition, a middleman, between God and man. Now that Jesus removed the middle wall or partition we can boldly enter into the Holy of Holies and make known our requests to God in prayer (Ephesians 2:14).

If Jehoshaphat could speak of the Lord in such a personal way and He only knew Him by His acts – the symbols of the Temple and priesthood and the shadows of the law of Moses – how much more should our faith be lifted up to believe God for the impossible.

Our relationship is no longer through an intermediary, nor is our salvation and the remission of our sins through the shedding of the blood of bulls and goats, but through the sacrifice of His Son, Jesus Christ. If Jehoshaphat was able to touch God with such a prayer, we have the potential for far greater victories for the Kingdom of God.

Jehoshaphat continues his prayer: *"Are not thou our God, who didst drive out the inhabitants of this land before thy people Israel, and gavest it to the seed of Abraham thy friend forever?"* (2 Chronicles 20:7) This is the turning point of this great prayer. Here is the transition from just the fact of Who God was to what the Lord God had done. Now we are moving from the static to the active. By remembering not only Who God was but what He had done, Jehoshaphat was moving from wishful thinking to positive action.

It is one thing to wait and another thing to move. Your faith can grow by recounting the historical acts of God, which we have already mentioned, but do not stop there because that is only the beginning.

Beloved, I believe this is the day when God wants you to move beyond recounting what He has done for you in the past to recognizing who He is in the present. You must move from the static to the active.

Make this declaration: "Today, I will move beyond recounting what God has done for me in the past to recognizing who He is in the present. I will move from the static to the active."

Recalling God's Provision

Scripture reading: Psalm 42:1-11

Take some time today to look at your own life and recall the mighty provision of God in spiritual and physical deliverance in your own life. Start at the point of your salvation. Stop for a moment and think what God has saved you from and the transformation that has come into your life since you have known Jesus as your personal Savior.

Can you remember something God did for you in the past – that answered prayer, that healing in your body, that financial miracle, the blessing in your home? Begin to recall the acts of physical healing and deliverance that you or your own immediate family have received.

If you are a new-born Christian and none of your family have ever had a real born-again experience, then read in the Bible the miracle stories of God's deliverance. Read what God has done in the past, but also be aware of the fact that God is still performing outstanding miracles in our day because *"Jesus Christ is the same yesterday and today and forever"* (Hebrews 13:8, NIV).

Although in the natural you may not have a family member that has had a true spiritual experience with God, you are part of the great family of God in the earth today and when you hear of a miracle someone else has received, you can rejoice just as if it were your own flesh and blood.

Realize that God is no respecter of persons and what He has done for anyone else, He will do for you in your life. Don't forget for one moment that the same nail-scarred hand that was laid upon your sin-fevered brow brought you perfect peace which is the greatest miracle in all the world. God is standing by your side right now to deliver you from the problem that is troubling you today. What God did for you in the past, He will do for you again. Remember what God did for you in the past and you will know what He is going to do for you in the future.

Why is there such a tremendous interest in the occult, and in particular, predictions of the future in the realm of psychic phenomena, fortune-telling, and astrology in this present hour? People have never been more insecure about the future. They want to know what is going to happen, whether it be good or bad. We don't need any crystal-ball gazer to tell us how to solve our problems and that the daily astrology chart is the guide of our life. Our faith stands on the Lord Jesus Christ, the Son of the Living God!

Make this declaration: "My faith stands in the Lord Jesus Christ, the Son of the Living God. He knows my future. He holds my today and tomorrow in His hands."

Standing On The Word

Scripture reading: Psalm 119:9,25,41,58,76,154,169,170

What has God taught us so far as we have mediated on the story of King Jehoshaphat?
One: We begin our petition with praise to God and a declaration of Who He really is.
Two: We begin to recall what He has done in the past, both in the Bible and through the living testimony of our lives, beginning with our salvation and the other mighty miracles He has done and is doing in this very hour.

Are you ready today for the next step to personal victory? The next step is to stand!

Jehoshaphat declared, *"If, when evil cometh upon us, as the sword, judgment, or pestilence, or famine, we stand before this house, and in thy presence, (for thy name is in this house,) and cry unto thee in our affliction, then thou wilt hear and help"* (2 Chronicles 20:9). After Jehoshaphat declared exactly Who he was praying to and recognized what God had done for him in times past, he then began to recite some examples of God's power.

We live in a very power-conscious world. Our television and newspapers remind us every day of the power of governments, of military might, and nuclear bombs. God is going to put such a spiritual foundation under you that when you pray you will not pray some formula or pre-printed prayer. You are not coming to a man, you are coming to God, the one that created the heavens and the earth, the one that rules over all the Kingdom. The authority in this world is in the hands of Almighty God and the greatest power in all the world is the power of the Word of the Living God.

Now we see Jehoshaphat lay the cornerstone of his prayer, declaring that the very basis of his approaching God is that he is standing before the house which contained His Name and His Word. He resorted to the highest authority known to man – the Word of God.

What Jehoshaphat is saying is, "Lord, the petition that I am about to make is not based on folklore, mythology, the pagan religions of our neighbors, superstition, witchcraft, or any form of demon power, but I stand here upon Your Word for You have declared Yourself to be the God of Israel." Jehoshaphat stood upon the Word!

You cannot make a wishful request if you are standing upon the Word of God. We only deal in wishful thinking in our prayer life when we are not actually sure of what we really want. If you really want your prayers answered, then make a positive request based upon the Word of God.

What are you trusting God for today? Perhaps you have prayed about the matter many times. Whether it is for healing, financial help, deliverance, or a lost loved one, pray for your need again today. This time, make your request based upon the Word of God.

Make this declaration: "I will make my requests to God today on the basis of His Word, knowing I will receive because God cannot lie. He will be true to His Word."

How The Word Becomes Effectual

Scripture reading: Acts 2:1-19

Why would anyone say prayers if they really didn't want their prayers answered? Simply because they are going through a form of self-imposed spiritual obligation.

"Lord, deliver me from cigarettes!" How many times that prayer has been prayed from an insincere heart that really doesn't want to give up cigarettes. Record the difference in the tone of their prayer from the time that they are just paying lip service to the habit until the time they pray after the doctor tells them that they have lung cancer and there is nothing medical science can do for them. Now we hear a prayer from the depths of their being. Now they begin to search out the Word of God to really stand upon.

We have often heard the phrase, "God answers prayer." It would be more correctly stated that "God answers fervent prayer that is based upon His Word." Take the example in the Book of Acts. The Apostles are all gathered together praying that Peter will be delivered from prison. Suddenly, a knock comes at the door.

"Who is there?"

"It's me, Peter! Let me in."

"Please don't disturb us right now. We are in the middle of a very important prayer meeting."

They had a hard time accepting the fact that their prayers were answered. I believe that Peter was delivered from prison as a result of his own prayers and petition to God, for with stocks on his feet in that cold, clammy dungeon his prayer had far more fire in it than the sequestered believers in their prayer closet.

The most powerful words in the Scriptures are, *"Thus saith the Lord."* There is a stamp of divine intervention when we stand upon the Word of God, claim the promises that it contains, and declare, *"Thus saith the Lord."* The Bible tells us that *"God is not a man, that he should lie; neither the son of man, that he should repent"* (Numbers 23:19).

What do we mean when we say "stand upon the Word?" Does it mean to take the Bible and place your hand upon it? Of course not! It means that the promises that are written in God's Word are also written in our heart, and only when it is transferred from the parchment to the tablets of our heart and mind does that Word become effectual for our lives.

Make this declaration: "The Word of God is being transferred from the parchment pages to the tablets of my heart and mind. The Word is becoming effectual in my life."

July 14

Depth In The Word

Scripture reading: Psalm 103:1-22

We must know what God has said in His Word. We must be able to recall His promises in order for them to be activated in our prayer life.

We must know that, *"He was wounded for our transgressions, he was bruised for our iniquities: the chastisement of our peace was upon him; and with his stripes we are healed"* (Isaiah 53:5).

We must know that David the psalmist declared under the inspiration of the Holy Spirit, *"Bless the LORD, Oh my soul, and forget not all his benefits: Who forgiveth all thine iniquities; who healeth all thy diseases"* (Psalm 103:2-3).

It is not enough in your hour of need to look at the Bible sitting on the shelf and then say, "Yes, Lord, I believe everything that is written in that book," when you really don't know what is in there because you have never taken the time to find out.

Does this mean that you have to be a Bible scholar to have answered prayer? No! In fact, here is where the standing – which is one of our steps to victory – comes in. You not only must know the Word of God, you must put it into action.

There is another very important reason for standing upon the Word of God as we make our petitions known to our Father in Heaven. We are living in an age of confusion where people are running about eagerly following any new thing that comes along and it seems that the farther that it is from the real truth of the Gospel, the more people are ready to follow.

Sad to say, confusion has taken hold of the hearts and minds of many of God's people. I can't tell you the number of people that have come to me in our Schools of Ministry and declared that they were in an utter state of confusion. They have told me that even in their Spirit-filled churches they are confronted with conflicting doctrines from one week to the next until they don't know where they are.

It is the devil's aim to bring confusion to God's people and it comes through a lack of knowledge of what the Bible really says. Sometimes teaching is only considered deep if a person cannot fully understand it. I call that confusing! I know that there is a teaching that the Holy Spirit wants to bring to every Christian that is beyond the basic foundations of our Christian life.

Depth, to me, is the degree of power of the Spirit in the life of the believer to live a more victorious, overcoming, meaningful life for the Kingdom of God. Teaching that motivates a person to greater holiness, not by separating himself from the world but by being able to live in the world, be a witness for Christ in the world and yet keep himself holy before God – that is what I call depth.

Make this declaration: "As I go deeper into the Word of God, the power of the Holy Spirit will be released in my life to enable me to live a more victorious, overcoming, meaningful life for the Kingdom of God."

Cigarettes In Your Pocket

Scripture reading: Isaiah 43:18-25

I often tell those that come to the School of Ministry that I am not interested if they get one blessing in the whole school. My objective is that their lives become infused with the power of God so that they may leave the school not just with a heavenly vision, but with the power to go out and defeat the work of the enemy.

Never let spiritual depth destroy your sense of purpose in your Christian life or you will know that you are on the wrong track. Let the Word of God have a place in your heart and when you approach the presence of the Lord, as Jehoshaphat, you will know exactly what to expect.

Only the Word of God can let you know exactly where you stand. The Bible says, *"The word of God is quick, and powerful, and sharper than any two-edged sword, piercing even to the dividing asunder of soul and spirit, and of the joints and the marrow, and is a discerner of the thoughts and intents of the heart"* (Hebrews 4:12).

It has been said, "Above all: to thine own self be true." Those are beautiful, commendable words from the pen of a great writer. The only problem is that it is not always possible. When we have allowed sin and compromise to creep into our lives, we no longer are able to judge what is really truth.

Isaiah speaks of a tragic condition where the judgment of God is upon a man, but he never really comprehends what has happened to him:

> *Therefore he hath poured upon him the fury of his anger, and the strength of battle; and it hath set him on fire around about, yet he knew not; and it burned him, yet he laid it not to heart.* *(Isaiah 42:25)*

Why didn't he know the meaning of it? I have a doctor friend in one of the largest medical centers in America. He told me that patients will come into his office for help, desperate because they have only been given weeks to live. Some of them already have a gaping hole in their throat and an electronic voice box, yet both of their vest pockets will be bulging with packs of cigarettes. When he asks them if they are not aware that those cigarettes are the source of their suffering, they reply that they don't think that the evidence of the link between cigarettes and cancer is really conclusive.

Hidden sins in our life are like cigarettes in the pocket. Sin brings forth death. It burns, but we have not yet laid it to heart. This is why we must stand upon the Word of God because it will reveal the real thoughts and intents of our hearts. It will restore tender flesh to the areas that have become calloused.

Make this declaration: "God will reveal the thoughts and intents of my heart. I will repent of my sin and He will replace the hardened areas of my heart with a heart that is tender towards God and His Word."

July 16

YOUR CONFIDENCE IS RISING

Scripture reading: Psalm 119:129-136

Beloved, today…

…If you need salvation, stand upon the Word of God which says, *"That if thou shalt confess with thy mouth the Lord Jesus, and shalt believe in thine heart that God hath raised him from the dead, thou shalt be saved"* (Romans 10:9).

…If you feel rejected, stand upon the Word of God which says, *"…him that cometh to me I will in no wise cast out"* (John 6:37) and *"But as many as received him to them gave he power to become the sons of God"* (John 1:12).

…If you need forgiveness, stand upon God's Word which says, *"If we confess our sins, he is faithful and just to forgive us our sins, and to cleanse us from all unrighteousness"* (1 John 1:9).

…If you are discouraged and feel that God has forsaken you, stand upon His Word that declares, *"I will never leave thee nor forsake thee"(Hebrews 13:5) and "Lo, I am with you always, even unto the end of the world"* (Matthew 28:20).

…If you are sick and need to be healed, stand upon God's Word. *"I am the Lord that healeth thee"* (Exodus 15:26) and the God *"Who forgiveth all thine iniquities: who healeth all thy diseases"* (Psalm 103:3)"…and *"with his stripes we are healed"* (Isaiah 53:5).

Right now take this third step to victory for your need: Stand upon God's Word! I know that there is coming over you the sense of God's presence – but more than that, something is beginning to happen in your spirit. A confidence is rising and an assurance is surfacing.

You have not even told God what your need is yet! But this is true prayer, communicating in the Spirit, your spirit with God's Spirit in communion. You are not so enraptured with your need that you have the spiritual "gimmies" – "Lord, gimme this, Lord, I need that. Oh God, don't you see the trouble, the heartache, the pain I am having?" Instead, you are entering into His courts with praise and thanksgiving already in your heart and upon your lips.

Simply recognize that you are calling upon God Almighty; acknowledge that in His hand there is power and might and nothing or no one is able to withstand Him; and stand firm in the absolute authority of His promises!

Can't you feel it? Your confidence is rising! Your fear is dissipating! You are ready to take the next step to personal victory.

Make this declaration: "My confidence is rising. My fear is dissipating. I am ready to take the next step to victory."

Utter Dependence Upon God

Scripture reading: 2 Corinthians 7:12-16

The fourth step to victory is perhaps the most critical step of all because it is so hard for us to admit that we can't do it ourselves and even more difficult to be willing to take our hands off.

Consider the greatness of this man Jehoshaphat. He was their king, the leader of the people. In their eyes he was their defender and deliverer. There have been many leaders in history who would rather drag their whole nation down to defeat than to openly admit that they were not strong enough for the battle.

You have an invisible shield that is established through self-sufficiency that is very difficult for even God to penetrate. As long as you think that you have everything in hand and can fight your own battles, the Lord will let you do just that.

No doubt at this climactic moment when Judah faced certain annihilation, Jehoshaphat remembered the words of the Lord to Solomon, his predecessor, as he dedicated this very temple that he stood before: *"If my people, which are called by my name, shall humble themselves, and pray, and seek my face and turn from their wicked ways; then will I hear from heaven, and will forgive their sin, and will heal their land"* (2 Chronicles 7:14). The words of King David may have also rung in Jehoshaphat's ears: *"He forgetteth not the cry of the humble"* (Psalm 9:12).

As King Jehoshaphat stood before the house of the Lord with all of his people gathered with him, he prayed: "God, I want you to know something; I am not standing here before you in my own strength. I want you to know that I completely recognize my own inability to cope with this situation. It does not take a military genius to understand that we are so outnumbered and outflanked that it is beyond human strategy, human experience, or natural understanding to resolve."

The Scripture is very succinct and we see the King going immediately before the people to pray; but you had better believe that there was a period of dying in the life of the King before this spirit of humility took over. I will guarantee you that if there was any way that he could have figured out how to do it himself, he would have done it because that is human nature.

We are a race of people who make up the proverb, "God helps those who help themselves." This simply is not true in the Word of God. God is teaching us to be utterly dependent on Him.

Make this declaration: "I am utterly dependent upon God. He will come through for me today in the circumstances of my life!"

July 18

Trusting God

Scripture reading: Psalm 31:1-5

One of the biggest problems God has with us is to wrestle our problems out of our hands. It is only when we come to the end of ourselves and are ready to die that we give up the reins so God can set His deliverance in motion.

One of the great missionary statesmen of our times, E. Stanley Jones, was broken in body and spirit as a very young missionary in India. One day he prayed, "Lord you know that I am at the end of the rope." God said to him, "If you will give that up, I will take care of everything." He quickly replied, "Yes, Lord, I gladly give it up to you right now." After that, his ministry touched millions through writing and preaching.

We want to hang on to something, no matter how small it is – and that tiny, insignificant thing can be the blockade to the power of God to meet our need. Only when we are at the end of self and are ready to die can God come in and take over.

Maybe the most frequent prayer that is ever prayed is for God to guide our lives. Usually we pray this most when problems start pressing in from more than one side. When you are asking for the guidance of the Lord, you must be willing to step down from the throne. As long as you are there ruling and reigning, it is impossible for God to work on your behalf.

When we ask God to lead us but still have the reins in our own hands, we are like the parent who gives his child a choice. If the child makes the right choice, the parent goes along with it; but if the child makes the wrong choice, the parent overrules. In other words, the child did not really have a choice at all.

When we ask God to help us and God directs in a way that is pleasing to us, then we will go along; but if it is contrary to our desires, then we rebel. Who is on the throne then? We are!

So many people say, "I want the will of God for my life!" What they are really saying is, "I hope that God's will for my life is the same as my will for my life." When you are on the throne and calling the shots, so to speak, then you have to be ready to take responsibility for the results- the sicknesses, the problems, and all of the other ramifications.

We live in a highly technological world where abilities are becoming more important every day. I am speaking of specialized skills that are required in order to make a living. More and more, as this age progresses, skills will become increasingly important for success in society. This is not true in spiritual power. Your spiritual power lies solely in your ability to trust God and throw yourself completely on Him.

Make this declaration: "I am throwing myself completely upon God. He will direct me and I will walk obediently in His way."

Leave Your Problems In God's Hand

Scripture reading: John 12:23-33

Oftentimes, when we ascend a staircase, we may skip one or two steps; but this does not work on the spiritual staircase that God has given to personal victory. Only as we place our spiritual feet firmly on one step are we prepared and qualified to move to the next step.

Most people will not put their trust completely in God as long as they feel there may be some other way out. I see them come to our crusades by the hundreds; people are "trying" God to see what will happen. These people would be very happy if God would heal them so that they would not have to go back to the doctor or incur the terrible expense of another operation, but by no means are they ready to put their trust in God completely to deliver and direct their life for them. If it works, fine; if not, they still have something else to fall back on.

This is really the critical step. You must humble yourself before God and recognize your insufficiency to cope with the situation. You can begin to recognize God. You can remember the things that He has done for you in the past, and you can stand upon His Word; but unless you are willing to step down and humble yourself, recognizing your own insufficiency, you will never reach the ultimate that God has for you. This step is really not a step up, but a step down – down from the throne of self.

At this great moment in the prayer of Jehoshaphat, "self" died. If it had not been for this statement, the prayer – as great as it was – would simply have been a beautiful piece of oratory. Here is where he moved the hand of God. Here is where he struck the heavenly chord that God could not resist.

Jesus said, *"Except a corn of wheat fall into the ground and die, it abideth alone"* (John 12:24). Until death to self happens, you are on your own. God will let you work it out yourself.

Now, the most important thing to learn from this step, other than the fact of dying to self, is that you do not have to come to your extremities or the end of your rope before you give up and let God do it. This is why the Bible gives examples of men and women who came to the end of themselves before they let God have His way in their lives. We see what the end result is, and therefore, if we allow the Holy Spirit to really speak to our hearts, we can let go of other possibilities and leave our problems completely in God's hands.

Make this declaration: "I am abandoning all of my own plans and strategies and leaving my problems in God's hands today. He will work in my behalf."

Humble Yourself

Scripture reading: Psalm 24:1-10

You do not have to wait till the end of the road when there is no way left to turn. The Holy Spirit can give you the power to die to self.

- When you get off the throne of your life, the responsibility for that physical body becomes God's instead of yours.
- When you get off the throne, the responsibility for choosing a life's mate is no longer yours.
- The responsibility for that unsaved son, daughter, husband or wife is no longer yours, it is the Lord's.

Why? Because, *"The earth is the Lord's and the fullness thereof; the world and they that dwell therein"* (Psalm 24:1). It is all His responsibility when you take your hands off. At the moment of total surrender it is His!

Humility is the answer:

> *The Lord is nigh unto them that are of a broken heart; and saveth such as be of a contrite spirit.*
> *(Psalm 34:18)*

> *Yea, all of you be subject one to another, and be clothed with humility: for God resisteth the proud, and giveth grace to the humble. Humble yourselves therefore under the mighty hand of God, that He may exalt you in due time.*
> *(1 Peter 5:5-6)*

What is so often prayed before God is, "Lord, make me humble." But God says, "Humble yourself."

You do not know how God will provide an answer to your prayer, but you have no might against the sickness, that personal problem, or that financial need. When you do not have the answer and there is no way out through man, you will find that God is still on the throne! He's waiting for you to humble yourself, to lean not to your own understanding, and to surrender yourself. That's when you will find Him right by your side.

Right now, take this marvelous step to victory. Humble yourself before your Heavenly Father. Put all preconceived ideas, all traditions, church denominations, doctrines and ways of man on the altar. Pray and surrender all to God.

Now, make today's declaration in power and authority…then you are ready to move on to the next step.

Make this declaration: "I surrender all. I have no power in myself to carry this burden, to fight this battle, to bear this pain. I am not able in myself."

Face Your Enemy

Scripture reading: 2 Chronicles 20:13-19

Look back over our daily readings for the past week and you will see that in Jehoshaphat's prayer the first four steps are taken by man toward God. In step five, however, we see the beginning of God's answer.

God's message through His prophet to the king and the people of Judah was, "I have heard your prayer." What a fantastic word of assurance! The most beautiful thing about the Word of God and the Holy Spirit is that when we follow the same path in prayer as Jehoshaphat, we too can hear in our innermost being: "I have heard your prayer!"

The answer came to Jehoshaphat, and the answer will come to you! As a servant of the living God, I can guarantee you that as you humble yourself before God, the answer will come. It will come with such unmistakable clarity so you will know beyond the shadow of a doubt that you have heard from God and your deliverance is on the way.

Listen to the assurance that the word of the Lord brings to the king and the people: *"Be not afraid nor dismayed by reason of this great multitude..."* (verse 15). Living in constant fear is a living hell. There is nothing worse except hell itself.

When humility came and the king got off the throne and, in essence, said, "Lord, you deliver the people because I can't," then the Spirit of God showed them exactly where the enemy was. You cannot deal with your problem until you know exactly what it is. You must be ready and willing to face it head on. You can't cover it over and pretend that it isn't there.

The element of trust is manifested in your life when you are ready to face your enemy completely unarmed in your own strength. This is a vital point. When the Spirit of God shows you exactly where your enemy is, there is a great temptation to do a little of your own warfare on the side. At least to have something in reserve to protect yourself when you stand face to face before your adversary.

God emphasized again to the king and the people: "...Fear not, nor be dismayed ... for the Lord will be with you." Trust the Holy Spirit as you face the enemy today.

Make this declaration: "I am not afraid. I am not dismayed. I am trusting the Holy Spirit as I face the enemy today."

Put The Battle Into His Hands

Scripture reading: Isaiah 41:8-13

When God sends His answer, receive it with a willing and open heart, even if it breaks every rule of human understanding. Men today want to put God in a test tube and dissect Him according to the rules of logic. Let me tell you something: God is not always logical according to our rules of logic, because the smartest man that ever lived (apart from Jesus) only "knew in part." The whole picture has never been revealed to man because finite minds are unable to contain it.

As long as we remain shackled to the natural world, there are things that God will ask us to do that appear unreasonable. This is where our trust comes to the forefront: our natural minds must be overruled, and we must put our complete trust in Him.

I am sure that you can see that each one of these steps we are meditating on leads to the next, and unless you are ready and willing to take each step as a part of your road to complete victory for body, soul and spirit, you cannot take the next.

Two words are repeated in 2 Chronicles 20:15 that are important for us to understand. We have mentioned the word "fear," but let us also close this step with a brief definition of the meaning of "dismayed." "Dismay" means to be filled with apprehension and lose our courage and confidence at the prospect of danger or trouble. You have heard the well-worn story about the day the devil decided to sell every tool and weapon and device that he possessed: that is everything with only one exception: The wedge of discouragement was not for sale.

If we have absolute trust, there is no room for discouragement. That wedge has no place to get a foothold between our faith and trust in God. Be ready to have your trust in God put to the test.

God said, "Here is where the enemy is coming. It is not necessary for you to take any weapons to the battle and fight. Just go and stand in the midst, in the face of the oncoming enemy, and trust that I will fight for you."

Right now, before you approach the next step, determine that you are going to put your trust in the Lord to bring you complete victory for body, soul, and spirit. Trust Him – He knows more about the problem and the need than you do. Trust Him enough to be willing to put the battle totally into His hands.

Make this declaration: "God knows more about my problems than I do. I am putting the battle totally into His hands."

ACTING IN FAITH

Scripture reading: 2 Chronicles 20:20-25

"And they...went forth" (verse 20). God's deliverance requires action on our part – action that is based on faith in the Word that He has sent. Jehoshaphat could have prayed the greatest prayer ever prayed, he could have humbled himself before God, and God could have sent His anointed Word through His prophet – but if that is where it stopped, God's people would have been eliminated from the face of the earth.

Jehoshaphat, at the moment, was concerned for his life and the lives of his people; but think of the ramifications for generations to come that hung upon his decision to put his faith into action and do what the Lord had commanded. Jehoshaphat could have said he trusted the Lord, and yet, refused to obey Him and go forth under His direction in the hour of testing.

How many times I have said that God never does anything to us or for us, but always through us. The battle is not ours, but we must remain the willing instrument in the hand of God that we might *"see the salvation of the LORD"* (2 Chronicles 20:17).

Jehoshaphat believed, but he also had to convince the people to follow him. He stood before the people and declared, *"Hear me, O Judah, and ye inhabitants of Jerusalem; Believe in the LORD your God, so shall ye be established; believe his prophets, so shall ye prosper"* (verse 20).

Here we can draw a spiritual parallel. We read in the Word of God that *"the spirit indeed is willing, but the flesh is weak"* (Matthew 26:41). In our hearts we can know that the Word of God is true, but our will must be brought into subjection to the Spirit of God.

Action must be the result!

Judah had to move out of the comfort of Jerusalem in order to confront the enemy face-to face. If they waited until the enemy reached the gates of the city, it would be too late to act.

Here is another great secret that the revelation of the New Anointing has taught us. We must put the enemy on the run. If we wait until the enemy is on our doorstep, then we are on the defensive and it is harder to gain the victory.

I am a firm believer in meeting and dealing with every problem the moment it rears its ugly head. Do not live under the cloud of illusion that problems will work themselves out; they will only work you out – out of peace of mind and out of the health, happiness, and prosperity God wants you to have.

When Jehoshaphat heard of the problem, he immediately took steps to deal with it. He did not wait until the enemy was at the city gates. Because he dealt with the problem immediately, the Spirit of God was able to establish him in an offensive posture. He was able to pursue the enemy, not in his own strength, but in the power of the Holy Spirit.

Make this declaration: "Today, I will pursue the enemy not in my own strength, but in the power of the Holy Spirit."

FAITH IS AN ACT

Scripture reading: Hebrews 11:1-40

Belief in the Word of God was not enough. It was a foundation upon which to build, but in itself was insufficient.

If you were to stop one thousand people on the streets of your city and take an opinion poll asking the following question, "Do you believe in God?" over ninety-five percent of them would answer "yes," but that is not enough because the Bible tells us that even demons believe and they tremble. The Apostle Paul declares:

> *Without faith it is impossible to please him: for he that cometh to God must believe that he is, and that he is a rewarder of them that diligently seek him.*
>
> *(Hebrews 11:6)*

If the Lord had told Judah, "I have heard your prayer and I will take care of everything. You just go back to the regular routine of business as usual," what measure of faith would that have required? Virtually none. They would have done so without a moment's hesitation.

Every blessing that we receive from God comes to us through faith. We are saved by faith: *"For by grace are ye saved through faith; and that not of yourselves: it is the gift of God"* (Ephesians 2:8). We are healed by faith, we live by faith – all things come to us by faith.

I have seen so many people struggle for faith. Some feel if they tell themselves often enough that they have faith, that eventually it will become a reality. You can say, "I have faith, I have faith" from now until your last breath and it will not increase the measure of your faith.

Early in my ministry God gave me a little saying that has been an eye-opener to tens of thousands of people around the world concerning faith. It is this: Faith is a fact, but faith is an act. If you want something to repeat over and over again, try that one until it really sinks in.

Jehoshaphat could have heard the Word of the Lord through the prophet and said, "We believe." But if they did not do what was told them, they would not be acting on their belief or their faith in the Word. They could hear his mighty prophecy and be greatly inspired knowing that the Lord had answered the King's prayer, but if they did not act it would all be for naught.

I believe in acting on the Word. The greatest revelation in the world is worthless unless it motivates you to do something about it. I am not interested in tickling people's ears. I want to see them so moved by the Holy Spirit that they cannot sit still, but they have to get out and put into practice what the Holy Spirit has taught them.

Make this declaration: "I will act upon the Word. I will put into practice what the Holy Spirit is teaching me."

Put Your Trust In God

Scripture reading: James 1:1-8

For some time now, God has led us to conduct what we call Schools of Ministry, in conjunction with every Victory Miracle Crusade. As we go into cities throughout the world, it has been our opportunity to sit in a classroom atmosphere and teach those that come the revelation called *The New Anointing.*

The New Anointing is a message of action. You cannot hear it and sit still. It compels you to begin to use the spiritual weapons that are placed in your hands. The unique thing about these schools is that you are given the opportunity immediately to see if the principles of spiritual warfare you are given really work. My Jewish heritage, I think, has given me the attitude that I am not so interested in how high-sounding something may be – what I want to know is, does it really work?

After the people come out of the day sessions of the School of Ministry, they are brought into the evening crusade rallies and come face to face with every type of sickness, disease, and demonic power. Right there they can begin to use the weapons of spiritual warfare that Christ has given His Church. They see what they have been taught come to pass right before their eyes, in many cases because they have exercised the faith God has placed in their hearts.

Faith is action! Faith is positive, *"For he that wavereth is like a wave of the sea driven with the wind and tossed. For let not that man think that he shall receive any thing of the Lord"* (James 1:6-7).

Now, I am a man that thanks God for what we receive in the natural. I thank God for doctors, but isn't it interesting that one day you're told to take certain medicine and after you have taken it for fifteen years they tell you, "Oh, you've been taking the wrong thing. This drug causes such and such kind of disease. In fact, it could kill you."

What am I trying to say? They are practicing physicians – they practice on you. Now I'm not being critical. There is nothing wrong with man learning from experience and advances in science, but I hope it will put the weakness and frailty of man into perspective.

You cannot trust in the arm of flesh…

> *…with him is an arm of flesh; but with us is the Lord our God to help us, and to fight our battles…*
>
> *(2 Chronicles 32:8)*

Put your trust today in Almighty God!

Make this declaration: "I will not trust in the arm of flesh. I will trust in Almighty God."

THE INFALLIBLE, IMPREGNABLE WORD

Scripture reading: John 1:1-14

One day I was sitting in my kitchen at home in San Diego, California. All of a sudden I felt a hand touch my shoulder. I turned around to look, but there was no one there. Then the presence of God engulfed me and God spoke. He said, "Son, tell My people that everything that can be shaken is going to be shaken. Tell them to fix their faith on that which cannot be shaken."

When you fix your faith on that which cannot be shaken, no theologian can destroy your position. No one can come along with a passing doctrine and sway you this way or that until you are in absolute confusion.

Receive this into your spirit: There is only one power in the world that cannot be shaken. It's not the power of governments nor the power of men's minds. It is the power of the Word of the Living God:

> *"In the beginning was the Word, and the Word was with God, and the Word was God...All things were made by him; and without him was not anything made that was made."*
>
> *(John 1:1,3)*

Put these two words into your spirit:

One: Infallible which means, "no margin for error."

Two: Impregnable, which means "impossible to be taken by assault."

The Word of God is infallible. That means it will never fail you. The Word of God makes you impregnable. That means the devil can throw anything and everything at you he wants to, but you can't be taken because your faith is fixed on that which will never fail.

When you hear someone say that it is not God's will for them to be well, you know they have bought the biggest lie the devil ever told. Don't let anyone ever tell you that it is God's will for you to be sick or afflicted. I have heard people go so far as to believe they are suffering their afflictions for Christ. This is worst yet, for then you are saying that what Christ suffered was not enough and you have been called to complete His work.

Did Jehoshaphat waver at the command of God? Did he sit down and say, "Now I wonder if it really is the will of God to deliver Judah from the hand of the enemy? Maybe we ought to really consider this for some time, because we certainly don't want to go out and defeat the enemy if it is not God's will."

Sound silly? Yes, but no more silly than a born-again, Spirit filled Christian saying, "I wonder if it really is the will of God for me to have victory over this sin in my life? I wonder if it really is the will of God to save my husband or wife or bring back that wayward, prodigal son or daughter?"

Make this declaration: "I will fix my faith to that which cannot be shaken: The infallible impregnable Word of God."

Praise God For The Victory

Scripture reading: Psalm 150:1-6

It is not only important to believe God, but Jehoshaphat said to the people that they must believe God's servants as well – "So shall ye prosper." The spirit of prophecy came upon a man named Jahaziel, who was the son of Zechariah. He told Jehoshaphat and Israel what to do, they believed it, and acted upon it.

Today, if someone gives a message in prophecy relative to persons and circumstances, people might brand him as a religious fanatic. If that is being a fanatic, I want to be used of God as one! We need more Jahaziels in our world today.

You can meditate on the steps to victory we are studying and begin to reason in your mind or try to apply human understanding to them. You can sit down and pick apart this point or that, but where will it get you? On the other hand, you can say, "Lord, I believe that you are giving me a progressive revelation each day through these devotions and though I may not fully comprehend every point, I am going to do just what the prophet of God has said and I am going to have victory in Jesus' Name."

Are you ready for the final step to victory? This is the capstone – don't miss it because you have gone too far to stop now! The next step is praise the Lord for the victory:

> *And when he had consulted with the people, he appointed singers unto the LORD, and that should praise the beauty of holiness, as they went out before the army, and to say, Praise the LORD; for his mercy endureth for ever.*
>
> *(2 Chronicles 20:21)*

The final and most powerful step of all to victory is to praise and worship the Lord. There might be a temptation to say, "This step is so wonderful, all I really need to do is start right here. I don't need to go through all the effort of those other six steps." Man has always been prone to look for spiritual short cuts, and this is where false cults and doctrines develop. We always want to find an easier way.

It is not God's intention to make the path difficult, but He knows what we are made of and what steps we must take if victory and abundant life are to be manifested in our lives. To start this step with no regard to the other steps would make our praise a "sounding brass and tinkling cymbal." A hollow, shallow noise that goes no further than the ceiling is the result.

Make this declaration: "I will praise the Lord for the victory, even before it comes. I will give Him the glory for what He is going to do!"

A Reason To Be Excited

Scripture reading: Revelation 3:14-22

It is easy to praise God after the fact, but that is not the key that is given here. Immediately after they had received the Word of the Lord from the Prophet Jahaziel, Jehoshaphat bowed his head until his face touched the ground. Jehoshaphat was not just going through a form. He was putting his heart and soul into his worship of God.

When the king set the example (verse 18), the Bible tells us that all of Judah and the inhabitants of Jerusalem followed suit. What a picture! We may look at this as a group of primitive people, but let me tell you they had as much protocol as we could ever imagine.

Let your imagination take hold of you for a moment and picture this in mind. Can you conceive the spiritual explosion that would take place in this nation if next Sunday morning every minister would face the congregation right at the opening moment of the service and declare that the order was going to be different, the printed program would not be followed, and the entire congregation were going to bow their heads and hearts before the Lord in worship and praise?

Do you know why that won't happen? Because of two things. The first six steps we have learned have not taken place because the church does not really feel it is in crisis. It has not been given the news that the enemy has amassed its strength and is about to attack. In many cases the message has come through, but has been disregarded. Remember what the Holy Spirit said to the Church that said, *"I am rich and increased with goods, and have need of nothing?"* (Revelation 3:17). Their real spiritual picture was that they were naked and blind.

Jehoshaphat and all the people bowed and worshiped the Lord, thanking Him for the Word He had given. They did not complain that they had to leave their homes and go out to meet the enemy, that they might miss a meal or a day's pay on the job. They began to thank Him for the victory when they were being asked to do what seemed to be an impossible task. God has asked them to go forth to meet the enemy without all of the regular preparation for war! God had asked them not only to go forth to meet the enemy, but told them to *"stand still, and see the salvation of the Lord."* They were to go out and confront the enemy, but the Lord said the battle was not theirs, but His.

As the king and all of Judah were on their faces praising and worshiping God, the priests rose up to *"praise the Lord God of Israel with a loud voice on high"* (verse 19). It sounds to me that they were getting a little emotional. When you know that the answer has come from the Lord, there is a good reason to get excited!

Make this declaration: "I know my answer is coming from the Lord. I will praise Him with a loud voice on high!" (...Now do what you just declared!)

Joy Is Your Strength

Scripture reading: Psalm 21:1-13

Certain doom had been changed into certain victory for Jehoshaphat in 2 Chronicles 20:17. The people had received a word from the Lord and they were expressing their feelings of joy and happiness by praising and worshipping God.

You do not need to wait until deliverance is accomplished; in fact, the deliverance may not come until you praise the Lord for the victory. Praise the Lord before the answer comes.

In this period of praise, immediately after they had heard from the Lord, Judah was strengthened to stand up and go into the wilderness to face the enemy. The Bible says, *"The joy of the Lord is your strength"* (Nehemiah 8:10). This is a simple, yet profound, statement of fact. Do not try to go out and face your enemy until you have the joy of the Lord in your heart. This strength of joy will come into your heart as you begin to praise and worship the Lord by faith.

In the natural, Judah had nothing for which to praise the Lord at this moment. The enemy was still a united front against them and at that very moment was marching toward Jerusalem. But after the day of prayer and praise, the entire nation was ready to march. Yes, I said the nation. There was no paid, professional army that was designated to go out and meet the enemy. It was the whole body of the nation that was going to face the battle together.

The Bible tells us that they rose up early (verse 20), and there is a tremendous lesson in that example. They gathered together in the wilderness, but before they met the enemy the king called a halt again. He wanted to find the strongest men to put in front of the weaker men, women, and children.

Is that what he did? No! He found those with the best and strongest voices to lead the people in praise and worship as they went forth into direct contact with the enemy. What a strange leader! Why didn't he find the most accurate marksman, those with the biggest muscles, or those with the best armor? Because the physical battle was not theirs and the victory did not depend upon the arm of flesh.

Remember this however: Just because the battle was not theirs but the Lord's does not mean that they did not do anything. The entire nation of Judah began to praise the Lord in the face of battle.

Think about your own situation...what might happen if you begin to praise the Lord today in the face of your enemy, in spite of your problem?

Make this declaration: "The joy of the Lord is my strength. I praise God today in the face of my enemy, in spite of my problem."

Confronting A Defeated Enemy

Scripture reading: Psalm 98:1-9

In 2 Chronicles 20:21-22 we find Jehoshaphat employing a very unusual battle strategy – praising and worshiping the Lord. What happened as a result of this unusual strategy?

The Bible says, *"And when they began to sing and praise, the Lord set ambushments against the children of Ammon, Moab, and mount Seir; which were come against Judah; and they were smitten"* (verse 22). Another translation of this verse says, *"The moment they began to praise the Lord, their enemy was defeated."* You see, our victory is in learning how to overcome an enemy that's already defeated.

When Israel began to praise the Lord, their enemies turned on each other and completely destroyed one another The enemy destroyed itself. Praise and worship of God from the depths of an honest and sincere heart is like pushing the self-destruct button for everything that opposes you, whether it be sickness, emotional problems, or financial bondage.

Do you want to fight your own battles, winning one here and losing two there, always losing ground, things always getting worse and seldom getting better? Or do you want to follow God's formula and be able to stand back and see the deliverance of the Lord?

This is so beautiful: *"And when Judah came toward the watchtower in the wilderness, they looked unto the multitude and, behold, they were dead bodies fallen to the earth, and none escaped"* (verse 24). I like that! The multitude – their enemies – were dead bodies and none of them escaped!

Jehoshaphat did not know by what means the victory was going to come, or how God was going to work; but he stood firm in the midst of the enemy, faced the battle line, and praised God for the victory, even before he saw it come to pass.

Let me tell you something. It's one thing to believe this intellectually, but it's another thing to stand face to face with the devil when you've got cancer, your children are on drugs, or your home is breaking up. When you stand face to face with the enemy, you are going to have to believe the Word of God and the word that falls from the lips of the prophet!

They didn't know how God was going to bring them the victory and neither do you. Why do you care how He brings you deliverance and victory? Though you may not know how the victory will happen or when, you know this according to the Word of God: It will happen!

You can praise God in the midst of the enemy. You can praise God right on the battle line with the adversary of your soul, your home, and your body. Praise God for the victory!

Yes, before and even during the raging of the battle – praise Him, for the victory!

Praise Him, for you shall overcome! The enemy is a defeated foe!

Make this declaration: "Praise God for the victory! The enemy is a defeated foe. I will overcome in the strength and power of the Lord."

Seven Steps To Victory

Scripture reading: James 1:22-25

Right now, take a pen in hand, sit down and think of your battles, your burdens, your problems, your sickness, any need you may have – write it all out on a piece of paper. Remember: faith is a fact, but it is also an act.

God's victory is complete, it is never halfway. If you have followed His formula in these seven steps to victory, your deliverance will be complete. As we conclude our devotions for this month, let's review the seven steps we have meditated on during the past few weeks:

Step One: Jehoshaphat recognized the power and the authority of the God he was serving.

Step Two: He remembered the acts of deliverance God had already accomplished for His people.

Step Three: He affirmed his position to stand completely on the Word of God, for he was standing in the presence of the Lord, before the House of the Lord where the Word of God resided.

Step Four: Next, he humbled himself before the mighty hand of God and acknowledged his utter dependence upon God for deliverance.

Step Five: He put his trust in the Lord and God answered his prayer with directives on how the victory was to be achieved.

Step Six: Jehoshaphat showed his faith in the direction he received not only by a verbal declaration, but also by putting his faith in motion and marching towards his enemy depending not on the arm of flesh, but solely on the Lord.

Step Seven: Jehoshaphat led the people in praise and worship of the Lord before the answer came and then they went forth to meet the enemy.

These steps are applicable to every problem of life, whether it be physical, material, or spiritual. I do not believe that there is a person under God's heaven that cannot apply these steps, taking them one at a time, and not come into victory.

These seven steps to victory were given to me by God. They were revealed to me as I was on my knees in prayer! The steps you have learned during this past month cannot fail. Act upon what you have learned, walk in new faith, and remember – God has no defeats planned for you!

Make this declaration: "I will act upon these seven steps to victory. I will walk in new faith and victory. God has no defeats planned for me!"

August

Morris Cerullo overlooks the valley of Armageddon where the great and final battle on earth will be staged. The victory was already won over 2,000 years ago on Calvary – all you must learn is how to enforce it. This month's devotional readings will give you the strategies needed to penetrate enemy lines and seize what is rightfully yours.

August

Penetrating Enemy Lines

The only ability Satan has is the ability to attack our minds and deceive us into believing a lie. God has given us power over all the purported power, lies, deceptions, and attacks of the already-defeated enemy.

Israel's enemies had no power at all over them. After the one defeat at Ai which resulted from sin, God's people were totally victorious. God had promised them, *"Every place that the sole of your foot shall tread upon, that have I given unto you..."* (Joshua 1:3). The enemy nations of Canaan had no power over God's people. God gave them divine strategies for penetrating enemy lines and defeating their foes – and that is exactly what God wants to do for you.

Satan wants you to believe that he has power over you. He wants you to think you are his victim and that he can do whatever he wants to you. During the days ahead, we are going to strip Satan of his power. We are going to take the mask off and reveal him for what he really is – a defeated foe.

It is time for the Body of Christ to recognize that we have the entire armor of God, the spiritual forces of Heaven, and the power of Christ at our disposal. It is time to walk in the heritage of our King. It is time to enter into the cycle of Victory Miracle Living.

You may be saying, "Brother Cerullo, I just don't know if I can do that.?" You may be a new believer or it may be that you have failed or have been knocked down so many times in the past that you find it hard to get up and fight again.

But you can do it! You will do it! And this time, you will be victorious because you are going to fight from a position of knowing. You are going to learn strategies for penetrating enemy lines – just as Israel did – and you are going to be victorious.

You may say, "But Brother Cerullo, I don't know how to do it. I want to enter in, conquer the enemy, and live a victorious life, but I don't know how."

Well, then let me show you...It all starts with acting in obedience upon the first reading for this month. So turn there, and let's begin...

Spiritual Ignorance

Scripture reading: Luke 10:17-19

Satan's strategy is to hold you helplessly in a state of spiritual ignorance. He will do everything possible to stop you from knowing the truth about him, and to stop you from knowing your position of power and authority over him as a child of God. This is exactly the strategy he used on the people of Israel to keep them out of the Promised Land. It is the strategy that caused their defeat at Ai and Satan still employs these same methods today.

As we enter the new millennium, Satan has planted fear in the hearts of many Christians causing them to think they are weak and powerless against his attacks on their bodies, their finances, and their loved ones. One of the powerful strategies Jesus used to defeat Satan and live His life in 100 percent victory was that He faced the enemy from a position of knowing! Jesus never wavered, faltered, or retreated. Jesus knew the power and arsenal of the enemy and that He had power and authority to defeat him in every circumstance: *"For this purpose the Son of God was manifested, that he might destroy the works of the devil"* (1 John 3:8).

When Jesus died on the cross, He took Satan's authority over the earth away from him and gave us power to overcome and defeat all of his attacks:

> *Forasmuch then as the children are partakers of flesh and blood, he also himself likewise took part of the same; that through death he might destroy him that had the power of death, that is, the devil.*
>
> *(Hebrews 2:14)*

Today Satan desperately tries to keep that defeat a secret. Satan does not want the Church to know that through the death of Jesus Christ on the cross they now have the power: *"Behold, I give unto you power to tread on serpents and scorpions, and over all the power of the enemy"* (Luke 10:19).

The devil's only hope to lure souls into his eternal hell is to keep the Church functioning in a state of perpetual spiritual ignorance and that tactic has become one of his major strategies. To implement this strategy of spiritual ignorance, one of the major lies Satan has planted in the Body of Christ is to perpetuate the false fear that many Christians have concerning the purported power of the devil. He has distorted and embellished on his horns and pitchfork image, causing many to see him as more powerful than he really is.

Much of the Body of Christ still clings to the Hollywood image of the devil as portrayed in *The Exorcist.* They believe he has a force so powerful that he can freely throw people around rooms, twist the innocent heads of little girls, and toss praying priests out of two-story windows. These pathetic and false depictions are all nothing more than lies.

Make this declaration: "Satan will not keep me in spiritual ignorance. The Holy Spirit, the Spirit of truth, lives within me."

The Devil Does Not Have Dominion

Scripture reading: Ephesians 1:19-23

The Scripture does not say Satan is a roaring lion, only that he makes the noise of one. He functions as a roaring lion, which is completely different from actually being a roaring lion.

It is time to strip the mask off Satan. He has no strength, power, or authority over a child of God. He can only growl, hoping your own fear will take over and you will then attribute to him powers he no longer has.

Here's the key spiritual fact you must grasp firmly in your mind: The devil does not have dominion over the earth. Satan told Jesus when he tempted Him in the wilderness: *"...all this power will I give thee, and the glory of them: for that is delivered unto me; and to whomsoever I will I give it. If thou therefore wilt worship me, all shall be thine"* (Luke 4:6-7). But Jesus knew that Satan was a liar, that God never relinquished control of the earth, and that He never will.

There is not one verse in the entire Word of God that indicates the earth was ever given into Satan's evil hands, yet, many Christians believe this lie of the devil. Satan is not in control of the earth. Jesus, through His death and resurrection, destroyed the works of the devil.

> *And what is the exceeding greatness of his power to us-ward who believe, according to the working of his mighty power, Which he wrought in Christ when he raised him from the dead, and set him at his own right hand in the heavenly places, Far above all principality, and power, and might, and dominion, and every name that is named, not only in this world, but also in that which is to come: And hath put all things under his feet, and gave him to be the head over all things to the church, Which is his body, the fullness of him that filleth all in all.*
>
> *(Ephesians 1:19-23)*

Israel was promised power over all the enemies they would encounter in their Promised Land. God's Word declares that you and I have been given total dominion over this world. The power that was once in Christ's hands has been transferred to us through our spiritual inheritance as His children. Now, His power is in your hands.

Make this declaration: "I have total dominion over this world. The power that was once in Christ's hands has been transferred on to me. His power is in my hands."

DELEGATED POWER

Scripture reading: John 12:31-37

Jesus compared Satan to a strong man guarding his palace. When Jesus came, a stronger One than Satan had come. Jesus broke the power of Satan, bound him, and cast him out. As Jesus prepared to lay down His life to redeem and restore man, He said: *"Now is the judgment of this world: now shall the prince of this world be cast out"* (John 12:31).

Today, the very destiny of our cities, nation, and world depends upon our understanding that Satan is a liar and that we now hold all of Christ's power in our hands. One of the most strategic lies Satan has perpetrated concerning himself is that the earth is his kingdom, and that he has unlimited power and control. But Satan is a liar and *"When he speaketh a lie, he speaketh of his own: for he is a liar, and the father of it"* (John 8:44).

Satan used lies to trick Eve into sin in the garden of Eden. When Adam and Eve yielded to Satan and sinned against God, they lost their position of power and dominion over the earth. Spiritual darkness came upon them and covered the earth. Because of their sin, Adam and Eve relinquished their power and authority to the devil. Satan became the ruler of the spiritual darkness of the world, but he was never in control of this earth because *"The earth is the LORD'S, and the fullness thereof; the world, and they that dwell therein"* (Psalm 24:1).

Satan never had unlimited power over the world. Jesus knew that the devil did not even have the power to deliver kingdoms, even though he falsely claimed he could when he tried to tempt Jesus (Luke 4:5-7). In this modern day, many Christians still believe the devil has unlimited power to control their lives, but all the devil can really use are the same lies and deceptions he used in the garden of Eden. Lies and deceptions are the only weapons he has available in his feeble arsenal to attack us on the battlefield of spiritual warfare. He has no other power.

Jesus broke the power of Satan over man and defeated him at Calvary. Before Jesus ascended into heaven, He told His disciples:

> *All power is given unto me in heaven and in earth. Go ye therefore, and teach all nations, baptizing them in the name of the Father, and of the Son, and of the Holy Ghost.*
>
> *(Matthew 28:18-19)*

Jesus has complete power over Satan. The really great news is, Jesus has delegated that same power to us!

Make this declaration: "Jesus has complete power over Satan and He has delegated that same power to me."

August 4

SATAN'S ONLY ABILITY

Scripture reading: Acts 5:1-11

The only ability Satan has is the ability to attack our minds and to deceive us into believing a lie. God has given us power over all the purported power, lies, deceptions, and attacks of the already-defeated enemy.

Jesus has given us power over all things:

> *Far above all principality, and power and might, and dominion, and every name that is named, not only in this world, but also in that which is to come: And hath put all things under his feet, and gave him to be head over all things to the church, Which is his body, the fullness of him that filleth all in all.*
>
> *(Ephesians 1:21-23)*

It is only when we accept the devil's lies, such as Ananias and Sapphira did in the days of the early apostles, that we become vulnerable to his wicked schemes. Satan lied to Ananias and Sapphira, convincing them that if they held back part of the price they received for their land, they would somehow be better off than giving it all to the disciples for the work of the Lord as had been agreed upon by the disciples (Acts 5:1-3). They believed the devil's lie, attributed to him power he did not have, and paid the price of this deception with their lives.

There is a very well-known scripture often quoted in the Body of Christ, but usually without the full context:

> *For the weapons of our warfare are not carnal, but mighty through God to the pulling down of strong holds...*

This is usually where people stop the quote. But look carefully at what comes next...

> *Casting down imaginations, and every high thing that exalteth itself against the knowledge of God, and bringing into captivity every thought to the obedience of Christ.*
>
> *(2 Corinthians 10:4,5)*

It is true that the weapons of our warfare are not carnal because the nature of our battle is spiritual. But – as we learned in our meditations during March – many Christians fail to recognize that the battlefield is in the mind. They imagine that Satan is strong, that he has unlimited power, and they become captive to his purported power. But the only weapon the devil has to get your thoughts to be his evil thoughts is to deceive you, just as he deceived Ananias and Sapphira, through his lies.

The only way Satan has any power in your life is if you do not bring your thoughts into captivity and you fall for his age-old lie that he actually has power over the earth. Don't believe it!

Make this declaration: "Today I will bring every thought into captivity. Satan will have no power over me."

No New Tricks

Scripture reading: Matthew 4:1-11

You give Satan power when you constantly acknowledge his alleged power and attribute every bad thing that happens in your life to him as though he had the ability to manipulate every circumstance to destroy you.

- You give Satan power when you backslide into sin, believing in your mind that sin will not harm your relationship with Jesus Christ.
- You give Satan power when you constantly give him attention.
- You give him power to perpetuate his lies in your mind through your sins and false perceptions of his powers.

How do you change this process?

Begin to recognize the devil as the pathetic liar he really is. He has no new tricks!

He is still using the same old lies and deceptions he used in the garden of Eden to try to lure God's people into a fiery hell. Get it clear in your mind, your heart, and your spirit, once and for all, that the devil does not operate in unlimited power. The only power he has is rooted in lying signs and wonders operating through deception in your mind.

- The devil wants you to believe he has power.
- The devil wants you to believe you are his victim.
- The devil wants you to believe he can do whatever he wants to you.

But let's take the mask off! It is time for the Body of Christ to put this nonsense aside once and for all and recognize that we have the entire armor of God and the power of Christ at our disposal. It is time to walk in the heritage of the King.

Many Christians honestly believe they are weak and powerless against the devil and that Satan can somehow overcome them with mysterious powers that they cannot resist. The devil wants to plant a false fear into the hearts of Christians. He wants you to believe you are no match for his power.

The strategy here is clear: If Satan can keep you believing that he has more power than you do as a Christian, then you will walk in fear, afraid to face him in any confrontation.

The devil will become the thing you fear most in life, and you will live each day wondering where his next attack will come from and how you can defend yourself against a foe that is stronger than you.

Satan has no power in your life until you give it to him by sinning or accepting his lies. We give Satan the advantage – he cannot take it himself. Sin gives Satan an advantage in our spiritual warfare with him. When we stay in the will of God, free from sin, and are not ignorant of the devil's devices, then we can win the war every single time.

Make this declaration: "I am not a victim of the devil. I am a victor through the power of God. I will win every battle every time."

Lest Satan Take Advantage

Scripture reading: Romans 6:16-19

The Apostle Paul tells us that we can know and overcome the schemes of the enemy. In his letter to the Corinthians, Paul explains that one way to overcome the devil is to forgive those who have offended you. Harboring unforgiveness is a sin which opens an opportunity for the devil to enter into your life and begin his evil deeds.

Paul tells us that it is important to forgive those who have hurt us *"Lest Satan should get an advantage of us: for we are not ignorant of his devices"* (2 Corinthians 2:11). Notice the progression here. If you keep unforgiveness (a sin) in your heart, then Satan gets an advantage.

When you stay free from sin, the devil cannot penetrate your spiritual armor because Jesus Christ has delivered you from any power of the devil: *"Who hath delivered us from the power of darkness, and hath translated us into the kingdom of his dear Son: In whom we have redemption through his blood, even the forgiveness of sins"* (Colossians 1:13-14). We are delivered from the power of the devil. The only power he has left in his feeble war with the saints of God is to use lies and temptations to try to deceive you.

The devil is the father of lies. He actually creates them (John 8:33). He uses lies to deceive and tempt you, and to step-by-step – very subtly – take you farther away from the reality of Jesus Christ:

> *But I fear, lest by any means, as the serpent beguiled Eve through his subtilty, so your minds should be corrupted from the simplicity that is in Christ.*
>
> *(2 Corinthians 11:3)*

Notice carefully the spiritual truths here. Satan is subtle: He knows that he cannot get the saints of God to start sinning with murder and adultery, so he breaks them gently, luring them into a life of evil step-by-step. First, may be a "small" lie. Then another. He takes you on a subtle progression of evil so that you barely notice the pollution these sins are causing in your spirit.

Our devotions in March centered on your mind. We learned that the mind is the main battle field. That's where the devil attacks, and that's where he gets his power. The devil is a spiritual parasite that feeds off your mind. He has no power except the power you give to him by accepting his invitations to sin by believing his lies and deceptions. Don't let him take advantage of you!

Make this declaration: "Satan, you will not take advantage of me today! I will not accept your invitation to sin or believe your lies and deceptions."

THE DEVIL'S DESTINY

Scripture reading: Revelation 20:1-10

Instead of the saints of God continuously dwelling on the devil's gloom and doom in our minds, we should dwell on what the Word of God says concerning the ultimate destiny of the devil:

> *Now is the judgment of this world: now shall the prince of this world be cast out. And I, if I be lifted up from the earth, will draw all men unto me.*
> *(John 12:31-32)*

Now is the judgment – that means the devil is already defeated. He no longer has power over the earth. Jesus Christ has the power and, if we lift him up, no fiery darts of the enemy can ever harm us.

In the book of Revelation, we find details regarding Satan's ultimate destiny:

> *And the devil that deceived them was cast into the lake of fire and brimstone, where the beast and the false prophet are, and shall be tormented day and night for ever and ever.* *(Revelation 20:10)*

If you are to ever receive God's promises, claim your inheritance, and take the spiritual territory that is rightfully yours, then you must get these truths down in your spirit:

- The enemy has been judged.
- His power has been destroyed.
- His ultimate destiny is assured.
- Jesus destroyed the works of the devil (1 John 3:8).

Surely, if the Son of God has destroyed the works of the devil, then the saints of God need to grasp that breakthrough revelation and begin to function in the full power and authority God intended.

You must know and act upon these facts as the nation of Israel did:

- The devil has no power over you except what you allow him to have.
- The devil cannot make you do anything that you do not want to do.
- He has no power over your will!

Each morning, start your day focusing not on your current circumstances, but on the ability of God to overcome every obstacle before you. Each day, give God the glory as the creator of the universe and the source of all your needs. Start each day with the knowledge that there is no circumstance that you cannot overcome.

Make this declaration: "The power God gave me is greater than the forces of Satan. As a child of God, I can overcome all the purported power of the enemy."

August 8

We Do Not Have To Be Defeated

Scripture reading: Matthew 24:4-35

Satan wants you to believe that if you are undergoing some sort of problem in your life today, it must mean that God does not love you or that you are out of His will. But these are old lies from the feeble arsenal of an already-defeated devil. He will continue to pile his pathetic lies and deceptions upon all who will listen until the final conflict between Satan and the saints.

Just because we can win every war against the devil does not mean we will not encounter problems, but these situations do not automatically mean you are out of God's will. The Word of God clearly says there will be times of trouble and persecution.

From Revelation 12:10-12 we learn...

1. The devil will accuse you day and night.
2. The devil will attack you.
3. He has only a short time.
4. Spiritual persecution is one of his tricks.

From this knowledge of how the enemy operates, you can then be armed through the power of Jesus Christ to defeat every circumstance.

We all experience persecution. What separates Christians from the unbelievers is that we can emerge from every circumstance in victory. Jesus warned us there would be a time of persecution:

> *Then shall they deliver you up to be afflicted, and shall kill you: and ye shall be hated of all nations for my name's sake And then shall many be offended, and shall betray one another, and shall hate one another.*
>
> *(Matthew 24:9-10)*

And in another place:

> *For then shall be great tribulation, such as was not since the beginning of the world to this time, no, nor ever shall be. And except those days should be shortened, there should no flesh be saved: but for the elect's sake those days shall be shortened.*
>
> *(Matthew 24:21-22)*

But remember, even during a time of great persecution, God is still in control. The saints living upon the earth during these persecutions are not weak, cowering believers who are subject to the wiles and whims of a supposedly powerful devil.

Through the blood of Jesus Christ, we can and will be completely victorious in every personal persecution. We do not have to be defeated!

Make this declaration: "If persecution comes today, I will not be defeated. Through the blood of Jesus, I will be victorious in every situation."

Why Are You Sleeping?

Scripture reading: Luke 22:39-46

God wants you to know that regardless of the persecution which will come into your life today or in the future, He is in control! There is no circumstance, there is no trial, there is no test, there is no persecution that Satan can bring into your life that can defeat you!

The devil is doomed for complete defeat and total destruction. You personally need to know, that you know, that you know, what God's Word says. When the enemy comes against you, you can be filled with spiritual knowledge because...

> *But the anointing which ye have received of him abideth in you, and ye need not that any man teach you: but as the same anointing teacheth you of all things, and is truth, and is no lie, and even as it hath taught you, ye shall abide in him.*
>
> *(1 John 2:27)*

Christians are not exempt from the attacks of the enemy, but we are the only ones prepared, equipped, and empowered to defeat him. Let the anointing of God burn this spiritual truth into your heart. Your position during every test, trial, and temptation should not be one of fear, discouragement, or weakness. Rather, it should be one of power, complete confidence, and victory!

Thousands of Christians have fallen victim to one of Satan's deadliest snares and are totally unaware of what has happened to them. Using one of his most subtle strategies, Satan has caused Christians to fall into a sleep of spiritual apathy and complacency. The devil tries to be so subtle in his lies so that many Christians are unaware of his movements and have a false sense of security concerning their own spiritual condition. They remain unaware of the lateness of the hour or the impending danger until it is too late. Satan's strategy is to lull Christians into a spiritual sleep so they are unable to hear the voice of the Lord as He speaks to them. Satan wants to give you a false sense of security concerning your own spiritual condition.

Today, the Church is filled with Christians like the early disciples who fell asleep when Jesus asked them to pray (Luke 22:39-46). When the soldiers came to arrest Jesus, the disciples fled in defeat. They had fallen into the apathetic snare of the enemy. Jesus had warned the disciples of the coming dangers. He knew that they should fortify themselves through prayer to prepare for the attacks and the temptations, but they failed the test.

The Church today is filled with Christians who, like the disciples, have fallen asleep spiritually. They are going to church regularly, paying their tithes, and working in the church, but they are unaware of what is happening in the spiritual realm. They are not prepared to recognize the signs of the times.

God's message to those who find themselves in this sleep is, "Why are you sleeping? Arise and pray, lest you enter into temptation."

Make this declaration: "I will not fall asleep spiritually. I will be alert and pray. I will not enter into temptation."

A Hardened Heart

Scripture reading: Hebrews 3:7-13

Satan's strategy to make Christians fall asleep spiritually is to alienate our hearts from God. His plan is to cause our love for God to grow cold. One of the ways he does this is to get our eyes off pleasing God and doing His will and get them on ourselves. Satan knows that if he can succeed in getting you to let other things take first place in your life, he gains a strong foothold.

Satan's strategy is to cause you to become so involved with providing for the needs of your family that you have little or no time left for the things of God that are necessary to strengthen your spiritual life. The only true foundation for your house – your life – is to build on the solid rock.

Many Christians feel that a hardened heart could never happen to them, but Paul warned the Church, not the unbeliever, *"...if ye will hear his voice, Harden not your hearts..."* (Hebrews 3:7 8). A hardened heart does not harken to the will of God, and is closed to the Holy Spirit. A hardened heart cannot be penetrated by the love and grace of God. A hardened heart feels no compassion for lost souls and thinks only of selfish things.

To avoid a hardened heart, Christians must understand what the Bible teaches that causes such a condition:

> *But exhort one another daily, while it is called To day; lest any of you be hardened through the deceitfulness of sin.* *(Hebrews 3:13)*

Many Christians today are in the process of hardening their hearts through sin and unbelief, therefore they are unable to see the fulfillment of God's covenant promises in their lives.

Like hardening of the arteries, hardening of the heart does not happen overnight. It happens gradually through a progression of sinful steps. To protect against this hardening process, each day you must examine yourself to see if you have allowed Satan to get a foothold in your life through sin and gradual unbelief.

Before you say it cannot happen to you, remember that the men who were closest to Jesus, those who sat under His teaching, who saw with their own eyes the blind eyes opened, the deaf ears unstopped, still battled unbelief and hardness in their hearts.

Satan's strategy is subtle. By keeping you so busy that you have no time alone with God, you gradually become unable to hear God's voice with clarity. Because you do not recognize His voice with the certainty you once did, you then fail to respond and act when you do hear it. As a result of this gradual spiritual dying process, when you ask God to heal your body and supply your needs, you no longer really expect to receive fulfillment because your heart has become hardened through unbelief.

Make this declaration: "My heart will not be hardened through unbelief. I will hear God's voice with clarity, recognize it, and act upon what He says."

Exercise Your Power

Scripture reading: Hebrews 3:14-19; 4:1

Many believers have allowed their hearts to become hardened through sin because they have placed their confidence in man and material possessions. Scripture warns that in the last days this will happen: *"...because lawlessness is increased, most people's love will grow cold"* (Matthew 24:12, NAS); that *"men shall be lovers of their own selves"* (2 Timothy 3:2); and that men will be *"lovers of pleasures more than lovers of God"* (2 Timothy 3:4). Their faith has been shaken because it has not been rooted daily in the unshakable, irrefutable, impregnable Word of God.

Many fail to recognize that a key strategy of Satan is to gradually weaken the Church through lies and deceptions until sin stagnates and hardens the heart of the believer.

To avoid this process, you must seek God daily because it is in your daily walk that you are fortified by the Word of God:

> *For we are made partakers of Christ, if we hold the beginning of our confidence steadfast unto the end; While it is said, Today if ye will hear his voice, harden not your hearts, as in the provocation.*
>
> *(Hebrews 3:14-15)*

For thousands of years, the devil has carefully disguised himself. He is an imposter, and one of his strategies is to deceive. Throughout the ages, Satan's objective has been to hinder man from knowing the truth through his deceptions.

The devil uses two basic strategies to deceive the children of God: He perverts the truth, and he lures us into sin. His non-changing strategy is to accuse, slander, act as adversary, deceive, tempt, destroy, and lie in order to bring down the Body of Christ.

Jesus knew and recognized the enemy's strategies. He never feared to address him directly as the demon of darkness and destruction. Jesus was never taken by surprise.

Through the power and anointing of the Holy Spirit, He took authority over Satan, addressed him, rebuked him, and cast him out.

The Church of Jesus Christ should never be reluctant to discuss the devil or speak directly to him because he is a real being. His strategies are real. His power is real, but the power within you is greater than Satan's power.

Exercise that power in your life as you face the challenges of today!

Make this declaration: "As I face the challenges of my life today, I will exercise the power God has given me over all of the power of the enemy."

August 12

Lord Over Circumstances

Scripture reading: 2 Corinthians 1:8-10

Prior to crossing into the Promised Land, the nation of Israel was warned by Moses regarding what they would face in Canaan. Moses knew they would fight many battles and that the enemy would try to create confusion, wear them down, and defeat them.

The same is true today. Using the adverse circumstances of everyday life, Satan will constantly try to wear you down physically, mentally, and spiritually until you are so weary, discouraged, and depressed that you are ready to throw your hands up in defeat. He will try to create havoc in the circumstances in your life – bringing affliction and disease upon your body, attacking your children and loved ones, causing strife in your marriage relationship, binding your finances, and creating problems on your job. He will try to grab hold of your mind in a vise-like grip through fear, worry and doubt, and try to keep your mind in a constant state of turmoil. His strategy is to get you so tired that you are no longer able to resist his attacks and you are caught in a vicious cycle that hinders you from releasing your faith.

Satan's objective in attacking you through your circumstances is to steal your health, tear apart families, destroy marriages, bind the minds of your children through alcohol and drugs, and keep you in financial bondage:

> *The thief cometh not, but for to steal, and to kill, and to destroy: I am come that they might have life, and that they might have it more abundantly.*
> *(John 10:10)*

The Apostle Paul told the Corinthians:

> *We are afflicted in every way, but not crushed; perplexed, but not despairing; persecuted, but not forsaken; struck down, but not destroyed.*
> *(2 Corinthians 4:8-9, NAS)*

Paul faced circumstances that were beyond his strength to endure, even to the point of despairing for his own life, but he knew what to do, and placed his faith and hope in God's hands. He told the believers in Corinth:

> *We do not want you to be uninformed, brothers, about the hardships we suffered in the province of Asia. We were under great pressure, far beyond our ability to endure, so that we despaired even of life. Indeed, in our hearts we felt the sentence of death. But this happened that we might not rely on ourselves but on God, who raises the dead. He has delivered us from such a deadly peril, and he will deliver us. On him we have set our hope that he will continue to deliver us.*
> *(2 Corinthians 1:8-10, NIV)*

No matter what circumstance you are facing, you do not have to be a victim of those circumstances. There is no situation, trial, or persecution that Satan can bring into your life that can defeat you! Paul knew that, and walked in that knowledge. So can you.

Make this declaration: "There is no circumstance in my life that I cannot defeat through God's strength and power. God is Lord over all of my circumstances."

THREE COMMON DECEPTIONS

Scripture reading: Ezekiel 36:25-27

All of the strategies of Satan are focused for one ultimate purpose: To weaken, divert, and gain control over man's will through disobedience. Since man's first encounter with the enemy in the garden of Eden until now, Satan has been lying, deceiving, tempting, and using every possible tactic to cause man to disobey God.

Because of their disobedience, the children of Israel initially failed to take possession of their inheritance in the Promised Land. Today, Satan's strategy is to deceive you into thinking that it is impossible for any man to walk in 100 percent obedience to God. Don't listen to his lies! God made it possible for you under the New Covenant. He promised:

> *A new heart also will I give you, and a new spirit will I put within you: and I will take away the stony heart out of your flesh, and I will give you an heart of flesh. And I will put my spirit within you, and cause you to walk in my statutes, and ye shall keep my judgments, and do them.*
> *(Ezekiel 36:26-27)*

God fully intends for us to walk in His Spirit, to obey His statutes, and to keep His judgments. He has given us His grace and ability to achieve that goal. Satan lies to you, however, and tells you it is not possible to stay pure in heart and deed. Three of the common deceptions he uses to try to move you into disobedience are:

Rebellion: Rebellion is failure or refusal to act upon the Word of God. Our churches are full of Christians walking in rebellion because they fail to act and become doers of the Word. You can do nothing in your own strength, but through the power of God inside of you, you have the power to obey all that God expects from you. Satan cannot force you to rebel. You must choose to disobey. It is an act of the free moral will God has given to you. You can also choose to obey God, knowing sin no longer has power over you.

Self-will: Since the beginning of time, Satan has tried to lead God's people into rebellion and disobedience by tempting them to reject God's voice and follow their own pathetic self-will. This strategy is one of the most dangerous because many people are so deceived they cannot recognize when they have fallen into the snare of the enemy.

Compromise: Satan wants to deceive you into compromising what God has told you to do and lure you into living your life as you please. Do not be deceived. Just because you are submissive and obedient to God in a few areas of your life it is not enough! You must be obedient in all areas.

Make this declaration: "I reject the power of rebellion, self-will, and compromise. They will not operate or exert control in my life."

A New Wave Of Holiness

Scripture reading: Malachi 3:1-3

Prior to his death, Moses warned Israel repeatedly of the importance of living holy, separated lives when they entered the Promised Land. God has revealed to me that a new wave of holiness is coming upon God's people. This coming wave of true holiness will be a deep inner working where Christ will do some deep cleaning in our lives to prepare us for His coming. There is coming a time of cleansing and purging to the Body of Christ! God's people are going to be tried as gold is tried in the fire, until all the impurities – everything in our lives that is unholy and displeasing to Him – are burned up.

This time of purging will not be easy. It will be painful. God is going to allow circumstances to come into our lives that will bring to the surface the hidden attitudes of our hearts, the impurities, and unholy thoughts and actions. Malachi prophesied:

> *But who may abide the day of his coming? And who shall stand when he appeareth? For he is like a refiner's fire, and like fullers' soap: And he shall sit as a refiner and purifier of silver: and he shall purify the sons of Levi, and purge them as gold and silver, that they may offer unto the LORD an offering in righteousness.* *(Malachi 3:2-3)*

As we enter into this new dimension of holiness, there will be no more hypocrisy – saying one thing and doing another. There will be a genuine manifestation of holiness in our conversation, thoughts, attitudes, and relationships.

Today and tomorrow we are going to meditate on ten principles that will enable you to develop and maintain integrity and live in this new dimension of holiness:

Principle One: Ask God to expose sin in your life: Cry out to God and ask Him to expose any sin in your life that may be hidden to you. Pray as King David did, *"Who can understand his errors? cleanse thou me from secret faults"* (Psalm 19:12).

Principle Two: Present yourself to God: Willingly submit yourself to God and permit Him to cut away everything that is ungodly or unholy. Don't try to measure yourself by the world's standards or the standards of man. Submit yourself to God's standards:

> *I beseech you therefore, brethren, by the mercies of God, that ye present your bodies a living sacrifice, holy, acceptable unto God, which is your reasonable service. And be not conformed to this world: but be ye transformed by the renewing of your mind, that ye may prove what is that good, and acceptable, and perfect, will of God.* *(Romans 12:1-2)*

Make this declaration: "God, expose any sin in my life. Cleanse me from secret faults. I submit myself to you as a living sacrifice, holy, and acceptable."

Principles For Holy Living

Scripture reading: Revelation 2:1-7

We are meditating on ten principles that will enable you to live in this new dimension of holiness to which God is calling us. The first two principles, which we meditated on yesterday, are to ask God to expose sin in your life and to present yourself to God.

Principle Three: Break unholy alliances: As God exposes sin in your life, you must get rid of it – this is part of the transformation and renewal process that will bring you into this new wave of holiness. If you have formed any unholy alliances with the world, you must break them!

Principle Four: Reject ungodly attitudes: If God reveals jealousy, hatred, bitterness or other ungodly attitudes, take authority over them, rebuke these attitudes, and get them out of your heart.

Principle Five:Cleanse yourself from an ungodly lifestyle: Paul told the Corinthians:

> *Having therefore these promises, dearly beloved, let us cleanse ourselves from all filthiness of the flesh and spirit, perfecting holiness in the fear of God.*
> *(2 Corinthians 7:1)*

Your righteousness was secured by the death of Jesus, but you must apply that blood in order for it to be effective in your life. That is why Paul says we must *"cleanse ourselves from all filthiness of the flesh and spirit."*

Principle Six: Make God's Word your standard: The Word of God is a lamp to walk by and a mirror to reflect your continual need of cleansing. Reject worldly standards and make the Word of God your standard for living.

Principle Seven: Maintain a sense of purpose: When you have a vision and understand your purpose, it helps you maintain proper conduct and attitudes.

Principle Eight: Be accountable: God holds you accountable for your conduct, relationships, thoughts, attitudes, and finances. Maintain your integrity by being accountable to God, your mate, your boss, your co-workers, and others with whom you relate.

Principle Nine: Renew your first love: Just as sin mars a marriage relationship when a couple loses their first love, so unholiness enters our lives when our love for God dims. Study Revelation 2:1-7 and heed God's call to the Church at Ephesus to return to their first love.

Principle Ten: Keep your focus on eternity: Weigh each decision, each challenge, and each temptation in light of eternity. Be like Moses, who chose to suffer affliction with the people of God rather than enjoy the pleasures of sin for a season (Hebrews 11:25).

Make this declaration: "In the Name of Jesus, I break unholy alliances in my life and pledge to cleanse myself from an ungodly lifestyle, make the Word of God my standard, maintain a sense of purpose, be accountable, renew my first love, and keep my focus on eternity."

Mixing The Word And Faith

Scripture reading: Psalm 56:1-3

In a world that is searching and struggling to find an answer for how to cope with stress, God has planned for us not to simply cope with stress, but to overcome it and allow Him to turn it around for our good.

There is no circumstance, no trial, no persecution that Satan can bring into your life that can defeat you. None of the circumstances, trials, and tribulations we face are responsible for creating stress. It is how we react to our circumstances, trials and tribulations that activate stress in our lives.

When Satan attacks us through our circumstances, it triggers a response within us. If we are fearful, worried, or doubtful that God will deliver us, we will be full of stress. However, if we face the circumstances Satan brings into our lives full of faith and confidence in God, knowing that He will deliver us in that situation, our hearts and minds will be filled with a peace that passes all understanding. There will be no room for stress.

Once Satan has attacked us through stressful circumstances he bombards our minds with fear, worry, and doubt. As long as we keep our eyes on the circumstances we are facing – the sickness, financial problems, family problems, problems on the job – we will continue to be fearful.

Satan will try to fill our minds full of fear concerning these circumstances, until we are so bound we are unable to release our faith. Once fear grips our minds, we begin to worry about our circumstances, our sickness, our children, and our overdue bills. Night and day, our minds are in a constant state of turmoil as we try to decide what we are going to do about the circumstances we are facing. The longer we keep eyes set on our circumstances the more we worry, the more we are fearful and the more stress we feel, until Satan has us in a deadly grip that paralyzes and hinders us from receiving what we need from God.

The moment we begin to worry, we open ourselves up for defeat. Satan begins to attack our minds with doubt. "If God loves you, why did He allow this problem to happen to you?" Satan will cause you to doubt God's Word and His covenant promises to you. If you do not see an immediate change in your circumstances, Satan will tell you: "It's not God's will to heal you. He doesn't care about your finances. God doesn't do miracles today."

Satan's strategy is to fill our minds so full of fear, worry, and doubt that the Word of God becomes ineffective in lives because it's not mixed with faith.

Make this declaration: "I am mixing my faith with the Word of God. The Word of God is living, active, and at work in me and my circumstances."

No Room For Stress

Scripture reading: Psalm 91:1-15

The adverse circumstances Satan uses against us can create stress which attacks our entire beings, including our hearts and minds as well as our bodies. Stress is one of Satan's deadliest strategies he uses to keep Christians living in a cycle of defeat.

The world we live in today is filled with stress. Stress is a sign of the times in which we live. No one is exempt from it! Large corporations have developed stress management programs to help their executives cope with the stress they must face day after day. There are books and seminars everywhere teaching various techniques of how to effectively cope with stress. There are even books written specifically to teach children how to deal with the stress in their lives.

Stress is considered public health enemy number one. It tears down our immune system and increases our susceptibility to almost every disease. It inhibits the body's production of cancer fighting cells. Fifty percent of the two million deaths in the United States each year – two deaths every minute – are attributed to hypertension that has been brought on by stress.

The majority of Christians today have become victims of stress. Their minds and bodies have become weakened and many are suffering from sicknesses that are brought on by stress. It is time for us to recognize stress as more than just a sign of the times or a major contributing factor of disease. We must recognize stress as a weapon Satan uses to defeat us. We must know what stress is and how he uses it to attack and destroy us.

When Satan attacks through your circumstances, it triggers a response within you. If you are fearful, worried, or doubtful that God will deliver you in the circumstances you are facing, you will be full of stress. However, if you face the circumstances, trials, and tribulations Satan brings into your life full of faith and confidence in God, knowing that He will deliver you, then your heart and mind will be filled with a peace that passes all understanding. There will be no room for stress.

Make this declaration: "There is no room for stress in my life – so stress, get out in the Name of Jesus!"

August 18

SEVEN SIMPLE KEYS

Scripture reading: Hebrews 12:1-3

Today and tomorrow, I want to give you seven simple keys that will help you break out of the cycle of defeat in your life – in your finances, your family relationships, your body, and in every area of your life – and begin to live in a cycle of victory:

1. Replace stress with peace. When Satan attacks in your circumstances, instead of being filled with stress, set your mind on God and His faithfulness, and allow Christ's peace to guard your heart and mind:

> *You will guard him and keep him in perfect and constant peace whose mind (both its inclination and its character) stayed on You, because he commits himself to You, leans on You and hopes confidently in You.*
> *(Isaiah 26:3, TAB)*

2. Rejoice. James said to count it all joy when you encounter trials and temptations (James 1:2). Peter said:

> *Beloved, think it not strange concerning the fiery trial which is to try you, as though some strange thing happened to you; But rejoice, inasmuch as ye are partakers of Christ's sufferings; that, when his glory shall be revealed, ye may be glad also with exceeding joy.* *(1 Peter 4:12-13)*

Regardless of whether or not you feel like praising God, offer up sacrifices of praise – not for the circumstances Satan has brought into your life, but because you know He is with you and using your circumstances for your good.

3. Take your eyes off your circumstances and fix them on Jesus and the Word.

> *Wherefore seeing we also are compassed about with so great a cloud of witnesses, let us lay aside every weight, and the sin which doth so easily beset us, and let us run with patience the race that is set before us . Looking unto Jesus the author and finisher of our faith; who for the joy that was set before him endured the cross, despising the shame, and is set down at the right hand of the throne of God. For consider him that endured such contradiction of sinners against himself, lest ye be wearied and faint in your minds.*
> *(Hebrews 12:1-3)*

4. Resist Satan. God has provided victory in every circumstance you face, but it is not automatic. God would not have provided us with His armor to protect us and have given us powerful weapons to take authority over Satan if He had not planned for us to fight. God does not want you to sit back and allow Satan to run all over you. You must pick up your weapons and go to battle. The moment Satan attacks your mind with fear, worry, and doubt, cast it out in the Name of Jesus. Bind him from working in your circumstances.

Make this declaration: "Stress is being replaced in my life with peace as I take my eyes off of my circumstances, resist Satan, and rejoice in the Lord."

BREAKING THE CYCLE OF DEFEAT

Scripture reading: Isaiah 40:28-31

Yesterday, I gave you the first four keys that will help you break out of the cycle of defeat in your life and begin to live in a cycle of victory:

1. Replace stress with peace.
2. Rejoice.
3. Take your eyes off your circumstances and fix them on Jesus and the Word.
4. Resist Satan.

Today, I want to give you three more additional keys:

5. Release Your Faith. In prayer, cast all your cares on Jesus, believe God's covenant promises to you, and release your faith to God. As you do, act on what you believe as if it were already done. Believe that you have received the answer even before you see any evidence.

> *But let him ask in faith without doubting, for the one who doubts is like the surf of the sea driven and tossed by the wind. For let not that man expect that he will receive anything from the LORD, being a double-minded man, unstable in all his ways.*
>
> *(James 1:6-8, NAS)*

6. Wait on the Lord. We are only defeated when we become weary and faint in our minds. The moment you feel that you are becoming weary – tired of resisting, tired of struggling, tired of casting out fear and doubt, tired of releasing your faith, tired of the circumstances – take time out to wait on the Lord in prayer. God is faithful, Who will strengthen and sustain you.

> *He giveth power to the faint; and to them that have no might he increaseth strength. Even the youths shall faint and be weary, and the young men shall utterly fall; but they that wait upon the Lord shall renew their strength; they shall mount up with wings as eagles; they shall run, and not be weary; and they shall walk and not faint.*
>
> *(Isaiah 40:29-31)*

7. Persevere. Regardless of how long you may have waited for deliverance in your circumstances, regardless of the pain or heartache you feel, persevere in the Spirit. Keep on pressing through in prayer. Keep on believing. Keep on releasing your faith and expecting God to deliver you.

Make this declaration: "I am releasing my faith in God for deliverance. I will wait on the Lord and persevere until He sets me free."

The Glory Of God

Scripture reading: 1 Samuel 4:1-22

In Old Testament times, the Ark of the Covenant was synonymous with and symbolic of the glory of God. When the word "glory" is used of God it is an expression of His divine essence – His honor, praise, splendor, radiance, power, worthiness, likeness, beauty, and rank.

God told Moses, *"If you make the tabernacle according to the pattern, my glory will dwell there"* (Exodus 25:8-9). The Ark was created according to God's pattern, His glory inhabited it, and it symbolized His presence in the midst of His people. The Ark preceded Israel as they travelled through the wilderness, it led the way as they crossed the Jordan River into Canaan, and it went before them in battle as a symbol of God's presence.

In 1 Samuel chapters 4-7 there is a sad story about how the Ark was lost to the enemy. First Samuel chapter 4 opens with the nation of Israel encamped at a place called Ebenezer battling with the Philistine army. Israel was losing the fight because of sin in their midst, especially in their leadership (see 1 Samuel chapter 3).

When 4,000 men died in a single battle – instead of asking the Lord to reveal the true problem – the Israeli army decided to send the Ark before them into battle. In previous battles, at God's command, the Ark had preceded their army and led them to victory.

Israel sent to Shiloh for the Ark and it was brought to Ebenezer. When it arrived in the camp, Israel gave a great shout of joy because they were sure that now they would be able to defeat the Philistines. But the Ark was only a symbol of God's presence and this time – because of sin- the presence of God was not with His people so the symbol was meaningless. When the battle resumed, 30,000 Israeli soldiers were killed and the Philistines seized the Ark of God.

When a messenger came to tell Israel's priest, Eli, he was so devastated that he fell off his chair, broke his neck, and died. The shock caused his pregnant daughter-in-law to go into labor and give birth to a son. She named the child "Ichabod" which means "the glory of God is departed" (1 Samuel 4:22).

In many instances, the Church has settled for symbols of the presence of God – crosses, doves, holy communion dishes, candles, and great church buildings. We have even had the emotion as Israel did when the Ark entered their camp. But whenever we settle for symbols, rituals, tradition, or mere emotion, the presence of God withdraws.

In a similar manner, man was originally created for the glory of God, but when Adam and Eve sinned, the pattern was changed and the glory was lost (Romans 3:23). The glory departs from God's people when sin separates them from His presence.

Make this declaration: "I was created for God's glory. He has forgiven my sin so I can once again reflect His glory to a lost and dying world."

A Name That Lives

Scripture reading: 1 Samuel 5:1-12

When the Philistines seized the Ark of God at Ebenezer, they took it first to a place called Aphek which means "strength and fortress." This was one area that Israel was never able to take from the enemy when they conquered Canaan and it eventually became a base from which they launched attacks against Israel (1 Samuel 4:1; 29:1).

In Aphek, the Philistines defiled and plundered the Ark. Any area of your life left unconquered becomes a stronghold and fortress of Satan. It will be used by the enemy to launch attacks against you to defile, violate, and plunder. The glory of God cannot remain in such an environment.

In 1 Samuel 5, the Bible records that the Philistines next moved the Ark to a place called Ashdod. Ashdod was one of the five chief cities of the Philistines, the location of one of the most famous temples of their national god, Dagon. At Ashdod, the Philistines tried to set up the Ark with other idols but their gods kept falling on their faces and finally were broken in pieces. The Philistines learned that you cannot set up the glory of God alongside worldly idols.

Beloved, you can't have the glory when you are trying to see just how much of the world you can hold on to and still have the presence of God. All other idols must fall before His glory, just as they did in Ashdod. The Ark could not coexist with idols, so it travelled on to Gath.

Next, the Ark was moved to a place called "Gath" which means "winepress." Whenever you compromise and allow the enemy to establish strongholds in your life (symbolized by Aphek and Ashdod) you will soon find yourself in his winepress. Satan will squeeze and press you until every last ounce of God's glory is drained out of your life.

At Gath, the repercussions of the lost glory finally began to be manifested and the wrath of God began to fall. God sent judgment on the Philistines. God will not allow His glory to remain amidst ritual, desecration, and compromise. So…the Ark moved on to Ekron.

Ekron means death. If we don't heed God's judgment, then spiritual death soon follows. We become like the Church at Sardis with a name that we live, yet we are dead (Revelation 3:1). Death came to the people of Ekron when the Ark came to their city, so they, too, sent it on its way.

Make this declaration: "I serve a God who lives. His life force flows in and through me today and equips me to face every circumstance of my life."

Potential For Power

Scripture reading: 1 Samuel 6:1-21; 7:1-2

It was at Ekron that the Ark of God changed directions and began its long and slow journey back to Israel, but God's presence did not immediately return to His people. It would take over 20 years to arrive in Jerusalem!

After what happened at Ekron, the Philistines decided they had better send the Ark back where it belonged (1 Samuel 6). They built a cart, placed the Ark on it, and used two milk cows to pull the load. These cows had calves shut up at home and their natural instinct when left to their own resources would have been to head towards their young ones. Instead, the cows headed straight towards Israel. (Even dumb animals know the glory belongs to God's people and not in the hands of the enemy!)

The cows and their precious cargo arrived first at a place called Bethshemesh where some people of Israel were reaping their harvest. When they saw the Ark, they began to rejoice. They took the cart and cows and offered them as an offering to the Lord.

But the men of Bethshemesh did not realize that holiness must be at the heart of any move of God. These men wanted God's presence among them, but they did not want it God's way. They looked into the Ark, which was forbidden by God, and 50,000 people died because of their sin (Exodus 19:21). Many times we rejoice when the presence of God begins to move in our midst, yet we refuse to walk in holiness. We want the presence of God, but we refuse to abide by His standards. Because of this, God's move passes us over.

Because the people of Bethshemesh refused God's standards, the Ark was sent on to Kirjath-Jearim where it remained for 20 years (1 Samuel 7:2). Although the Ark itself was in the midst of the people of Kirjath-Jearim, they rejected its power. They had the symbol of the presence of God in their midst, yet they and all of Israel "lamented" after the Lord. This means they longed for His power, hungering and thirsting for His presence. Jerusalem was void of the presence of God – and all the time the Ark of God was only eight miles away!

For years, the Ark of God's glory was shifted from city to city with little or no regard, respect, or understanding of its significance. The Ark carried within it the potential of all the mighty power of God, but nobody understood, nobody cared, and no one paid much attention to it. That is how many of us today treat the power and presence of God that dwells within us, dormant until we release its potential.

Make this declaration: "I am releasing the power of God's presence which has laid dormant within me."

Bring Back The Glory

Scripture reading: 1 Chronicles 13:1-7

We have traced the Old Testament account of the journey of the Ark of God during the past few days, drawing spiritual significance to its movement. For over 20 years, the Ark – which symbolized God's presence – was lost to God's people.

There finally came a day when the Lord found a man after His own heart who would not rest until the Ark of God's glory was fully restored to His people. That man was King David. In 1 Chronicles 13:1-7 we read of David's first attempt to bring back the Ark. Because the Philistines had carried the Ark on a cart he thought Israel might move it in a similar way, so he placed it on an oxcart. When the oxen stumbled, two men reached up to steady it and they were stricken dead. Judgment fell because the Ark was not moved the way God had commanded. David later recalls this experience in 1 Chronicles 15:1-24. He said the first attempt to bring back the glory was *"not after the due order."* The glory of God could not return on a man-made cart. It had to return according to God's plan.

We have tried to bring the glory back through man-made carts. We have tried programs and rituals and called for unity and brotherhood. Interesting enough, the names of the two men who died for touching the ark mean *strength* (Uzza) and *brotherly* (Ahio). Our man-made carts may look strong and brotherly, but the end result of man-made carts is stumbling, slipping, sliding, and eventually…death. The best ox-carts of our own making will never be good enough to carry God's glory. If the glory is to return, it must return God's way.

Turn to 1 Chronicles 13-17 and read about the successful return of the Ark. David didn't let past failure defeat him. He held on to his dream. He set an objective – *"Let us bring again the ark of our God to us"* (1 Chronicles 13:3) – then he rose up to make the dream a reality. This time, David brought the Ark back God's way.

It is interesting to note that on the way to Jerusalem the procession bearing the Ark stopped every six paces to sacrifice an ox and a fatling. It is estimated that it is between 12 and 15 kilometers from the ancient site of Obed-edom's house where the Ark was to Mt. Zion. It would probably take a man about 30,000 paces to cover that distance, which means the procession stopped 3,500 times to make sacrifices enroute to Mt. Zion! It wasn't easy. The trip was long, expensive, and bloody. But there was no sacrifice that David wasn't willing to make to get back the glory.

Now let me ask you…Is your desire that great for God's glory? Just how much do we want the Ark of God's presence among us? We've talked about it. We've longed for it. We've tried to bring it on man-made carts. Now the Ark of God – His glory and presence is near – it is only a short distance away. God is just waiting for men and women with enough vision – enough longing for His presence – to rise up and lead the way!

Make this declaration: "God's manifested presence is within me – all I need to do is rise up in power and authority and release it."

August 24

THE LATTER GLORY

Scripture reading: John 4:10-14

When Joshua led Israel across the Jordan River from the wilderness to the Promised Land, the waters of Jordan rolled back all the way to a city called Adam (Joshua 3:16) and the waters flowing into the Dead Sea were cut off.

This was symbolic of what happened when Jesus died on the cross. When Jesus was crucified, blood and water flowed from His side (John 19:34). It may have looked like a trickle in the natural world, but in the spirit world it became a mighty flood that has cleansed millions from sin. His death shut off the river of death which issued from Adam and flowed down through the centuries into the sea (which is symbolic of nations) of dying humanity.

At Calvary, Jesus released the flow of living water which He had promised:

> *In the last day, that great day of the feast, Jesus stood and cried, saying, If any man thirst, let him come unto me, and drink. He that believeth on me, as the scripture hath said, out of his belly shall flow rivers of living water. (But this spake he of the Spirit, which they that believe on him should receive: for the Holy Ghost was not yet given; because that Jesus was not yet glorified.)*
> *(John 7:37-39)*

Jesus made this declaration during the Jewish Feast of Tabernacles which was also called the Feast of Ingathering because it represented the final great harvest of the Jewish year. Specifically, He made this declaration on what was called the "Great Day" of the feast when a ritual called the outpouring of waters occurred. On this day, water was taken from Siloam Spring and poured on an altar while Isaiah 12:3 was quoted... *"Therefore with joy shall ye draw water out of the wells of salvation."* (The name "Siloam" means "sent". See John 9:7.)

The two verbs "come" and "drink" used in this passage are in the present imperative tense of the Greek language. This means they are commands, not just invitations, and they denote continuous action. Jesus was saying, "Keep coming and keep drinking." The verb "flow" is also in the present tense. If we keep coming and drinking, we will keep being filled and the power and glory of God will keep flowing through us.

The Prophet Haggai prophesied of a greater glory coming to the latter Temple (Haggai 2:1,9). This prophecy was given when Israel was rebuilding their Temple which had been destroyed by the enemy. The new temple was where, years later, Jesus Christ preached and worked miracles. Indeed – the latter glory was greater than the former.

The glory of God is returning to His people. It won't be in outward show or splendor, but the latter glory will be greater than the former because it will flow from within. Are you ready to receive it?

Make this declaration: "I am ready to receive God's manifested glory. My latter end will be greater than the former, today better than my yesterdays."

Borderline Christianity

Scripture reading: Numbers 32:1-19

All of Israel did not settle in the Promised Land. There were two and a half tribes, Reuben, Gad, and half of the tribe of Manasseh, that settled east of Jordan. When they saw the riches, houses, pastures, lands, wells, and cattle, they looked at these things and thought, "It would be nice to settle down right here. Let's just remain on this side of Jordan." The leaders of these tribes came to Moses and said, "We want to stay right here and settle down."

Numbers 32, tells the story of what happened to Reuben, Gad, and half of the tribe of Manasseh. They came and told Moses, "We want to remain on this side of Jordan." Moses was rather upset when they made this request (Numbers 32:6-8). Moses reminded these men that their fathers had made a similar error which resulted in forty years of wilderness wanderings.

Eventually, the leaders of this tribe made a compromise with Moses:

> *But we ourselves will go ready armed before the children of Israel, until we have brought them unto their place: and our little ones shall dwell in the fenced cities because of the inhabitants of the land. We will not return unto our houses, until the children of Israel have inherited every man his inheritance. For we will not inherit with them on yonder side Jordan, or forward; because our inheritance is fallen to us on this side Jordan eastward.* *(Numbers 32:17-19)*

Moses agreed to this, and the two and a half tribes began to settle down, divide up the territory, and possess the land on the east bank of Jordan.

The question is: Were they right in what they did? Were they right in settling down east of Jordan?

The answer is: They were wrong. This is an example of borderline Christianity. These men were moved by sight instead of faith. The promise was clear – it was the west side of Jordan that Israel was to possess:

> *Moses my servant is dead; now therefore arise, go over this Jordan, thou, and all this people, unto the land which I do give to them, even to the children of Israel. Every place that the sole of your foot shall tread upon, that have I given unto you, as I said unto Moses.* *(Joshua 1:2-3)*

Examine your own heart today. Are you living a life of borderline Christianity or are you moving ahead to claim all that God has for you?

Make this declaration: "I won't settle for borderline Christianity. I will move out in faith to claim all that God has for me."

The Settled For Life

Scripture reading: Matthew 13:1-9;18-23

The Israeli tribes of Reuben, Gad, and the half tribe of Manasseh asked Moses to allow them to settle east of Jordan instead of crossing over into the Promised Land. Psalm 106:15 says: *"And he gave them their request; but sent leanness into their soul."* An answered request – an answered prayer – is not always a blessing. Sometimes God gives you an answer to your prayers because it's what you desire so much, but your soul will become spiritually lean.

Why did Moses agree to their request? It is a picture of God, who will never force His will upon anybody. He is not going to coerce you to do anything. He is patient and loving and will always leave the door open to provide opportunities for you to inherit His best, but it is your choice.

Borderline Christianity is a picture of the believer who has come out of Egypt and the wilderness, but who refuses to enter the Promised Land. In his heart, he doesn't want to go back to the slavery and bondage of Egypt, which is a type of sin. He doesn't want the wilderness – it is a baron, dry waste land. But he is willing to settle for just a taste of victory without all the problems that go along with possessing a Promised Land that is inhabited by giants and fortified cities.

The parable of the sower in Matthew 13 relates the response of those who hear the Word of God. There are those who remain in Egypt (sin) verse 19; those who remain in the stony places of the wilderness (verse 20); and those who receive seed among the thorns and the environment chokes out the seed (verse 22). The latter is a picture of borderline Christianity:

> *He also that received seed among the thorns is he that heareth the word; and the care of this world, and the deceitfulness of riches, choke the word, and he becometh unfruitful.*
>
> *(Matthew 13:22)*

Will you be a borderline Christian, or will you act upon the promises of God's Word and reap the blessings of the Promised Land?

> *But he that received seed into the good ground is he that heareth the word, and understandeth it; which also beareth fruit, and bringeth forth, some and hundredfold, some sixty, some thirty.*
>
> *(Matthew 13:23)*

The victorious Christian is one who is always moving forward, always pushing ahead, always hungering and thirsting for all that God has for them.

Those who settled on the eastern side of the Jordan wanted so much to look like those who were on the other side of the Jordan that they later built a religious structure to imitate the one which was constructed in Jerusalem and it caused division in Israel. That's a picture of borderline Christians...they select the good instead of the best, they build their own religious structures, and they cause division among God's people. They live the "settled for" life.

Make this declaration: "I will not live the settled for life. I will be all God intended me to be and do all He designed for me to do."

THE WEAPON OF PRAYER

Scripture reading: Psalm 94:1-23

There is a tremendous weapon God has given us which is powerful enough to penetrate into the spirit, break through enemy territory, bind the power of the enemy, and destroy Satan's strongholds. That weapon is prayer!

We are wrestling *"against principalities, against powers, against the rulers of darkness of this world, against spiritual wickedness in high places"* (Ephesians 6:12). There are organized demonic forces, spiritual powers, and rulers of darkness – spirits of violence, confusion, greed, hate, murder, rebellion, fear, lust, perversion – that are hovering over our cities, communities, and nations. They were in the Promised Land. They are everywhere in your land – on the streets, in your place of work, and in your home.

- Their objective: to tear down and destroy all that is holy, all that is pure, all morality, all sense of decency.
- Their objective: to destroy and tear down governments, pervert justice, and establish atheistic, immoral leaders.
- Their objective: to discredit Christian leaders, to weaken their godly influence in our cities and countries, to silence their voices.
- Their objective: to divide the Body of Christ – to bring division and strife within churches – to get the Church sidetracked from its true purpose of winning souls into the Kingdom of God.
- Their objective: to destroy families – turn husbands and wives against one another; sow seeds of rebellion in the hearts of the teenagers; cause feelings of rejection, mistrust, bitterness, and resentment.
- Their objective: to prevent you from growing up into the full stature of Jesus, to stop you from praying, to hinder God's answers to your prayers, to stop you from reading and studying the Word, to hinder you from applying the Word to your life, to stop you from doing God's will.

These evil powers and principalities have established strongholds in our homes, communities, and nations that no atomic bomb, nuclear weapon, or carnal weapon can ever destroy. The only power strong enough to penetrate these evil forces and pull down these strongholds in the nations of the earth are the prayers of God's people!

Every stronghold in your life, every bondage in your family, finances, job, or ministry can be penetrated by this powerful spiritual force. The last great end-time anointing will be the prayer anointing. God is releasing it right now, all over the world...and He is releasing it in you.

Make this declaration: "I am receiving the end time prayer anointing which God is releasing all over this world. Every stronghold in my life, every bondage in my family, finances, job, and ministry will be penetrated by this powerful spiritual force."

August 28

Penetrating Enemy Lines

Scripture reading: Psalm 13:1-6

Stop for a moment and think about the battles you are facing right now. Think about the strongholds Satan has tried to build in your life, your home, your community, and your nation. You may be facing situations and problems that are causing you to be confused and afraid. You don't even know how to pray concerning them. You may be facing Satan's attacks in your finances. Your debts keep piling up and, regardless of how hard you try, there is never enough money to go around. Your mind has been in constant turmoil. You have been worrying about what you are going to do and where you are going to get the money you need.

You may have been attacked physically and mentally. Your strength seems to be gone and discouragement and doubt are setting in. You may have been seeking God for direction – to hear His voice and know His will concerning important decisions you need to make. You have done everything you know to do. You have wept. You have pleaded. You have cried out to God. Yet, it seems as if the heavens are as brass and that somehow your prayers aren't getting through. Like the great warrior King David, from the depths of your spirit you cry out, *"How long, Oh Lord? Will you forget me forever? How long will you hide your face from me?"* (Psalm 13:1, NIV).

God has not left you alone. He has not forgotten you. The very first moment you called upon Him, He heard you. You are fighting against powers and principalities that have built a strong line of defense, a barrier to hinder you from receiving the answers to your prayers. God is not withholding Himself from you.

Before you can experience the breakthroughs you need in your life, you must penetrate deep into the Spirit and break through that barrier the enemy has built. To be able to penetrate the enemy lines where the forces of evil are positioned, locate, and bind the enemy, you must pick up your most powerful weapon – your weapon of prayer – and go beyond the surface into the spiritual realm.

The word "penetrate" means "to pierce; to enter by overcoming resistance." Through prayer Jesus penetrated enemy lines and tore down the strongholds of Satan. Using His mighty weapon of prayer Jesus opened blind eyes, unstopped deaf ears, restored crippled limbs, cleansed lepers, loosed tongues, cast out demons, and raised the dead.

Like a conquering soldier, Jesus broke through the enemy's line of defense. He penetrated Satan's territory and took authority over him. Out on the front lines He did not pray for demons to leave, He commanded them to go. When He saw those who were blind, deaf, or mute, He did not pray asking God to open their eyes, unstop their ears, or loose their tongues. He commanded them to see, hear, and talk. This was because He fought and won the battle before going out to the front lines.

Make this declaration: "I will fight the battle before going out to the front lines, then like a conquering soldier, I will break through the enemy's line of defense."

BEFORE YOU MOVE

Scripture reading: Matthew 26:36-46

The great victories Jesus experienced out on the front lines as He ministered to the needs of the people – teaching, opening blind eyes, unstopping deaf ears, healing all manner of diseases, and raising the dead – were not automatic. They were won in prayer before He went out on the front lines.

Although He was the Son of God, Jesus did not win these victories in His own strength. He had stripped Himself of His divine abilities and was subject to the same human limitations as you and I. He did absolutely nothing independent of the Father. None of the tremendous miracles and healings He did were automatic. Jesus said, *"By myself I can do nothing.* (John 5:30, NIV) and *" ...the Son can do nothing by himself; he can do only what he sees his Father doing, because whatever the Father does the Son also does"* (John 5:19, NIV). He was totally dependent upon the Father.

Everything Jesus did was a result of what the Father revealed to Him while He was in His presence in prayer. Before He went out onto the battlefield, Jesus spent time in His Father's presence where God revealed His will to Him. Through prayer, Jesus penetrated the enemy lines, fought the battle, and won it!

- Before Jesus began His ministry, He spent forty days in fasting and prayer and returned in the power of the Spirit into Galilee (Luke 4:14).
- Before feeding the five thousand by multiplying five loaves of bread and two fish, He went by boat out to a solitary place where He prayed (Matthew 14:13).
- Before choosing the twelve apostles, Jesus went to the mountain and spent the whole night in prayer to God (Luke 6:12, NAS).
- Before a great healing campaign in Gennesaret where the people brought the sick to Him, Jesus went up on a mountainside by Himself where He prayed (Matthew 12:22-23).
- Before He began His ministry throughout Galilee where He preached and cast out demons, Jesus prayed (Mark 1:35-39, NIV).
- Before going to the cross, Jesus went up to a mountain where He prayed (Luke 9:28).
- Before He was arrested, beaten, mocked, and crucified, Jesus went into the garden of Gethsemane where He agonized in prayer (Matthew 26:36-46).
- Before you make any move that will affect your life or family, shouldn't you enter into spiritual warfare through prayer, penetrate enemy lines, and win the battle?

Make this declaration: "Before making any move that will affect my life or family, I will penetrate enemy lines and win the battle in prayer."

August 30

Breaking The Barrier

Scripture reading: 1 Chronicles 14:8-11

To become a powerful spiritual warrior like Jesus, you must pick up your mighty weapon of prayer and use it to penetrate the barrier the enemy has built that is hindering you from experiencing the victories you need in your life.

The type of prayer we are talking about that is capable of piercing enemy lines is not the average five or ten minute prayer. It is not based upon a specific formula. It is a deeper dimension of prayer where you are travailing, agonizing, and groaning before God. This type of prayer involves engaging the enemy in one-to-one combat, battling the powers, principalities, and evil spirits that are at work, and not letting go until you have overcome their resistance and have broken their hold.

This type of spiritual warfare during prayer is like a soldier out on the battlefield. The enemy has formed a strong line of defense, a barrier that seems almost impossible to penetrate. There is no way of escape. The soldier has been struggling to overcome the enemy's resistance, but before the battle can be won he must penetrate enemy lines and overcome the enemy.

I have experienced this type of spiritual battle during prayer many times. In Madras, India, I faced a battle where the presence of demonic forces, spiritual powers, and rulers of darkness were so prevalent I could literally feel them wherever I went. The evil powers and principalities over that city had built a strong line of defense. They didn't want the Word of God to go forth and the power of God to be released which would break their hold on the people, so they stirred up the Hindu priests to organize a plot to kill me. I knew that before Madras could experience a breakthrough in the Spirit, it was necessary to penetrate the barrier the enemy had built. Before the meeting began, I closed myself in my room and used the powerful weapons of fasting and prayer. For twenty-four hours I agonized and wrestled with spiritual wickedness and rulers of darkness. I had known before what it was to pray, but I began to travail in prayer in a way I had never experienced before.

In your life there are going to be times when you face battles where evil forces, powers and principalities have built strongholds in the lives of your loved ones. There are going to be times when you face problems that are so overwhelming that you just don't know what to do and you must hear from God.

The only way you are going to break through the barrier and tear down those strongholds is to get on your face before God in times of prayer and fasting and stay there until you know that you know that you know Satan's power has been broken. You must enter into a time of intensified spiritual warfare where you are weeping and travailing before the Lord.

Make this declaration: "I will be spiritually aggressive in prayer today. I will enter into a new dimension of intensified spiritual warfare."

YOU ARE NOT ALONE

Scripture reading: 2 Kings 6:8-17

As you pick up your mighty weapon of prayer and begin to struggle against the evil forces that are attacking you, the hosts of heaven join with you in combat!

In the battles you are fighting right now in your life, you are not fighting alone. The hosts of heaven are fighting with you. The Almighty God, Jehovah is fighting for you!

The promise Moses gave the children of Israel belongs to you: *"The LORD your God who goes before you will Himself fight on your behalf..."* (Deuteronomy 1: 30, NAS).

The angels are fighting for you! *"The angel of the Lord encamps around those who fear him, and he delivers them"* (Psalm 34:7, NIV) .

When Elisha was surrounded by the Syrian armies he told his servant, *"...Do not fear, for those who are with us are more than those who are with them"* (2 Kings 6:16, NAS). Then God opened the eyes of Elisha's servant and allowed him to see that the mountain was full of a heavenly host of horses and chariots all around them (2 Kings 6:17).

The Holy Spirit is fighting for you! As you begin to pray, *"...the Spirit Himself goes to meet our (your) supplication and pleads in our (your) behalf with unspeakable yearnings and groanings too deep for utterance"* (Romans 8:26, TAB).

Jesus Christ, your Great High Priest, is fighting for you! While He was on the earth, He prayed for His disciples. He told Peter, *"But I have prayed for thee, that thy faith fail not..."* (Luke 22:32). He prayed for all believers (John 17:20-21) and He is at the right hand of the Father where He is making intercession for you right now (Romans 8:34; Hebrews 7:25).

The reason why many Christians live in defeat is because they are still relying upon their own strength and understanding. They are using carnal weapons instead of the mighty weapons of prayer and fasting. To become the same powerful spiritual warrior as Jesus and live in the same dimension of revelation, you must use the same weapon He used – prayer. You must strip yourself of all self-righteousness and spiritual pride. You must come to a place in your life where you are willing to say as Jesus did, "By Myself I can do nothing." You must come to a place where you are no longer relying upon your own strength or understanding.

You must be willing to forsake your way of doing things and develop a day-to-day dependence upon Christ where you are willing to spend time alone with Him, even if it means getting up at two or three in the morning or spending all night in prayer.

As you come to know, rely, and totally depend upon Jesus, in prayer, as He did the Father, you will able to face your circumstances from the same powerful position of knowing as Jesus. You will be able to face every circumstance, trial, and test knowing that you will not be defeated. Remember...you are not alone!

Make this declaration: "God has not abandoned me. I am not alone. He goes before me into battle to fight for me."

September

A knowledge of God's Word is your greatest protection against the snare of the enemy. This month, we will learn biblical strategies for dealing with deception.

September
Dealing With Deception

In Joshua chapter 9 we have an interesting account of a major deception of the enemy perpetrated against God's chosen people, Israel. Prior to entering the Promised Land, Israel was warned to eliminate all of the evil, heathen nations that resided there. When news of what happened at Jericho and Ai spread throughout the land, the Gibeonites came up with a deceptive plan:

> *And when the inhabitants of Gibeon heard what Joshua had done unto Jericho and to Ai, They did work wilily, and went and made as if they had been ambassadors, and took old sacks upon their asses, and wine bottles, old, and rent, and bound up; And old shoes and clouted upon their feet, and old garments upon them; and all the bread of their provision was dry and mouldy. And they went to Joshua unto the camp at Gilgal, and said unto him, and to the men of Israel, We be come from a far country: now therefore make ye a league with us. And the men of Israel said unto the Hivites, Peradventure ye dwell among us; and how shall we make a league with you? And they said unto Joshua, We are thy servants. And Joshua said unto them, Who are ye? And from whence come ye? And they said unto him, From a very far country thy servants are come because of the name of the LORD thy God: for we have heard the fame of him, and all that he did in Egypt, And all that he did to the two kings of the Amorites, that were beyond Jordan, to Sihon king of Heshbon, and to Og king of Bashan, which was at Ashtaroth. Wherefore our elders and all the inhabitants of our country spake to us, saying, Take victuals with you for the journey, and go to meet them, and say unto them, We are your servants: therefore now make ye a league with us. This our bread we took hot for our provision out of our houses on the day we came forth to go unto you; but now, behold, it is dry, and it is mouldy: And these bottles of wine, which we filled, were new; and, behold, they be rent: and these our garments and our shoes are become old by reason of the very long journey. And the men took of their victuals, and asked not counsel at the mouth of the Lord. And Joshua made peace with them, and made a league with them, to let them live: and the princes of the congregation sware unto them.*
>
> *(Joshua 9:3-15)*

The deceptive plan of the Gibeonites worked. Joshua failed to consult the LORD regarding the matter and made a league with them instead of destroying them as God had commanded.

Deception is one of the major strategies that the enemy uses. It started in the Garden of Eden when he deceived Eve and it has continued down to the new millennium. In these final days before the return of Jesus Christ, there is going to be a massive wave of satanic deception that will flood the earth. The Bible warns us that Satan is so deceptive that he can be transformed into an angel of light (2 Corinthians 11:14). If we are to experience Victory Miracle Living in the new millennium, then we must seek counsel of the Lord and prepare to face this tremendous onslaught from the enemy.

The strategic mistake which many Christians make is that they are willing to settle for a compromise with the devil. To them a compromise is a victory. There is no such thing as a stalemate with the enemy of men's souls. There is no negotiable truce line when you have a compromise agreement and you say to the devil, "You stay over there, and I'll stay over here." You either win or you lose and to compromise is to lose. To accommodate is to lose. You are only a winner if you overcome the forces that are trying to defeat you.

Pause for just a moment and consider your present standards of Christian conduct. Line them up against what you stood for and believed ten years ago, five years ago, or even one year ago. Does the comparison show a drastic difference? Could it be that you have been deceived into a compromising position?

Satan wants you to be lulled into a false sense of security in thinking that it is not possible that you could ever be deceived. I warn you, this is one of the most dangerous spiritual positions a Christian can be in. If you fail to recognize that there is the possibility of your being deceived, you will become a sitting target for Satan.

Our devotions this month will expose some of Satan's most diabolical strategies. Just as the Gibeonites deceived Israel, Satan is still working through deception but he is an enemy who has been defeated, stripped of his power over us, and doomed for destruction. He is a deceiver, imposter, the father of lies – and he is about to be exposed for what he is…

Just keep on reading…

We Must Have A Revelation

Scripture reading: Psalm 124:1-8

During our devotions this month, we will rip the mask off Satan and expose some of his most diabolical strategies! Just as the Gibeonites deceived Israel, Satan is still working through deception, but he is an enemy who has been defeated, stripped of his power over us, and doomed for destruction.

For years, the Church has been running from Satan, afraid to talk about him and afraid to acknowledge his existence. Our perception of him has been distorted. The majority of Christians have perceived Satan as a strong, formidable being possessing authority and dominion over the earth and over man. They have looked upon the horrors of sin, sickness, death, perversion, and murder that he has brought upon the world and have felt powerless against him. Because of their lack of knowledge, they have become victims instead of victors.

"Where there is no vision, the people perish..." (Proverbs 29:18). More than likely, you have heard this familiar verse quoted many times. The Hebrew word for "vision" is "revelation." For a lack of revelation concerning our adversary, Satan, the Church has been almost at a standstill, while our cities and nations remain filled with people who have been taken captive and are bound by Satan's power.

The world today is full of sin, sickness, death, and destruction because God's people have not had a revelation of the position of victory God has given them in this world over Satan. Satan's strategy is to keep us in spiritual ignorance. Our counterattack must not only be to expose Satan's lies concerning himself, but to know the truth and act on it.

We must have a revelation – a drawing away of the veil of darkness – concerning Satan's power, his strategies, and our power over him and his evil forces. Head knowledge is not enough. We must have a revelation within our spirit where we are able to exercise the power and authority God has given us over Satan to destroy his works in the lives of our loved ones, friends, neighbors, and co-workers.

We must have a revelation within our spirits where we will not be satisfied until we see Satan and the powers of darkness pushed back and his power broken over our nations. We must have a revelation within our spirits of ourselves as God intends for us to be – strong, powerful, full of the Holy Ghost and power – victorious over all the power of the enemy. As sons of the living God, we must take the position God has planned for us in these final, closing hours of time before Jesus returns.

Make this declaration: "God is giving me a revelation of how He intends for me to be – strong, powerful, full of the Holy Ghost and power – victorious over all the power of the enemy."

We Are At War

Scripture reading: Ephesians 6:10-12

We are at war! We are involved in a deadly one-on-one combat with a merciless enemy who is driven by an all, consuming, burning hatred of God, His saints and all that is pure and holy.

Paul told the Corinthians:

> *For we are not wrestling with flesh and blood – contending only with physical opponents-but against the despotisms, against the powers, against (the master spirits who are) the world rulers of this present darkness, against the spirit forces of wickedness in the heavenly (supernatural) sphere.*
>
> *(Ephesians 6:12, TAB)*

Paul compared our conflict with the enemy to wrestling. Wrestling is close combat which involves two enemies, each exerting their whole force and strength against the other. In the natural world during a wrestling match, each opponent enters the arena. The clang of the bell sounds at the beginning of the match. The wrestler's grab hold of one another, each striving to get the best advantage over his opponent. They struggle. They fight. They choke. They throw one another to the floor. Round after round they wrestle until one is declared the winner.

The moment you were born into the Kingdom of God you entered the arena of spiritual warfare and the wrestling match began. Unlike wrestler's in the natural world who can see their opponent face to face and feel his flesh as they wrestle, you are wrestling with spiritual forces that you cannot see, touch, or feel with your natural senses.

Paul told the Ephesians, *"For we are not wrestling with flesh and blood."* Every born-again child of God is involved in an intense struggle with Satan and his forces of darkness. This is a very simple, basic truth concerning spiritual warfare that I have been preaching and teaching for more than twenty years, yet the majority of Christians still have failed to grasp hold of it.

There are Christians today who are not even aware that a war is going on! They are unable to detect the works and strategies of Satan; they are failing to pick up the powerful weapons God has given them to fight the enemy; and, as a result, they are living in defeat.

You and I are involved in an unseen war that is raging all around us. Satan and the forces of darkness are not figments of our imaginations, they are real! In this ongoing battle we are fighting, Satan has developed strategies to discourage, tempt, deceive and defeat us. The major goal of his strategies are to rob us of our inheritance as children of God and destroy our spirit, soul, and body in hell. Behind all of his strategies, all of his temptations, all of his deceptions, this is his one all consuming desire.

Make this declaration: "I am not wrestling with flesh and blood. I am wrestling against principalities, powers, and spiritual wickedness. I will prevail, for I am supernaturally equipped by the power of God."

Will You Use It?

Scripture reading: Luke 10:1-20

We are wrestling, fighting, and struggling against Satan and the forces of darkness. We have no choice! We are in the arena of conflict and we must go into battle as Jesus did, knowing the strategies of Satan and recognizing our power and authority over him,

Satan wants to deceive you and keep you in a place of spiritual ignorance. One of the powerful strategies Jesus used to defeat Satan was that he faced the enemy from a position of knowing! Jesus never wavered. He never faltered. He never retreated. He knew the power of the enemy and He knew He had been given power over him. He faced the power of the enemy knowing the will of God, that He had been sent by God to destroy the works of the devil (1 John 3:8).

God has planned for you to face your trials, temptations, and circumstances – every attack of the enemy – from this same powerful position of knowing. Satan's deceptive strategy is to keep you in a position of spiritual ignorance. He will do everything possible to stop you from knowing the truth about him; to stop you from knowing your position of power and authority as a child of God; to keep you from knowing God's will for your life.

Jesus gave His twelve disciples power and authority over "all devils" (Luke 9:1). He appointed seventy disciples and gave them power over all the power of the enemy (Luke 10:19). The seventy returned rejoicing, saying, *"Lord, even the devils are subject unto us through thy name"* (Luke 10:17). They were amazed that they had power over evil spirits.

Jesus told the seventy disciples: *"Behold, I give unto you power to tread on serpents and scorpions, and over all the over of the enemy…"* (Luke 10:19). Satan was not in control. Jesus had been anointed by the Holy Spirit and was given power and authority over Satan. He came to destroy the works of the devil. He had all power over the enemy and gave it to His disciples.

Through the power of the Holy Spirit, Jesus took authority over the enemy. He rebuked the power of Satan and cast out devils simply by speaking the Word. He spoke, and devils trembled and fled. The people did not understand how a man could have such power and authority. They accused Jesus of casting out devils by Beelzebub, the chief of devils.

Jesus has given you that same power and authority. The only question is, will you use it today?

Make this declaration: "I will use the power and authority God has given me over the enemy."

The Earth Is The Lord's

Scripture reading: Luke 4:1-13

The position God has planned for His chosen people to take is not that of weak, cowardly soldiers running in defeat. As we approach the coming Day of the Lord, the position God has planned for us to take is marching forward, conquering an enemy that has already been defeated.

There are four basic facts you must know about Satan in order to be prepared to face his deceptions and seize the victory. These are very basic truths, but until you have a revelation within your spirit where you know that you know that you know, you will not be able to exercise the power and authority God has given you to defeat Satan in the circumstances in your life.

1. Satan is not in control over the earth.
2. Satan's power is limited.
3. Satan has no power over you, except what you allow him to have.
4. Christians are not exempt from persecution.

Satan's strategy has been to lie and to keep you from knowing these truths. Our counterattack is to expose Satan's lies.

The first lie Satan perpetrates is that he has power and dominion over the earth. There are many Christians today who have been deceived by this lie. They believe that when Adam and Eve yielded their wills to Satan and sinned against God, that God turned the earth over to Satan to rule. They base their belief on Satan's claim to power during the time he tempted Jesus in the wilderness.

Satan took Jesus up to a high mountain and showed Him the kingdoms of the world. He told Jesus: *"...All this power will I give thee, and the glory of them: for that is delivered unto me; and to whomsoever I will I give it. If thou therefore wilt worship me, all shall be thine"* (Luke 4:6-7).

Satan is a liar and he is the father of lies (John 8:44). Satan claimed that all the kingdoms of the world had been delivered into his hands. This is a lie! God has never relinquished control of the earth to Satan and He never will! Satan is not in control over the earth. It does not belong to him; it belongs to God: *"The earth is the Lord's, and the fullness thereof; the world, and they that dwell therein"* (Psalm 24:1).

Make this declaration: "This earth and the fullness therein belongs to God – including my friends and family, my possessions, and me!"

Satan Is Limited

Scripture reading: Matthew 26:47-56

Satan wants us to believe that he has been given unlimited power in the world, but he is a liar. Satan is not a free agent. God has not turned this world over to him to do whatever he desires to do.

God is in control, not Satan. God has a Master Plan that consummates in the establishment of a new heaven and a new earth where all things are brought together in one under Christ (Ephesians 1:10). Within God's plan and purpose for man, He has given Satan limited power and a limited amount of time to work.

Satan is not omnipresent. He cannot be in more than one place at a time. He is not all knowing. He does not know the thoughts and intents of our hearts. However, he does have the ability to attack our minds, as he did Ananias (Acts 5:1-3) and Judas Iscariot (John 13:27).

When Satan tempted Jesus in the wilderness, he did not do it without God's knowledge. It was not a chance meeting. Jesus was led by the Spirit into the wilderness. God allowed Jesus to be tempted in all points as we are so that we would have a faithful High Priest Who is touched by the feeling of our infirmities (Hebrews 4:15).

Satan did not have the power to take Jesus' life from Him because before the earth was created, God planned to give His only begotten Son as a sacrifice for the sins of the world. God allowed Satan, working through men, to crucify Jesus. Satan was not in control. God allowed them to work at that particular time, knowing that Jesus would rise from the dead, break the power of sin and death, and return to Heaven just as He had planned.

When the Jewish leaders and soldiers came to arrest Him, Jesus told them, *"When I was daily with you in the temple, ye stretched forth no hands against me: but this is your hour, and the power of darkness"* (Luke 22:53). This was the hour God permitted Satan to work through these men to bring Jesus to this point of suffering and death. It was not because Jesus was not powerful enough to prevent this from happening.

Jesus told them, *"Thinkest thou that I cannot now pray to my Father, and he shall presently give me more than twelve legions of angels? But how then shall the scriptures be fulfilled, that thus it must be?"* (Matthew 26:53-54). At Jesus' trial, when He refused to answer Pilate's questions, Pilate asked Him, *"Don't you realize I have power either to free you or to crucify you?" Jesus answered, "You would have no power over me if it were not given to you from above"* (John 19:10-11, NIV).

Satan's power is limited. He has been given certain boundaries by God and he cannot go beyond these boundaries. Satan is working upon this earth today, but God has placed limits upon him concerning his attacks on believers.

Make this declaration: "Satan's power against me is limited by the almighty power of God."

A Powerful Force

Scripture reading: Romans 6:16-19

Satan deceives believers by making them think he is more powerful than they are and that they are weak and powerless against him and his attacks. You are no match for Satan in your own strength, but through the power and authority of the Holy Spirit you have been given power over all the power of the enemy.

Satan has no power over you except what you allow him to have. Satan cannot make you sin. He can lie, deceive, and tempt you to sin; but, he has no power over your will! One of the most powerful things God has given you is your will. Unless you yield your will to Satan, he cannot have any power over you. He cannot make you lie. He cannot make you steal. He cannot make you commit adultery. He cannot make you sin against God. He cannot blot your name out of the Book of Life.

When you were born again, the power of sin was broken in your life. You have been delivered out of Satan's dominion and control and are now under Christ's dominion in the Kingdom of God. God has *"delivered us from the power of darkness, and hath translated us into the kingdom of his dear Son"* (Colossians 1:13).

Satan, as the ruler of darkness and the god of this world, has control over unbelievers and they are taken captive by him at his will (2 Timothy 2:26). He blinds their minds to the Gospel: *"...the god of this world hath blinded the minds of them which believe not, lest the light of the glorious gospel of Christ, who is the image of God, should shine unto them"* (1 Corinthians 4:4).

But Satan does not have power over you to make you sin. You are no longer under Satan's dominion and he has no power over you unless you yield your will to him. Paul told the Romans:

> *Know ye not, that to whom ye yield yourselves servants to obey, his servants ye are to whom ye obey; whether of sin unto death, or of obedience unto righteousness? But God be thanked, that ye were the servants of sin, but ye have obeyed from the heart that form of doctrine which was delivered you. Being then made free from sin, ye became the servants of righteousness...even so now yield your members servants to righteousness unto holiness.* *(Romans 6:16-19)*

Your will is a strong force that Satan has no power over. One of his major strategies is to attack your will, trick you into yielding to him, and then cause you to follow after your own will instead of doing God's will.

Make this declaration: "I am no longer under Satan's dominion. He has no power over me. I will not yield my will to him."

A Defeated Enemy

Scripture reading: Revelation 12:7-12

One of the greatest lies Satan has fed the Church is concerning a time that is coming before Christ's Second Coming when he will unleash all his fury upon the saints. It will be a time of great travail and persecution, greater than the world has ever known. Satan wants to keep the Church in spiritual ignorance concerning what will occur. He wants us to be confused and unprepared.

For years, there has been a veil over our eyes. There are many Christians who know very little about this time of great distress that is coming. Many have not studied or searched the Word to see for themselves exactly what is going to take place. Either they have dismissed it from their minds because what they have heard has been too negative or frightening, or they have accepted what they have heard without confirming it with the Word.

The Spirit of God is calling us to prepare for this time that is coming. Those who are not prepared will be shaken. Many will fall if they don't realize that God is still in control. The Church of the living God will not be defeated!

In Revelation 12, God reveals to us Satan's activities during the days before Christ's return. There is a final end time struggle in the heavens. Satan will be cast out of the heavenlies to earth and a warning is issued to those living upon the earth: *"...Woe to the inhabitants of the earth and of the sea! for the devil is come down unto you, having great wrath, because he knoweth that he hath but a short time"* (Revelation 12:12). Satan, knowing his time is short, will pour out his wrath upon the saints who are living upon the earth at that time. Daniel warned us of this persecution (Daniel 7:25), as did Jesus (Matthew 24:9-10, 21-22).

God has revealed to us His end-time plan for a purpose. He wants us to be prepared. Not only has He shown us how Satan will be working upon the earth in the last days, He has shown us that we will be victorious. We will overcome by the blood of the Lamb and by the word of our testimony (Revelation 12:11). Satan will be destroyed forever (Revelation 19:20).

You are facing an enemy who is defeated and doomed for destruction. Once you have a revelation within your spirit:

... that Satan has been defeated

... that his power is limited upon the earth today

... that God is in control

... that you have been given power and authority over him...

You are prepared to face every attack, every testing, every trial in total victory.

Make this declaration: "Satan is defeated. His power is limited. God is in control. I have power and authority over Satan and I am prepared to face every attack, every testing, and every trial in total victory."

The Knowledge You Need

Scripture reading: 1 Corinthians 2:11-16

Paul told the Ephesians not to *"give place to the devil"* (Ephesians 4:27). In the New International Version it is translated, *"do not give the devil a foothold."* A "foothold" is a position usable as a base for further advance.

Our position is not one of running or hiding from the enemy. It is standing firm, not giving Satan a place in our lives that will enable him to advance and eventually defeat us.

Allowing spiritual ignorance to remain in your life regarding Satan, his power, and the plan and purpose of God for your life allows him to get a strong foothold. It gives Satan an opportunity to operate and attack you. Your counterattack must be to put on the mind of Christ. Paul told the Philippians, *"Let this mind be in you, which was also in Christ Jesus"* (Philippians 2:5). We must not allow our lives to be ruled by our natural minds. Our natural minds cannot comprehend the things of the Spirit.

Paul told the Corinthians:

> *But the natural, nonspiritual man does not accept or welcome or admit into his heart the gifts and teachings and revelations of the Spirit of God, for they are folly (meaningless nonsense) to him; and he is incapable of knowing them – of progressively recognizing, understanding and becoming better acquainted with them because they are spiritually discerned and estimated and appreciated. But the spiritual man tries all things-(that is,) he examines, investigates, inquires into, questions, and discerns all things…*
>
> *(1 Corinthians 2:14-15, TAB).*

We must fill our minds with the knowledge of the Lord. We must know Christ intimately so that His thoughts and purposes become our thoughts and purposes. There is no excuse for spiritual ignorance! There is no excuse for not knowing and understanding the Word of God. There is no excuse for not knowing the truth concerning the enemy and your position of power and authority over him. There is no excuse for not knowing how to defeat the enemy.

God has equipped you with the all-powerful Word of God which gives you detailed instructions on how to fight and win. God is not sending you out into battle empty-handed. He has given you all the knowledge you need to win the war. Use it!

Make this declaration: "God is not sending me out into battle empty-handed. He has given me all the knowledge I need to win the war."

The Word Of God

Scripture reading: Psalm 119:89-96

Satan will try to deceive you and make you think that God's Word is too hard to understand. He will try to make you think that you aren't smart enough or that you need someone else to interpret it for you. He wants you to become dependent upon other Christians, your pastor, or great preachers and teachers.

Now, don't misunderstand me. God has placed pastors, teachers, evangelists, and preachers within the Body of Christ to train, teach, and equip you with a knowledge of the Word of God. They are God's gifts to you to help you. But God does not want you to build your life on what other people know and believe. You can't survive on what other people know; you must know that you know that you know.

Can you imagine a soldier out on the battlefield fighting one-on-one combat telling the enemy, "You wait right here while I go ask my sergeant what to do to win this battle." The enemy would get him the minute he turned his back! During a battle, when you are face-to-face with the enemy, is not the time to try to find someone who can tell you what to do. You need to know what God's Word says and how to use it to defeat the enemy.

Every time the enemy comes against you, God doesn't want you running to someone else. He wants you to be strong and immovable. He wants you to be filled with spiritual knowledge. He has placed the Holy Spirit within you to reveal the Word of God to you:

> *But the anointing which ye have received of him abideth in you, and ye need not that any man teach you: but as the same anointing teacheth you of all things, and is truth, and is no lie, and even as it hath taught you, ye shall abide in him.*
>
> *(1 John 2:27)*

If there are things that you do not understand in the Word of God, ask Him to reveal it to you. Head knowledge is not enough. You can't win the battle using head knowledge. Begin to exercise spiritual discernment. When you hear the Word of God coming forth, search the Word to see if what you hear is true. As you do this, you will grow in knowledge and Satan will not be able to get a foothold in your life.

Confusion, apathy, spiritual pride, fear, worry, compromise, and discouragement are also methods Satan will use to get a foothold in your life. Rise up in the Spirit and make yourself a fortress against these strategies of the enemy.

I feel sorry for the devil when he comes up against me. Honestly, I really do. I look at him and say, "You don't have a chance. Go bother someone else!"

Make this declaration: "Satan, you don't have a chance. Go bother someone else!"

September 10

Standing And Opposing

Scripture reading: 1 Peter 5:8-11

There are so many Christians today running around complaining, grumbling and crying, "The devil's after me; I just don't know what I'm going to do!"

Have you ever been in a place in your life when it seemed as if you were being attacked on every side and you began whining and complaining about how Satan was attacking you? Did you know that is exactly what he wants you to say? He wants you to keep on believing the lie that he is more powerful than you.

Those words have no place in your mouth! The devil has no power and authority over your life and your circumstances except the place you allow him to have. If you don't give him a place, he doesn't have it. He cannot take it from you.

Can you imagine Jesus as He was being tempted and tried by Satan, saying "The devil has been after me. He's been attacking me from all sides. I don't know what to do." No, of course not! Don't give Satan a foothold in your life. God has given you the power to keep Satan on his side of the line. Do it!

Our position in the days to come is standing with our feet firmly planted, rooted in the infallible, impregnable Word of God, unmoving and steadfast – regardless of the trials, testing, or persecution the enemy may bring our way.

Peter told the believers in the Early Church:

> *Be self-controlled and alert . Your enemy the devil prowls around like a roaring lion looking for someone to devour. Resist him, standing firm in the faith, because you know that your brothers throughout the world are undergoing the same kind of sufferings. And the God of all grace, who called you to his eternal glory in Christ, after you have suffered a little while, will himself restore you and make you strong, firm, and steadfast.*
> *(1 Peter 5:8-10, NIV)*

In these verses, Peter was writing to Christians who were at that time facing severe persecution. They were being tortured, imprisoned, thrown to the lions, and burned at the stake. Satan thought he could destroy them through persecution.

Peter warned them to be alert, to be aware of their enemy, and to resist him. The position of the early Church in the midst of this severe persecution was not retreat but resistance. The word "resist" in the original Greek is *"lianthistemi"* which means "to stand against and to oppose."

This is the position the Church must take today.

Make this declaration: "I will resist Satan and his evil forces. I will stand against him and oppose him in the strength of the Holy Spirit."

Restored, Strong, Firm, and Steadfast

Scripture reading: John 15:19-25

Don't be deceived – in this world we are going to face persecution. Jesus told His disciples:

> *A servant is not greater than his master – is not superior to him. If they persecuted Me, they will also persecute you; if they kept My word and obeyed My teachings, they will also keep and obey yours. But the enemy will do all this to you – inflict all this suffering on you – because of (your bearing) my name, and on My account, for they do not know or understand the One Who sent Me.*
>
> *(John 15:20-21, TAB)*

Jesus also warned His disciples:

> *In the world you have tribulation and trials and distress and frustration; but be of good cheer – take courage, be confident, certain, undaunted – for I have overcome the world. I have deprived it of power to harm, have conquered it (for you).*
>
> *(John 16:33, TAB)*

Christians are not exempt from the attacks of the enemy. We are going to be tried and tested. Many of you have already faced times of severe testing and your faith has been tried through the circumstances Satan has brought into your life. You may be going through a time of testing right now. Some of you may be facing persecution.

Your position during this time of testing or trial is not one of fear, panic, discouragement, or retreat. You are to resist – to stand against and oppose Satan – to stand firm knowing that Christ will make you victorious.

Peter said:

> *And the God of all grace, who called you to his eternal glory in Christ, after you have suffered a little while, will himself restore you and make you strong, firm and steadfast.*
>
> *(1 Peter 5:10, NIV)*

These four words – "restore," "strong," "firm," and "steadfast" – describe what God has promised to do for you during times of trial or persecution as you resist Satan.

The word "restore" means to "perfect" or "make complete." God allows trials, testings, and persecution to come for the purpose of perfecting you. Regardless of the circumstances you face, as you resist Satan, God will use them to strengthen and perfect you. As you resist Satan, God will "make you strong." The original Greek translation of this phrase means "to strengthen one so he can stand fast in persecution." As you resist Satan, God will make you "firm and steadfast," which means "to be put on a firm foundation."

Make this declaration: "The God of all grace, who called me to His eternal glory in Christ, after I have suffered a little while, will restore me and make me strong, firm, and steadfast."

September 12

SIFTED LIKE WHEAT

Scripture reading: Luke 22:31-34

The apostle Peter was not deceived concerning Satan's power. He knew what it meant to wrestle with the powers of darkness. He knew what it was like to be tested and tried. He knew what it was like to be involved in one-on-one combat with the enemy. Jesus had warned him, *"Simon, Simon, behold, Satan has demanded permission to sift you like wheat; but I have prayed for you, that your faith may not fail; and you, when once you have turned again, strengthen your brothers"* (Luke 22:31-32, NAS).

Notice Jesus told Peter, *"Satan has demanded permission to sift you like wheat."* Before Satan can come against God's people, he must have God's permission. God allows these times of testing to perfect and strengthen you so that you will stand fast in persecution.

Although Peter had been forewarned, he was not prepared for this encounter with the enemy. He was filled with spiritual pride and overconfident in his own strength. He told Jesus, *"Lord, I am ready to go with you to prison and to death." Jesus answered, "I tell you, Peter, before the rooster crows today, you will deny three times that you know me"* (Luke 22:33-34, NIV).

Filling our hearts with spiritual pride is one of the strategies Satan will use to defeat us. You know the story, Peter was unprepared. He failed to resist and stand firm against Satan so he suffered a setback. He lost a battle but he did not lose the war.

There are some of you reading this who are being "sifted like wheat." Your faith is being tested. Today, you may feel weak and on the verge of giving up. I tell you, in the Name of Jesus, resist Satan. You have the power and authority. Don't rely on your own natural strength. God has placed the Holy Spirit within you, giving you power over all the power of the enemy. Draw upon that Holy Ghost power!

Your position during every test, trial, and temptation is not fear, discouragement, or weakness; it is one of power, confidence, and victory knowing that as you resist Satan, he will flee from you (James 4:7). Just as Jesus, full of the Holy Ghost, resisted Satan in the wilderness and was victorious, through that same power you can say to the enemy, "Get thee behind me, Satan!" and he will run from you.

My prayer today is that the Church will begin to receive a new revelation of the powerful position God has planned for us in these last days during the trials, testing, and persecutions that are to come. We are not to be fearful, hiding, or running from the enemy; the enemy is supposed to be running from us!

Make this declaration: "I will not fear the enemy. I will not run or hide from him. He will run from me!"

Satan's Deadliest Snare

Scripture reading: Luke 22:39-46

There are literally untold thousands of Christians today who have fallen victim to one of Satan's deadliest snares, yet, they are totally unaware of what has happened to them. Using one of his most subtle strategies, he has caused Christians to fall into a sleep of death. In this dangerous position, they are spiritually asleep and are unable to hear the voice of the Lord as He speaks to them. They are unaware of the movements of the enemy. They have a false sense of security concerning their own spiritual condition. They are unaware of the lateness of the hour, and the impending danger. They are deceived by spiritual apathy.

In Luke 22:39-46, we see the disciples asleep, unaware that the enemy is lurking in the shadows. Jesus had told them, *"Pray that ye enter not into temptation"* (Luke 22:40). Jesus knew He was going to be crucified and the disciples were going to be tested and tempted to stop following Him. The disciples' hearts were heavy with sorrow. They were physically and mentally exhausted. They were, no doubt, discouraged and afraid.

Think about the battles you are facing this week in your own life. Are you apathetic about them? Do you really realize that the battles you are facing are not against flesh and blood? Are you facing the enemy from a position of knowing the power of the enemy or have you listened to Satan's lies? You must realize that you are in one-on-one combat with Satan and the powers of darkness.

Receive a new revelation of the position of victory God has planned for you to have in these circumstances! By faith, lift your hands to the Father and begin praising Him for making you victorious! Praise Him for giving you power and authority over all the power of the enemy.

Now, in the Name of Jesus, take the authority God has given to you and use it. Get Satan on his side of the line. Say aloud:

> *"Satan, in the Name of Jesus, be rebuked from my mind. Be rebuked from my spirit. Be rebuked from my circumstances! I have all power over you. You are defeated. I will not allow you to have a foothold in my life, in my children, or in my family. I resist you in the Name of Jesus. I pull down your strongholds. I cast out all fear and discouragement. Satan, get out of my way, in the Name of Jesus."*

Now, walk in that victory!

Make this declaration: "Praise God! In the name of Jesus, I have pulled down every stronghold over my life and cast out all fear and discouragement."

September 14

SATAN'S PROPAGANDA

Scripture reading: Psalm 11:1-7

During the Yom Kippur War in 1973, Israel's armies were the victims of a surprise attack. The major strategy Egypt used against Israel was to feed them misinformation about themselves. Egypt began a "misinformation" campaign. Through a series of foreign press interviews in Egypt, they confirmed the misconceptions the Israelis had concerning the Egyptian lack of preparedness for war. A newspaper article in the London Daily Telegraph with a Cairo dateline described the poor maintenance of equipment in the Egyptian army and their lack of preparedness. A special staff was assembled to make sure this misinformation was confirmed.

In mid-September, major exercises were under way in Egypt, but Israeli military personnel were deceived into thinking they were nothing more than simple maneuvers. These repeated exercises were actually part of the Egyptian plan of deception and were so successful they not only deceived Israel but many other countries, including the United States.

Early in the morning of October 6, 1973, as the Israeli Cabinet was assembling to discuss a report that an attack would be launched against them, the Egyptian and Syrian armies struck. The Egyptian army had chosen October 6 because it was one of Israel's holy days – Yom Kippur (the Jewish Day of Atonement) – and they felt Israel's preparedness would be at its lowest on that day.

Egypt's strategy is considered the most outstanding plan of deception in the course of military history. Had the Israeli armies been prepared – had they known the power of their enemy, had the defense system been fully manned with the necessary artillery, armor, and air support – the enemy would have been prevented from making meaningful gains and many lives would have been spared.

All truth is parallel. For hundreds of years, the Church has been fed propaganda concerning Satan, his power and how he operates upon the earth today. Satan has lied and deceived many Christians into believing that as "prince" of this world, he in control. He has planted false fear in the hearts of many Christians, causing them to think they are weak and powerless against his attacks on their bodies, their finances, and their loved ones. He has succeeded in getting many to deny his existence. He has distorted his image, causing many to perceive him as having horns, a tail, carrying a pitchfork, and confined to the flames of hell.

This is God's hour for His people to rise up in power! It's time for us to strip the mask off Satan, to expose him and his strategies.

Make this declaration: "It is time for me to rise up in power. It is time for me to strip the mask off Satan and expose him and his strategies."

The Strong Man

Scripture reading: Luke 11:18-26

Jesus, speaking of Satan and His power over him, said: *"When a strong man armed keepeth his palace, his goods are in peace: But when a stronger than he shall come upon him, and overcome him, he taketh from him all his armor wherein he trusted, and divideth his spoils"* (Luke 11:21-22).

In these verses, Jesus compared Satan to a strong man guarding his palace. From Adam and Eve until Jesus came, Satan – the strong man – had held man in his power, bound by chains of sin. But now, a stronger One than Satan had come. A stronger One than Satan was standing before them who could bind the strong man, set men free from his power over them, and divide his spoils.

Jesus broke the power of Satan, bound him, and cast him out. As Jesus was preparing to lay down His life to redeem and restore man, He said: *"Now is the judgment of this world: now shall the Prince of this world be cast out"* (John 12:31). In this Scripture, Jesus was referring to the immediate results that would take place as a result of His death and resurrection. The very moment Jesus rose from the dead, the "prince of this world" was cast out.

Satan's power over man has been broken! Jesus has destroyed the works of the devil! The power of sin, sickness, and death has been broken! Satan is a liar! He never has been and never will be in control of the kingdoms of this earth. Jesus cast him out. His power has been broken!

Before Jesus ascended into Heaven He told His disciples: *"All power is given unto me in heaven and in earth. Go ye therefore, and teach all nations, baptizing them in the name of the Father, and of the Son, and of the Holy Ghost"* (Matthew 28:18-19).

Jesus said that all power is given unto you. He said, *"Go … teach all nations."* Satan has been cast out. His power has been broken. Satan is not in control of the nations of the earth. Jesus has given you power over all the power of the enemy!

Like a mighty conqueror, Jesus ascended into Heaven in great triumph: *"Therefore it is said, When He ascended on high, He led captivity captive – He led a train of vanquished foes – and He bestowed gifts on men"* (Ephesians 4:8, TAB).

Throughout the Word, the phrase "led captivity captive" refers to a victory procession after a great battle has been won, where the conquering king leads a procession of captives he has taken, along with the spoils of war, back to his own kingdom. Jesus faced Satan and defeated him. He broke the power of sin, sickness, and death Satan had over man, *"And having disarmed the powers and authorities, he made a spectacle of them, triumphing over them by the cross"* (Colossians 2:15, NIV).

Make this declaration: "At Calvary, Christ disarmed every power and authority of the enemy to operate in my life. I will walk in the freedom He purchased with His blood."

Taking Captivity Captive

Scripture reading: Philippians 2:5-11

When Jesus arose from the grave and ascended into heaven, He disarmed powers and authorities, including Satan and all the evil spiritual forces of this world. The Greek word for "disarmed" means "stripped." Like a conquered army being stripped of its weapons and armor, Satan and his horde of demons and spiritual rulers of darkness were stripped and put to shame. Then, having disarmed Satan and his evil forces, Jesus made a public spectacle of them. He exposed them to the universe as His captives.

Like a general leading a procession of victory, Jesus ascended into Heaven leading behind Him the defeated enemy, Satan, and all the powers and rulers of spiritual darkness, displayed before the entire universe as the spoils of battle. Jesus led captivity captive! Those powers that once held man captive – sin, sickness, and death – were destroyed.

Today, Jesus is seated in heaven in a position of supreme power where He has been exalted and given a Name that is above every other name.

> *God has highly exalted Him and has freely bestowed on Him the name that is above every name, That in (at) the name of Jesus every knee should (must) bow, in heaven and on earth and under the earth, And every tongue (frankly and openly) confess and acknowledge that Jesus Christ is Lord, to the glory of God the Father.*
>
> *(Philippians 2:9-11, TAB)*

Satan and all his demons must bow at the Name of Jesus! Every bondage, every sin, every sickness, and every disease must bow at the Name of Jesus!

In this exalted position, Jesus has been given power over powers, principalities, dominions in heaven, on earth and under the earth. God has exalted Him: *"Far above all principality, and power, and might, and dominion, and every name that is named, not only in this world, but also that which is to come: And hath put all things under his feet, and gave him to be the head over all things to the church, which is his body, the fullness of him that filleth all in all"* (Ephesians 1:21-23).

Satan does not have power and dominion over this earth. Jesus has given us power and authority over the power of the enemy. The truth is, Satan is not in control. The works of Satan- sin, sickness, and death – have been destroyed. Satan has been bound, cast out, defeated, and stripped of his power over us!

Satan's strategy against the Church has been to deceive us and keep us in spiritual ignorance concerning these truths. The Church of Jesus Christ must rise up in the power of the Holy Spirit and fight! We must enter into the arena of spiritual warfare and begin to bind the power of Satan. We must take the power and authority that God has placed in our hands and use it to tear down Satan's strongholds.

Make this declaration: "As I face the battles of life today, I will take the power and authority God has placed in my hands and use it to tear down Satan's strongholds."

A STRUGGLE OF PURPOSE

Scripture reading: Judges 13:1-23

There was a man in Old Testament times who experienced a tragic defeat due to the deception of the enemy. His name was Samson. His dramatic and true story is found in the Bible in Judges 13-16.

Samson was born at a time when Israel needed a deliverer. In fact, he was born for that very purpose. His birth itself was a miracle in several ways: It was a miraculous answer to prayer by his mother, who was barren. It also was an answer to the groanings of the people who had been under the rule of the Philistines for forty years!

From the beginning, this man was a chosen instrument of the Lord and his parents were instructed to raise him as one separated unto the Lord. Not a razor was to touch his head all of his life as a sign of his separation unto God.

As Samson grew and God blessed him, he began to be moved by the Holy Spirit and did many mighty works of God! When a lion came against him, Samson took it and slew it with his bare hands! On another occasion, Samson outran and captured three-hundred foxes, tied firebrands to their tails, and turned them loose to set fire to the fields of the Philistines.

Once, when the Philistines thought they had Samson securely locked in the city of Gaza, this superman of God took hold of the gates of the city and the two posts and carried them off to the top of a hill. Another time, he slew one-thousand Philistines using the jawbone of a donkey as his only weapon.

This was God's deliverer! This was God's judge of His people! This was God's chosen instrument! But, even in this man – anointed and used of God, separated unto God before his birth – a battle raged. Satan began to speak to Samson through his physical passions. Samson's eyes fell on a beautiful woman of the Philistines, heathen people God had warned His people to avoid.

Soon Samson began to desire that woman. Satan teased Samson with his fleshly appetites until we see this spiritual giant disregarding his spiritual heritage and going to the camp of the Philistines to indulge himself in sinful pleasures.

God had set a standard for Samson to follow. He wanted Samson to keep himself pure, to obey the laws of God, and to remain in the place of blessing. God wanted him to be a servant of God, a man who could hold his head high and walk down the street anointed with the power of the Spirit of God. Satan's purpose was to destroy Samson, destroy his life and effectiveness for God, and, in fact, reduce Samson to nothing! A zero!

Don't be deceived...the battles you are facing today are struggles between the purposes God has for you and the purposes the devil has for your life.

Make this declaration: "The battles I am facing today are struggles between the purposes God has for me and the purposes the devil has for my life. God's purposes will be fulfilled!"

September 18

YOUR WEAKEST POINT

Scripture reading: Judges 14:1-7

God's will is the most important element in a person's life. God has a purpose for every life. God had a purpose for Samson's life, but in spite of God's warnings, Samson's sinful appetites took over. His rebellious nature caused him to turn his back on God. When God said, "Don't do it!" Samson stubbornly declared, "I will!" The battle was on!

This is what the battle of life is all about: the struggle between the law of God and the law of man's sinful nature which rises up and says "I will!" This is what sin really is – the exercising of our will power to purposely disregard God and His will for our lives.

Sex was Samson's weakness. Let me tell you this about Satan. He never attacks us where we are strong. He will use our weakest point. It is at their weak points that Satan attacks many people today. Your weak point could be the same as Samson's or it could be one of many other areas – the drug scene, the occult, stealing, lying, cheating. Satan used Samson's weak point and caused him to lust after Delilah. What weak points might he use for you?

The true story of Samson illustrates another important point. A person does not usually turn his back on God all at once. A girl does not suddenly decide to become a prostitute. Young people experimenting with drugs don't become addicts all at once. People play with sin. They enjoy what the world calls "kicks" until they are suddenly no longer playing with sin, but sin is playing with them – only it is deadly serious play. Sin gets hold of them. Sin rules them. They are in bondage and cannot escape!

Samson committed fornication with Delilah until she had him right under her thumb. The Philistines sized up the situation and came to Delilah. "You must find out the secret of Samson's strength so we can defeat him once and for all," they told her.

"I'll find out," she promised. "He is so bound to me that he has forgotten that he ever was a servant of God! He is my servant now!"

Perhaps you are one who has dabbled in sin. Perhaps you are in deep bondage to habits and cannot seem to break away. You did not get in that condition all at once. First it entered your mind and you began to meditate on it. It absorbed more and more of your thinking, until you began to yield to temptation more and more.

Today is the day to break the cycle. If you are dabbling in sin or living in bondage, repent, ask God to forgive you, and then rebuke and bind the enemy from further operations in your life.

Make this declaration: "In the name of Jesus, I break the power of every bondage and sin in my life and I rebuke and bind the enemy from further operation."

Playing With Sin

Scripture reading: Judges 14:8-20; 16:1-20

The story of how Delilah got Samson's secret of spiritual strength is told in Judges 16. The first time she asked about the secret of his strength, Samson lied to her. "Bind me with seven fresh cords which have been dried," he told her. The next time she tried, he lied again, "Tie me with new rope." The third time he lied, saying, "If you take the locks of my hair and weave them all together, I will be as weak as any other man." Delilah finally said, "Samson, you have lied to me too many times. Now get out of my house and don't come back!"

The Bible says, *"To whom you yield yourselves servants to obey, his servants ye are to whom ye obey"* (Romans 6:16). If you play with sin, you break the spiritual law of God and you become a servant and slave to sin. That is what happened to Samson. Samson had gone down there for a few kicks, but now he was hooked. He had become a slave to his own passion.

"Please don't send me away, Delilah!" Samson begs, "I'll tell you what you want to know. Just don't make me leave!"

God had a purpose for his life, but Samson had let Satan cross-circuit that purpose. He was no longer the anointed judge and leader. Instead, he groveled at the feet of sin. Look at the humiliation of this champion who had become completely defeated. Samson gave in to the demands of the enemy and surrendered the secret of his strength. If this had been just a natural secret, it would have been sad enough. But, to give away the secret of spiritual power and betray to God, shows the lowest level to which sin can take a person.

Samson finally admitted, "If you were to cut off all my hair, I would be as weak as any other man." You see, Samson had been a Nazarite since his birth and a razor had never touched his head. This was his secret. Spiritually he had been set aside by God for a purpose.

As he slept with his head on her lap, Delilah called for a man to cut Samson's hair off. Then she said, *"The Philistines be upon thee, Samson. And he awoke out of his sleep and said, I will go out as at other times before and shake myself. And he wist not that the Lord was departed from him"* (Judges 16:20).

The Bible says "as at other times" Samson rose and shook himself, but something was different this time. Samson felt for the power of God, but it was no longer there! The power of God that had come upon him when he battled with the Philistines; the power of God that was upon him when he carried away the gates of the city; the power of God that was upon him when he was attacked by a lion; the anointing of God to do great exploits – it was all gone.

Make this declaration: "I will not play with sin because I do not want to lose God's precious anointing in my life."

God Means What He Says

Scripture reading: Judges 16:21-25

Samson played with sin. Now the time had come when Samson needed the power of God, and it was gone. How about you? Have you been playing with sin, continuing bad habits, and walking in rebellion against God, yet your sin has not been exposed?

Let me warn you right now that this is the mercy of God. If you continue on that path, there will come a time when you will reach out for the touch of God and you won't be able to find it. You will cry out for the presence of His Spirit, but find that the sense of conviction of the Spirit has departed! You will find your heavens have turned to brass.

Proverbs describes a dreadful condition:

> *I also will laugh at your calamity; I will mock when your fear cometh; When your fear cometh as desolation, and your destruction cometh as a whirlwind; when distress and anguish cometh upon you. Then shall they call upon me, but I will not answer; they shall seek me early, but they shall not find me: For that they hated knowledge, and did not choose the fear of the Lord: They would none of my counsel: they despised all my reproof. Therefore shall they eat of the fruit of their own way, and be filled with their own devices.*
>
> *(Proverbs 1:26-31)*

What a dreadful picture! The devil will try to tell you that this will not happen, that you are safe and that you will never be found out. Do not believe him! God says, *"Be sure your sin will find you out"* (Numbers 32:23). He says, *"For whatsoever a man soweth, that shall he also reap"* (Galatians 6:7).

The devil's purpose is to make you think that God does not mean what He says. That is how he convinced Eve in the garden. Belshazzar, the king, thought God did not mean what He said until, during a lavish feast, a hand appeared seemingly from nowhere and began writing on the wall, *"Thou art weighed in the balances, and art found wanting"* (Daniel 5:27). That night Belshazzar died. Ananias and Sapphira, in the fifth chapter of the book of Acts, did not think God meant what He said either. They lied to the Holy Ghost and were carried out dead! God means what He says!

Don't wait another day. Ask God to deal with your hidden sins and bad habits today. Repent for your rebellion and ask Him to place within you the spirit of obedience.

Make this declaration: "I repent of all rebellion and ask God to place within me the spirit of obedience."

The God Of A Second Chance

Scripture reading: Judges 17:26-31

Galatians 6:7 says, *"Whatsoever a man soweth, that shall he also reap!"* God means what He says. That is why Samson was unable to feel the power of God after he played with sin for so long. Judges 16.20 says that Samson *"wist not (or knew not) that the Lord was departed from him."* But he soon found out! For the Philistines took him, put out his eyes, bound him, brought him down to Gaza, and caused him to grind at the millstone.

One of the first things sin does is to blind you. Sin looks good at the beginning, but all of a sudden, you are blind. You don't know where you are going. You don't have any direction at all.

They brought Samson down to Gaza and put him in the prison house to grind, binding him to a stone wheel. There he is, blind, bound, and grinding day in and day out, week in and week out, month in and month out. Around, and around, and around he trudges.

Samson has plenty of time to think now. He has plenty of time to say, "If only I had listened to God!" What thundering, pulsating words! If only! He knows now that the pleasures of sin for a season are not worth the price he is now paying.

When you sin, you not only hurt yourself, but you hurt those around you, even those you love the most. Looking at Samson now, you would never recognize what God wanted him to be – a prophet of God, a ruler, a judge. Satan has reduced him to a nothing: A big zero!

And this is Satan's purpose for your life, to reduce you to a nothing, to so humiliate you through disobedience to God that he can hold your life up before the world as a washed-out nothing – a zero!

Perhaps you have watched people fall into this trap of Satan. Thankfully, God has an answer for everyone in this tragic position. God still has a purpose for your life, a high and noble purpose. He does not want to leave you blinded, bound, and grinding day after day in the despair of defeat, humiliation, and the inability to overcome. God wants you to be able to hold your head up high in this world, to be an overcomer, to be someone who radiates the power of the life of Jesus Christ!

Sin may bind, blind, and grind; but God is able to deliver. Whatever Satan has done to you, God can undo from this day forward! He is the God of a second chance.

Make this declaration: "From this day forward, God is undoing everything Satan has done in my life. Praise God – He is a God of a second chance."

September 22

Admitting Your Need

Scripture reading: Luke 15:11-32

Despite the fact that Samson was bound by the enemy, God had a plan for Him and God has a plan for you. If you want to be freed from the bondage of Satan, there are six important steps you must take which we will meditate and act upon during the next few days.

First, recognize the power of confession: First of all, you must face up to your need. This is not the first step by accident. All of the power God possesses cannot do one thing for you until you recognize that God has placed within you His power to lead you out of defeat and despair.

I believe in positive thinking. I believe a man can mentally take himself out of the doldrums and utilize his natural mental powers in a greater way, but with God's power we can lift ourselves from sin by our spiritual bootstraps. *"But as many as received him, to them gave the power to become the sons of God, even to them that believe on his name"* (John 1:12).

Romans 10:9-10 describes the importance of confession:

> *That if thou shalt confess with thy mouth the Lord Jesus, and shalt believe in thine heart that God hath raised him from the dead, thou shalt be saved. For with the heart man believeth unto righteousness; and with the mouth confession is made unto salvation.*
>
> *(Romans 10:9-10)*

When you confess Jesus, you are in essence also confessing your own inability to win the battle of life alone. There is only one person Who has ever lived Who won the battle by Himself, and that is Jesus. Because He won the battle and defeated the enemy, you can also share in that victory and be a winner. You can appropriate that same power in your life when you confess that you need Him.

Think again of Samson. He thought he could play with sin and not eventually pay. He felt he had it in control and that he was the master of his fate.

Think of the prodigal son whose escapades are described in Luke 15. "Confidence" was his name. His inheritance was a well that would never run dry. Payday came early and he sat in that barnyard sharing a meal with pigs. Likely he had been sitting there for many days and thought of home many times, but his foolish pride kept him from confessing his sin.

Finally, there was one last "pig meal" that broke the back of his spiritual pride. The Scriptures say, "he came to himself." It means he was finally willing to admit that he had made a mess of things. He stopped blaming luck or his friends and said, "It is me, Oh Lord, standing in the need of prayer."

Today, come to yourself. Be honest with yourself and admit your need. Acknowledge that you cannot help yourself. You have already tried and look where it has gotten you!

Make this declaration: "I cannot help myself. I rely this day upon God and Him alone."

More Steps To Freedom

Scripture reading: John 10:10-15

We are meditating and acting upon six important steps to free us from the bondage and deception of Satan. The first step was to recognize the power of confession.

Second, know who really loves you. People are so blinded by Satan that they think Satan is the one who loves them and has their happiness at heart. They think he can give them a good time with no chafing restrictions. In this blinded condition they believe God is a hard taskmaster who wants to deprive them of joy and pleasure.

In reality, it is Satan who destroys and brings unhappiness and grief, while communion with God brings joy and happiness.

> *The thief cometh not, but for to steal, and to kill, and to destroy: I am come that they might have life, and that they might have it more abundantly.*
> *(John 10:10)*

> *For God so loved the world, that he gave His only begotten Son, that whosoever believeth in him should not perish, but have everlasting life.*
> *(John 3:16)*

God does not act out of hatred or anger because of our sin, but out of great love and compassion in order that we might receive life more abundantly. To those who accept Christ into their lives, God's promises are for peace, love, joy, prosperity, and health – all that is good.

He promises restoration of that which Satan has destroyed. In Christ, communion between God and man is restored. The precious promise of Jesus is, *"Lo, I am with you always, even unto the end of the world"* (Matthew 28:20) and *"I will never leave thee, nor forsake thee"* (Hebrews 13:5). *The Psalmist says that God is "a very present help in trouble"* (Psalm 46:1).

Third, break sins' barriers. Receive and accept the work of atonement which Jesus accomplished for us by His death on the cross. If you are not a saved, born-again Christian, you can bow your head right where you are and invite Jesus Christ into your heart right now:

> *But as many as received him, to them gave he power to become the sons of God, even to them that believe on his name.* *(John 1:12)*

The work is done in your heart and in your mind through prayer as you turn your life over to God. This is a business transaction, so to speak, between you and God. You turn your sins over to God and He takes them.

> *All we like sheep have gone astray; we have turned every one to his own way; and the Lord hath laid on him the iniquity of us all.*
> *(Isaiah 53:6)*

Whether you are unsaved, a new Christian, or a Christian for many years, be sure to make today's declaration...

Make this declaration: "I am repenting of my sins and God is giving me power to live as His child."

Accepting The Rules

Scripture reading: Psalm 119:57-64

We are meditating and acting upon six important steps to free us from the bondage and deception of Satan. The first step was to recognize the power of confession; second was to know who really loves you; and third was to break sins barriers.

Before we take the next step, let me counsel you that a halfway commitment to God is no commitment at all. God is a jealous God. He wants all of you! Satan, on the other hand, tries to appear more liberal. "You can serve God," he will tell you, "Only let me have just this one little part of your life." He tries to make that arrangement sound reasonable, then he enlarges the little part you have given him until he has all of you. In this sense, God is not "reasonable." He plainly states that He wants all of you – a one hundred percent commitment.

> *And thou shalt love the Lord thy God with all thy heart, and with all thy soul, and with all thy mind, and with all thy strength: this is the first commandment.*
>
> *(Mark 12:30)*

Now, here is the next step…

Fourth, accept the rules. God wants you to play strictly by His rules. You must accept His Word as the rule for your life. Two basic types of law govern this world – natural laws and spiritual laws. An example of the natural laws that govern our physical world is the law of gravity. This law declares that "what goes up, must come down." Now I could say I don't believe in that law, but my not believing doesn't change the law. The law of gravity exists, regardless of my attitude toward it. To prove that I don't believe in the law of gravity, I might climb to the top of a building and step off. No matter how much I disbelieve the law of gravity, there is going to be a bloody mess where I fall. That proves the law of gravity is true. My not believing in the law did not nullify it in the least. I could say the same for many other basic natural laws governing our universe.

The same holds true for the spiritual laws of God. God's spiritual law is the law of the Spirit against sin, God's "Thou shalt nots." You could say, "I don't believe in God. I don't believe in the spiritual laws of God, especially the sin law." But, like natural laws, these laws are reality. Their existence does not depend upon our accepting them or believing in them. All you have to do to prove their existence (as is true with the law of gravity) is to violate these laws. Break them, and as surely as night follows day, there will be a price to pay.

Once we accept Christ's death as the atonement for past sin, we need to accept God's Word as the rule for our present life without hedging, without compromise, and in full faith that His Word will lead us in the way that we ought to go. This is an important step to overcoming the deception of the enemy and living in total victory.

Make this declaration: "I accept God's Word as the rule for my life without hedging, without compromise, and in full faith that His Word will lead me in the way that I should go."

Your Secret Weapon

Scripture reading: Psalm 28:1-9

There are six important steps to free us from the bondage and deception of Satan. The first step is to recognize the power of confession; second is to know who really loves you; third is to break sins' barriers; and fourth is to accept the rules. Today, we will take the last two steps:

Fifth, locate your enemy. We cannot only take God's Son as our salvation and God's Word as our rule of life, but we can take God's Spirit as our great weapon and the key to our greatest victories.

Many years ago when I was in India, I was wrestling in prayer against the evil spirits of that country (the spirits of idolatry and many other kinds of evil spirits). God drew my attention to the needs of America, and He said to me, "Son, the same thing you are battling here is what has to be battled in the other nations and in the lives of people everywhere. In America you are not just dealing with the drug culture or people trying to tear down the structure of society. You are dealing with spirits! "

The Apostle Paul declared, *"For we wrestle not against flesh and blood, but against principalities, against powers, against the rulers of the darkness of this world, against spiritual wickedness in high places"* (Ephesians 6:12). I know there are many things which you are battling against – bad habits, problems, family disorders, spiritual conflicts, and sicknesses – but no matter what, we have the key to victory:

> *And I will give unto thee the keys of the kingdom of heaven: and whatsoever thou shalt bind on earth shall be bound in heaven: and whatsoever thou shalt loose on earth shall be loosed in heaven.*
>
> *(Matthew 16:19)*

Sixth, learn how to use our secret weapon. God will not send us into battle against armored tanks with water pistols as weapons. He will, so to speak, arm us with missiles against the enemy. He knows that in ourselves we are no match for the devil's spirits, but He has taken care of this inequality by arming us with His own blessed Holy Spirit. Jesus told His disciples, *"But ye shall receive power after that the Holy Ghost is come upon you"* (Acts 1:8).

Writing under the inspiration of the Holy Spirit, the Apostle John penned these mighty words: *"Ye are of God little children and have overcome them: because greater is he that is in you, than he that is in the world!"* (1 John 4:4).

Think about it! How could anyone fail with a promise like that to hold onto? Armed with this tremendous Spirit of God, we have the right and authority to come against any evil spirit in the Name of Jesus. We can come against any invasion of Satan into our lives.

Make this declaration: "I am a child of God and can overcome all enemy forces because greater is He that is in me than he that is in the world."

September 26

YOU HAVE THE CHOICE

Scripture reading: 1 Kings 18:21-39

We need not wait for victories any longer! Through Christ Jesus we can do something about it, whether it's problems in our own lives or in society around us. We don't have to take anything Satan dishes out. He is already a defeated foe. We are already conquerors if we just put on the authority and power God has given to us: *"In all these things we are more than conquerors through him that loved us"* (Romans 8:37).

The time has come for you and me to begin a march on the devil. Today is a good day to start! We must march onto the battlefield of life and wage offensive warfare against Satan and all his powers of darkness. We cannot be victorious just through defensive measures. This is war, and you cannot sit idle. You must become a soldier in God's army!

We know that Satan is defeated and the time will come, and it is not far off, when the whole world will know. Isaiah 14:12,15-16 says, *"How art thou fallen from heaven, O Lucifer, son of the morning! How art thou cut down to the ground, which didst weaken the nations! Yet thou shalt be brought down to hell, to the sides of the pit. They that see thee shall narrowly look upon thee, and consider thee, saying, Is this the man that made the earth to tremble?"*

Don't be deceived...Satan is the loser. Those who are on his side will lose with him. God is the winner, and those on God's side will win with Him! You are the chooser! You can play for either side. God has given you that right of choice. He will not violate your will, although He will seek by His Holy Spirit to influence you and draw you to Himself.

Deuteronomy 11:26-28 maps out the battlefield for the battle of life in this way:

> *Behold, I set before you this day a blessing and a curse; A blessing if ye obey the commandments of the Lord your God, which I command you this day: And a curse, if ye will not obey the commandments of the Lord your God, but turn aside out of the way which I command you this day.*

This matter of choosing sides was put clearly to the people of Israel by Joshua when he said, *"Choose you this day whom ye will serve!"* (Joshua 24:15). Elijah presented the same choice in his mighty battle of the spirit with the prophets of Baal. *"How long halt ye between two opinions?"* he demanded of the people (1 Kings 18:21).

God is putting that same challenge to you today. Choose you this day who you will serve. How long halt ye between two opinions?

Make this declaration: "I will chose the way of blessing. I choose God's way."

Have The Gibeonites Infiltrated?

Scripture reading: 1 Peter 1:13-25

In the midst of this evil generation, God has called us to walk in holiness before Him. He called Israel to walk in holiness also. That is why they were not to make any covenant with the evil nations around them, yet, despite God's warnings, the Gibeonites deceived them.

As God's chosen people, we are commanded to be holy even as God is holy:

> *But as he which hath called you is holy, so be ye holy in all manner of conversation; Because it is written, Be ye holy; for I am holy.*
> *(1 Peter 1:15-16)*

The word "conversation" actually means "conduct," which means that as God's chosen people, we are to be holy in all manner of conduct.

At the beginning of the decade of the 1990's, the Lord showed me that a new spirit of true holiness, righteousness, and consecration was coming to the Body of Christ to prepare us for Christ's coming. Do not be deceived! Jesus is not coming to receive a harlot bride that is committing spiritual adultery with the world. He is not coming for a bride that has become polluted with the sinful lusts and desires of this world. He is coming for a holy end time remnant – a people who have been cleansed and purified – without spot or blemish.

Jesus came to earth, suffered the shame and agony of the cross, died and was resurrected so we could be holy and without blemish. Through the blood of Jesus, He made it possible for us to be clothed with His righteousness and enabled us to live a life that is holy and blameless before Him:

> *...Christ also loved the church, and gave himself for it; That he might sanctify and cleanse it with the washing of water by the word, That he might present it to himself a glorious church, not having spot, or wrinkle, or any such thing; but that it should be holy and without blemish.*
> *(Ephesians 5:26-27)*

God has planned for us to set a standard of holiness that will reflect the light of His gospel to the world. Paul told the Philippians:

> *That ye may be blameless and harmless, the sons of God, without rebuke, in the midst of a crooked and perverse nation, among whom ye shine as lights in the world.*
> *(Philippians 2:15)*

Instead of following God's standard of holiness, however, the Church has become polluted by the world. Instead of speaking out and rebuking the evil that surrounds us, we have compromised with the world and become comfortable with the sin that surrounds us!

Examine your own life. Have the "Gibeonites" infiltrated your life? Are you deceived and comfortable with sin?

Make this declaration: "I will not be comfortable with sin. I will not compromise. I will eliminate every 'Gibeonite' from my life."

September 28

Getting Rid Of Gibeonites

Scripture reading: Titus 2:11-15

When we take the mask off and take an honest look at the Church today, we can see that sin has so infiltrated the Church that it is difficult to distinguish Christians from the world. The Church is full of hypocrites, so-called Christians, who are professing to live a holy life but who are fulfilling the lusts of their flesh. Gibeonites!

There are those professing to be Christians who go to church, dance, shout, and praise the Lord, but the minute they walk out the door they begin to gossip, criticize, and tear down the pastor or other members of the Body of Christ.

There are men and women who are bound by lust and involved in adulterous relationships, yet see nothing wrong with it. They believe they can continue illicit relationships and still be Christians.

There are so-called Christians today who think nothing of lying, stealing, or cheating on their jobs. They take days off and call in sick, take supplies home from the office, take extra time at coffee breaks, and extend their lunch hours.

There are Christian businessmen who think nothing of lying on their income taxes, cheating their customers, purposely misrepresenting the facts, and becoming involved in questionable business deals with unbelievers.

There are Christians who allow their hearts and minds to be filled with all sorts of evil imaginations and fantasies. They watch R-rated movies filled with foul language, sex, and violence. They tell dirty jokes on the job, read cheap romance novels, and watch the "soaps" on television.

In many churches today worldliness is no longer condemned but is encouraged through doctrines that are being preached which cause Christians to set their affections on things of this world and teach them to seek material possessions.

In many churches, there are those who have an outward form of godliness, but who are filled with hatred, bitterness, unforgiveness, pride, greed, jealousy, and other ungodly attitudes.

Paul told Titus:

> *For the grace of God that bringeth salvation hath appeared to all men, Teaching us that, denying ungodliness and worldly lusts, we should live soberly, righteously, and godly, in this present world.*
> *(Titus 2:11-12)*

Peter told believers to *"...rid yourselves of all malice, and all deceit, hypocrisy, envy, and slander of every kind"* (1 Peter 2:1, NIV).

Examine our own life today. Are there some Gibeonites you need to identify and get rid of?

Make this declaration: "In the Name of Jesus I reject malice, deceit, hypocrisy, envy, and slander from operating in my life."

Unholy Alliances

Scripture reading: 2 Corinthians 2:14-18

Just as Israel was called to be separate from the evil nations around them, believers must not be unequally yoked together with unbelievers. There must not be any fellowship, communion or agreement made with unbelievers. Paul told believers in the Corinthian church:

> *Do not be bound together with unbelievers; for what partnership have righteousness and lawlessness, or what fellowship has light with darkness? Or what harmony has Christ with Belial, or what has a believer in common with an unbeliever? Or what agreement has the temple of God with idols? For we are the temple of the living God; just as God said, "I will dwell in them and walk among them; and I will be their God, and they shall be my people. Therefore, come out from their midst and be separate," says the Lord, "and do not touch what is unclean."*
>
> *(2 Corinthians 6:14-17, NAS)*

There are Christians today who have formed social and business relationships and partnerships with unbelievers. Christians who form such relationships with unbelievers are acting contrary to the Word of God. They are exposing themselves to worldliness and entering into friendship with the world.

Friendship with the world is *"enmity against God"* (James 4:4). The moment you enter into such a relationship with unbelievers, whether it be a business relationship or a close personal relationship, you become the enemy of God.

God does not want us to make unholy alliances with the world – any type of agreement that links us together with them. This separation and new dimension of holiness will be necessary if we are to be able to stand before Christ clothed in righteousness at His coming. Peter said,

> *Beloved, seeing that ye look for such things, be diligent that ye may be found of him in peace, without spot, and blameless.* *(2 Peter 3:14)*

Christ's message to lukewarm church members with one foot in the world and one foot in the Church, is... *"I counsel thee to buy of me gold tried in the fire, that thou mayest be rich; and white raiment that thou mayest be clothed and that the shame of thy nakedness do not appear..."* (Revelation 3:18).

There can be no truce with Satan. No compromise with sin. Watch out for Satan's deceptions. Make no covenants with the Gibeonites.

Make this declaration: "I will make no truce with Satan, no compromise with sin, no covenant with the Gibeonites."

Satan's Strategy Of Delusion

Scripture reading: John 8:42-51

Satan has commissioned legions of spirits to infiltrate and attack the Body of Christ by deception and delusion. Satan has released spirits of deception, delusion, lying, and seducing spirits that are attacking even Spirit-filled, on-fire, sold-out Christians!

One of the biggest lies of Satan is, "It doesn't really matter what you believe as long as you're sincere. After all, we are all serving the same God." Have you ever heard anyone say that? Don't listen to them. It does matter what you believe!

Satan wants you to be lulled into a false sense of security thinking it is not possible you could ever be deceived. I warn you, this is one of the most dangerous spiritual positions a Christian can be in. If you fail to recognize that there is the possibility of you being deceived, you will become a sitting target for Satan.

You must recognize Satan for who he is – an imposter. An imposter is "one who assumes identity or title not his own for the purpose of deception." As an imposter, his major strategy is to deceive. Jesus described his true character when he called him "the father of lies" (John 8:44). Just as Jesus manifests truth, Satan manifests all that is false.

Here are five major ways Satan will bring delusion:

1. Trying to weaken the authority of God's Word.
2. Working through men and women who are deceived.
3. Working through false teachers, adding to the Scriptures theories and heresies.
4. Keeping Christians bound by a lack of spiritual knowledge.
5. Doctrines of demons directly attacking the minds of Christians by injecting and transmitting lies into their thoughts so they will be deluded and deceived.

To wage war and defeat Satan and his army of deceiving spirits, we must…

1. Grow up to full maturity in our minds: You do this by continually filling your mind with the truth of God's Word.
2. Recognize and discern the spirits of delusion, deception, and the seducing, lying spirits as they attack your mind: You must learn to try the spirits to see if they are of God (1 John 4:1)
3. Hold fast to the truth: Once you have saturated your heart and mind with truth and have tested the spirits and proven the teachings which you have received with the Word of God, hold fast to them. Don't allow yourself to be swayed by every new thing that comes along. Live your life according to the things you have tested and proven to be true.

Make this declaration: "I am filling my mind with God's Word. I am holding fast to His truth so I will be able to recognize the spirits of delusion, deception, and seducing lying spirits."

October

Morris Cerullo Believers' Meeting, Moscow, Russia, 1964: Since God's supernatural call at the age of fifteen, Morris Cerullo has been on the cutting edge of God's open door for ministry in the nations of the world.

OCTOBER

PUTTING THE ENEMY UNDER YOUR FEET

In Joshua chapter 10, we find the account of a powerful enemy alliance which was forming. The kings of Jerusalem, Hebron, Jarmuth, Lachish, and Eglon were angry about the treaty the Gibeonites made with Israel, so they decided to go to war against Gibeon (Joshua 3:4-5). Fearful of their lives, the Gibeonites sent word to Joshua:

> *And the men of Gibeon sent unto Joshua to the camp to Gilgal, saying, Slack not thy hand from thy servants; come up to us quickly, and save us, and help us: for all the kings of the Amorites that dwell in the mountains are gathered together against us.*
>
> *(Joshua 10:6)*

At the direction of the Lord, and because of Israel's covenant with Gibeon, Joshua responded by launching a night-time attack against the enemy alliance. As the enemy fled before Israel, their armies were supernaturally slain by great hailstones that God rained down upon them and the sun and moon stood still until God's people had adequate time to avenge themselves upon their enemies.

The five kings of these enemy nations hid themselves in a cave, but Joshua sent Israelite soldiers to capture them:

> *Then said Joshua, Open the mouth of the cave, and bring out those five kings unto me out of the cave. And they did so, and brought forth those five kings unto him out of the cave, the king of Jerusalem, the king of Hebron, the king of Jarmuth, the king of Lachish, and the king of Eglon. And it came to pass, when they brought out those kings unto Joshua, that Joshua called for all the men of Israel, and said unto the captains of the men of war which went with him, Come near, put your feet upon the necks of these kings. And they came near, and put their feet upon the necks of them. And Joshua said unto them, Fear not, nor be dismayed, be strong and of good courage: for thus shall the LORD do to all your enemies against whom ye fight. And afterward Joshua smote them, and slew them, and hanged them on five trees: and they were hanging upon the trees until the evening.*
>
> *(Joshua 10:22 26)*

"Put your foot on the enemy's neck," instructed Joshua, "Because this is symbolic of what God wants to do to every enemy that rises up against you!"

Before Israel could advance to claim everything God had for them, they had to learn some important lessons and the story in Joshua 10 is one of them. God is saying the same thing to us as Joshua said to Israel. We must bring the enemy out of hiding, identify him, then put our feet upon his neck, and conquer him!

During the coming month you will learn strategies for winning the battle of life and dynamic tactics for defeating Satan drawn from over fifty years of my personal study and teaching of spiritual warfare principles. We will also spend time meditating on one of the greatest behind-the scene revelations concerning spiritual warfare, the book of Job.

As we begin this exciting month of spiritual revelation, God's promise to us is the same as that given to Israel by Joshua: *"Fear not, nor be dismayed, be strong and of good courage: for thus shall the LORD do to all your enemies against whom ye fight" (Joshua 10:25).*

The Enemy Is Under Your Feet

Scripture reading: Joshua 10:1-25

These are the words Joshua spoke to the captains of his army following a great victory over the Amorites (Joshua 10:24-25). During this battle God had supernaturally delivered their enemies into their hands. He rained down hailstones upon them and caused the sun to stand still until they were able to defeat their enemies. Joshua and his army slaughtered the Amorites and those who had survived were running from them in fear.

The five kings of the Amorites were hiding in a cave and Joshua sealed it up and posted guards outside. Knowing the battle would not be won until they had pursued and totally annihilated their enemies, Joshua told his soldiers: *"But don't stop! Pursue your enemies, attack them from the rear and don't let them reach their cities, for the Lord your God has given them into your hand"* (Joshua 10:19, NIV).

After they had totally consumed the army, the soldiers returned to Joshua and he instructed them to open the cave and bring the five kings to him. When they did this, he called all the men of Israel together and said to the captains of the army: *"Come near, put your feet upon the necks of these kings."* This act was a sign of triumph over the conquered enemy. Joshua wanted the Israelites to see that God had caused them to be victorious and made their enemies subject unto them.

There, in the sight of all the men of Israel, the captains came forward and put their feet upon the necks of the five conquered kings. Joshua said to them: *"Fear not, nor be dismayed, be strong and of good courage: for thus shall the Lord do to all your enemies against whom ye fight"* (Joshua 10:25).

Jesus has already conquered Satan and his principalities. At Calvary, He destroyed the works of Satan, stripped him of his power, and put His foot on his neck. He bruised Satan's head in fulfillment of prophecy: *"I will put enmity between thee and the woman, and between thy seed and her seed; it shall bruise thy head, and thou shalt bruise his heel"* (Genesis 3:15).

Through this victory God gave us authority over all the power of the enemy and, like the captains in Joshua's army who put their feet on the necks of their enemies, Christ wants us to put our feet on the necks of all our enemies. Jesus said, *"Behold, I give unto you power to tread on serpents and scorpions, and over all the power of the enemy"* (Luke 10:19).

Make this declaration: "I have power to tread on serpents and scorpions and over all the power of the enemy."

October 2

What Is God Saying To You?

Scripture reading: Psalm 31:1-24

Paul told the Romans, *"The God of peace will soon crush Satan under your feet..."* (Romans 16:20, NIV). To whose feet was Paul referring? The Church!

Through Christ you have been given the "dunamis" power of God, and it is your foot God wants to use to tread on and crush Satan! God wants you to get spiritually violent – pursue Satan and the evil spirits that are attacking you and fight against them until they are totally destroyed. He wants you to expose the forces coming against you, one by one, and put your foot on their necks.

One of the major forces we are exposing that is attacking, tormenting, and binding the majority of Christians today is fear. As we go out onto the battlefield, I want you to see yourself as the captain of Joshua's army and by faith take the victory Christ has already won over Satan and his demon spirits. Put your foot where Christ's is – on the neck of every enemy – and conquer every fear in your life.

As "Commander in Chief" over God's army, Christ's command to you as you go into battle today is the same as Joshua's was to his soldiers: *"Fear not!"* God is going to have a people today who are strong, courageous, and fearless even in the midst of great opposition, trial, and danger – even during the most fierce onslaughts of the enemy and in the face of death. *"Fear not, be strong and of good courage"* are not simply words of encouragement. They are given as a command by God to His people:

- God told Abram: *"...Fear not, Abram: I am thy shield, and thy exceeding great reward"* (Genesis 15:1).
- He told Isaac: *"Fear not, for I am with thee, and will bless thee, and multiply thy seed for my servant Abraham's sake"* (Genesis 26:24).
- God said to Jacob: *"...Fear not to go down into Egypt; for I will there make of thee a great nation"* (Genesis 46:3).
- He said to Moses: *"...Fear ye not, stand still, and see the salvation of the Lord..."* (Exodus 14:13).
- God told Joshua: *"Have not I commanded thee? Be strong and of a good courage; be not afraid, neither be thou dismayed: for the Lord thy God, is with thee whithersoever thou goest"* (Joshua 1:9).

Now...let me ask you a question. What do you think God is saying to you today?

Make this declaration: "God is telling me to be strong and of good courage, not to be afraid or dismayed, because He is with me wherever I go."

Fear Not

Scripture reading: Isaiah 43:1-3

If you are fearful, meditate on the following verses:

- God told Jehoshaphat and the children of Israel as they went to fight their enemies: *"Ye shall not need to fight in this battle: set yourselves, stand ye still, and see the salvation of the Lord with you, O Judah and Jerusalem: fear not, nor be dismayed; tomorrow go out against them: for the Lord will be with you"* (2 Chronicles 20:17).
- God said to Israel: *"...Fear not, for I have redeemed thee, I have called thee by thy name; thou art mine. When thou passest through the waters, I will be with thee; and through the rivers, they shall not overflow thee; when thou walkest through the fire, thou shalt not be burned; neither shall the flame kindle upon thee. For I am the Lord thy God, the Holy One of Israel..."* (Isaiah 43:1-3).
- He told Daniel: *"...Fear not. Peace be unto thee, be strong, yea, be strong..."* (Daniel 10:19).
- He spoke through Isaiah: *"Say to them that are of a fearful heart, Be strong, fear not: behold, your God will come with vengeance, even God with a recompense; he will come and save you"* (Isaiah 35:4).
- Jesus told His disciples concerning those who would persecute and try to kill them: *"Fear them not..."* (Matthew 10:26).
- Jesus said: *"But even the very hairs of your head are all numbered. Fear not then: ye are of more value than many sparrows"* (Matthew 10:31, TAB).
- Jesus told Jairus whose daughter had died: *"...Fear not. Believe only, and she shall be made whole"* (Luke 8:50).
- He told His disciples they are not to fear want: *"Fear not, little flock; for it is your Father's good pleasure to give you the kingdom"* (Luke 12:32).
- God spoke through an angel to Paul as his ship was ready to sink in a great storm: *"Fear not, Paul; thou must be brought before Caesar: and, lo, God hath given thee all them that sail with thee"* (Acts 27:24).
- Jesus appeared to John and said to him: "Fear not; I am the first and the last" (Revelation 1: 17).
- ...And God is saying to you – FEAR NOT!

Make this declaration: "I will not fear because God said I should not!"

October 4

God Has A People

Scripture reading: 1 Samuel 17:1-51

God has always had a people who were strong and fearless through whom He could fulfill His will upon the earth. He has always had men and women who would boldly face danger, oppose entire kingdoms, face hunger, pain, persecution, and death without fear.

When the armies of Israel saw the size and fierceness of the giant and heard his threats against the armies of Israel, they all ran in fear. Even when all the seasoned warriors were trembling in fear, God had a man who was not afraid. A young boy named David said, *"...Who is this uncircumcised Philistine, that he should defy the armies of the living God?"* (1 Samuel 17:26). He told Saul, *"...Let no man's heart fail because of him; thy servant will go and fight with this Philistine"* (1 Samuel 17:32). Without any armor protecting him and armed only with a slingshot and five stones, David fearlessly went out to face Goliath (1 Samuel 17:45-46).

Later, as a great warrior who fearlessly faced his enemies in battle, he said, *"The LORD is my light and my salvation; whom shall I fear? The LORD is the strength of my life; of whom shall I be afraid? ...Though an host should encamp against me, my heart shall not fear..."* (Psalm 27:1, 3).

In ancient Babylon when all the people bowed down and worshiped the golden idol built by King Nebuchadnezzar, there were three young Jewish boys – Shadrach, Meshach and Abednego – who fearlessly faced death in the fiery furnace rather than compromise their loyalty and faithfulness to God Almighty (Daniel 3:16-18).

A young man named Daniel fearlessly faced being thrown into the lions' den, rather than obey the king's decree that no one was to pray or ask a petition of any god or man for thirty days. He refused to compromise his relationship with God (Daniel 6:10).

The Apostle Paul wrote of men and women of faith who conquered kingdoms, administered justice, and gained what was promised; who shut the mouths of lions, quenched the fury of the flames, and escaped the edge of the sword: whose weakness was turned to strength; and who became powerful in battle and routed foreign armies (Hebrews 11:33-34, NIV).

He spoke of men and women who fearlessly faced persecution, temptation, cruel torture, and death rather than turn their backs on God:

> *Others had to suffer the trial of mocking and scourging, and even chains and imprisonment. They were stoned to death; they were lured with tempting offers (to renounce their faith); they were sawn asunder; they were slaughtered by the sword; (while they were alive) they had to go about wrapped in the skins of sheep and goats, utterly destitute, oppressed, cruelly treated.*
>
> *(Hebrews 11:36-37, TAB)*

Make this declaration: "I will not fear. I will be mighty in the Lord and have fearless faith."

A Death Blow To Fear

Scripture reading: Acts 4:1-31

After Pentecost, the Apostle Paul and the other disciples fearlessly faced severe persecution, imprisonment, beatings, stoning or death, rather than deny Christ. They were not intimidated by the threats of the Pharisees who demanded that they stop preaching the Gospel. They assembled the believers together and called upon God to fill them with boldness.

They prayed:

> *And now, Lord, behold their threatenings: and grant unto thy servants, that with all boldness they may speak thy word, By stretching forth thine hand to heal; and that signs and wonders may be done by the name of thy holy child Jesus.*
>
> *(Acts 4:29-30)*

When they were finished praying, the place where they were meeting was shaken, they were filled with the Holy Ghost, and they began to speak the Word of God with boldness – without fear!

Paul was not fearful concerning the afflictions and persecutions he endured. He did not fear what man could do to him. Not knowing what he would face in Jerusalem, he boldly said: *"...for I am ready not to be bound only, but also to die at Jerusalem for the name of the Lord Jesus"* (Acts 21:13).

Paul wasn't running or shrinking back in fear. He wasn't afraid of death. He said, *"I am ready!"* He said, *"None of these things move me!"* He was ready to face anything and everything Satan could bring against him.

What was it that enabled these men and women to be so strong, bold, and fearless in facing their enemies? What was it that enabled David to face Goliath with only a slingshot and five smooth stones; enabled Shadrach, Meshach and Abednego to go into the fiery furnace unafraid; empowered Daniel to march into the lions' den and face the ferocious lions; and helped believers in the Early Church face beatings, imprisonment, burning at the stake, and being thrown to lions while unafraid and singing praises to God?

Their fearlessness was not based upon anything they possessed or upon their natural strength or abilities. It was based upon the fact that they knew God was with them.

You will not conquer fear by therapy, education, or rationalization. You may be helped temporarily, but the root of fear will still be there. A death blow is only rendered to the root of fear when you realize God is with you. Your victory isn't in you – it is in Him.

Make this declaration: "I am rendering a death blow to the spirit of fear."

No Defeats

Scripture reading: Deuteronomy 31:6-8

Jesus' disciples did not possess boldness and fearlessness in themselves. In the natural, they were fearful and unbelieving. When they were out on a boat on the sea of Galilee being tossed by the pounding waves, they were afraid and panicked. They went to Jesus, shook Him from His sleep and said, *"...Lord, save us: we perish."* Jesus looked at them and said, *"...Why are ye fearful, Oh ye of little faith?"* (Matthew 8:25-26).

After Christ had risen from the dead, the disciples were running and hiding in fear. Jesus appeared to them and upbraided them for the unbelief which was at the root of their fear. It was not until after Pentecost that the disciples were bold, strong, and fearless.

Their fearlessness was not based upon their natural understanding. It was a divine characteristic imparted to them by God. The boldness and fearlessness believers in the early Church experienced was a result of being filled within their innermost being with the Holy Spirit ... the *dunamis* miracle power of God.

God originally planned to have a people who were invincible. He entered into a covenant with Israel and promised them: *"The Lord shall cause thine enemies that rise up against thee to be smitten before thy face: they shall come out against thee one way, and flee before thee seven ways"* (Deuteronomy 28:7).

The Israelites' fearlessness as they went out into battle against their enemies was based upon the fact that God's presence was with them to fight their battles and give them victory over their enemies. Moses told Joshua:

> *Be strong and of a good courage: for thou must go with this people unto the land which the Lord hath sworn unto their fathers to give them; and thou shalt cause them to inherit it. And the Lord, he it is that doth go before thee. He will be with thee, he will not fail thee, neither forsake thee. Fear not, neither be dismayed.*
>
> *(Deuteronomy 31:7-8)*

As long as Israel was obedient to God, God was with them and delivered their enemies into their hands. They were invincible! God told Joshua: *"There shall not any man be able to stand before thee all the days of thy life: as I was with Moses, so I will be with thee: I will not fail thee, or forsake thee"* (Joshua 1:5). The surrounding nations feared the Israelites because God was with them. God promised Israel, *"I will send my fear before thee, and will destroy all the people to whom thou wilt come, and I will make all thine unto enemies turn their backs thee"* (Exodus 23:27).

God did not plan any defeats for His people. He has not planned any defeats for you.

Make this declaration: "God has not planned any defeats for His people, therefore, He has not planned any defeats for me."

Fear Is Under Your Feet

Scripture reading: Deuteronomy 20:1-4

God's purpose and plan for His people has not changed. His command to us today as we face Satan's counterattacks is: *"Fear not, be strong and of good courage!"* God has planned for the soldiers in His end-time army to be strong, bold, fearless, and invincible!

Regardless of who you are, or how weak you may feel, God intends you to be fearless. He has made it possible for you to live your life free from fear. When you face sickness or disease, financial problems, family problems, trials, tribulations, or circumstances which seem impossible and you just don't know what to do, He expects you to fear not!

This fearlessness is based upon your faith and trust in God and His faithfulness to His promises. He has entered into a covenant with you, providing for all your needs. He has promised to deliver you out of the hands of all of your enemies, protect you, heal your body, and strengthen you. As you release your faith and trust God to fulfill His promises, there is no fear.

The fearlessness and boldness which God has planned to be manifested in your life is through the power of the Holy Spirit. It is imparted to you by the Spirit. You cannot work it up. Jesus said, *"But ye shall receive power, after that the Holy Ghost is come upon you"* (Acts 1:8). What was the purpose of this *dunamis* power? To make the disciples witnesses. To enable them to boldly and fearlessly face the powers of hell and not be moved. It is through the *dunamis* power of God within you that you can be fearless and bold in the face of persecution, danger, trials, affliction, and death.

You have not received a spirit of fear, but of power! Paul told Timothy, *"God hath not given us the spirit of fear; but of power, and of love, and of a sound mind"* (2 Timothy 1:7). The word "fear" in this verse means cowardice. Paul was exhorting Timothy not to be fearful or cowardly in preaching the Gospel or in facing the persecution which might come as a result of his witness and testimony of the Lord. God does not intend you to be bound by a spirit of fear. He does not intend your mind to be tormented, oppressed, or obsessed by a spirit of fear. He has given you a spirit of power.

Fear is a demon spirit that Satan is using today to torment, control, dominate, oppress, and bind Christians. But, through the power of the Holy Spirit, fear is under your feet!

Make this declaration: "I will not receive a spirit of fear. Fear is under my feet!"

Set Free From Fear

Scripture reading: Romans 8:1-9

Before we can be the strong, bold, fearless people God intended us to be, we must recognize fear as a demon spirit and understand how Satan is using it to attack us. If you allow it, the spirit of fear will spiritually immobilize you until you are unable to release your faith and receive what you need from God. If you let it gain a stronghold, you will be in terrible bondage.

Today there are millions of people suffering from phobias – fears which cause people to lose control of their wills. One example is the story of a woman who suffered from agoraphobia which is the fear of open spaces. She was so controlled by this fear that she left her apartment only three times in thirty years!

The spirit of fear will try to torment you until you no longer have any peace. If you fail to resist and take authority over it, it will gain entrance, first into your thoughts, then your emotions, and finally you will be under its control.

Through Christ, we have already been set free and released from the tormenting bondage of fear. Fear has no place in our lives! Paul told the Romans:

> *For as many as are led by the Spirit of God, they are the sons of God. For ye have not received the spirit of bondage again to fear; but ye have received the Spirit of adoption, whereby we cry, Abba, Father.*
>
> *(Romans 8:14-15)*

We have not received a spirit of bondage to fear. There is no reason why we should ever be fearful. The spirit of fear came upon man as a result of his rebellion and sin against God. When Adam and Eve were perfect, they did not possess fear. As children of God, Adam and Eve enjoyed the loving provision and protection of their heavenly Father. They were not afraid of the lions, bears, or any other animals. They were not afraid of anything. They enjoyed perfect peace.

After they rebelled against God by yielding to Satan, Adam and Eve realized what they had done and they experienced a new feeling – guilt. Along with guilt, a spirit of fear came upon them and they hid from God. That day in the garden of Eden a spirit of fear was loosed upon this earth, and it has tormented and oppressed countless millions through the ages.

Under the old covenant, God's people lived in bondage to the law and it brought great fear to their hearts. They feared the wrath of God. Today, in the covenant relationship we have with God, this fear has been abolished. We do not serve God as slaves bound by fear. Through Christ we have been set free from slavish fear and have entered into a new relationship as sons of the living God!

Make this declaration: "Fear has been abolished. I am living in new covenant relationship with God, my Father."

A Spirit Of Sonship

Scripture reading: Romans 8:10-17

As children of Almighty God, we have not been given a spirit of bondage which causes us to live our lives cringing in fear and dread of His wrath. We have received the spirit of adoption. Through the Holy Spirit we have been begotten to divine sonship whereby we can come to God in complete trust and confidence and cry out "Abba, Father." We are no longer slaves to fear, we are children of God! We have been adopted into the family of God.

The word "adoption" has changed in meaning throughout the years. The Greeks and Romans had a custom of adoption, but it was not taking a child into a family that wanted it. It involved the transfer of a child from one family to another for the purpose of inheritance. The one who was adopted became the legal heir of the one who adopted him. He acquired, as part of his inheritance, the property and name of the one adopting him.

The Greek word for "adoption" means much more than it does in our English language. The Amplified Version translation of Romans 8:15 gives us a clearer understanding of the original meaning:

> *For (the Spirit which) you have now received (is) not a spirit of slavery to put you once more in bondage to fear, but you have received the Spirit of adoption-the spirit producing sonship – in (the bliss of) which we cry, Abba! (That is,) Father!*
>
> *(Romans 8:15, TAB)*

The Greek word for "adoption" is ***huiothesia.*** The first half of this word is *"hios,"* which is the word used for an adult son. The second half of the word is *thesia,* which means "the placement of a person or a thing in its place." The whole word means not so much adoption, but rather the placing of a son.

We have received a spirit of sonship! Through the Holy Spirit we have been placed as a son or daughter into the family of God and have become legal heirs, possessing all rights and privileges as His children. This verse is not referring to the adoption of a child into another family, but welcoming as a fully matured man, the one that has been born again. In this covenant relationship, as full-grown sons and daughters of God, not only have we received all the covenant rights and privileges as joint heirs with Christ, but we have full access to God whereby we can call upon Him as our "Abba Father."

In this relationship with God, we can come to Him in complete faith and confidence knowing that He receives us as one of His own and that He will meet every need in our lives. We enter into a relationship with Him whereby we come to know, through experience, His real love toward us. And, as we walk in this relationship with God where His love is being manifested in our lives, we are set free from fear!

Make this declaration: "I am walking in a new relationship with God where His love is being manifested in my life, setting me free from fear."

The Fear Of Death

Scripture reading: 1 John 4:16-18

As children of God, we have been set free from a fear of God's wrath and the fear of death. Paul told the Hebrews:

> *Forasmuch then as the children are partakers of flesh and blood, he (Christ) also himself likewise took part of the same; that through death he might destroy him that had the power of death, that is, the devil; And deliver them who through fear of death were all their lifetime subject to bondage.*
>
> *(Hebrews 2:14-15)*

Christ has destroyed Satan, broken the power of death, and set us free from the fear of death. There is no reason why any of God's people today should fear death. You have been set free from a fear of death!

There is a place we can reach in our relationship with God where there is no fear! When we come into a relationship with God where we are one with Him – Christ is in us and we are in Him- we come to know and experience God's love. As we walk in His love, believing, trusting, and relying upon His love for us, we are set free from all fear:

> *And we know (understand, recognize, are conscious of, by observation and by experience), and believe (adhere to and put faith in and rely on) the love God cherishes for us. God is love, and he who dwells and continues in love dwells and continues in God, and God dwells and continues in Him. In this (union and communion with Him), love is brought to completion and attains perfection with us, that we may have confidence for the day of judgment – with assurance and boldness to face Him because as He is, so are we in this world. There is no fear in love – dread does not exist; but full-grown (complete, perfect) love turns fear out of doors and expels every trace of terror! For fear brings with it the thought of punishment, and (so) he who is afraid has not reached the full maturity of love – is not yet grown into love's complete perfection.*
>
> *(1 John 4:16-18, TAB)*

It is God's will that we have a full manifestation of His love in our lives where we are living in freedom from fear. He has planned his love to be perfected, to grow to full maturity in our lives, until every trace of fear is cast out. Once we understand and recognize the power of God's love and walk in it, we will be able to face every circumstance, trial, persecution, Satan, all the demons of hell, and even death without fear!

Make this declaration: "I can face every circumstance, trial, persecution, Satan, the demons of hell, and even death without fear!"

Spoils Of War

Scripture reading: Romans 8:35-39

Paul told the Romans:

> *Who shall separate us from the love of Christ? shall tribulation, or distress, or persecution, or famine, or nakedness, or peril, or sword? ... Nay, in all these things we are more than conquerors through him that loved us. For I am persuaded, that neither death, nor life, nor angels, nor principalities, nor powers, nor things present, nor things to come, nor height, nor depth, nor any other creature, shall be able to separate us from the love of God, which is in Christ Jesus our Lord.*
>
> *(Romans 8:35, 37-39)*

We are set free from a fear of death...set free from a fear of persecution...set free from a fear of calamity...set free from a fear of want...set free from a fear of hunger...set free from a fear of sickness ... set free from a fear of danger...set free from a fear of things to come...set free from a fear of Satan and his principalities...we are set free from all bondage to fear.

God has planned us as soldiers in His end-time army to be bold and fearless as we march forward through the nations of the world, pursuing the enemy, and tearing down his strongholds in a manifestation of His power. God expects us today to boldly and fearlessly wage war against Satan and his principalities. Regardless of what strategies he may use to oppose, hinder and stop us – regardless of the dangers, the persecution, afflictions, or suffering we may face in fulfilling the will of God – we must not be weak or cowardly. We must cast off the spirit of fear and march forward, knowing God will deliver our enemies into our hands and give us the victory.

God has set us free from fear. He has given us *dunamis* power to boldly proclaim God's Word in a demonstration of power to the world. He has given us powerful weapons that enable us to demolish Satan's strongholds. He has given us His spiritual armor that makes us invulnerable and invincible.

What is holding back the army of the living God in the United States, Canada, Britain, Europe, the Philippines, Trinidad, Indonesia, Africa, Malaysia, Mexico, India and the other nations of the earth, from moving out in God's power to take their villages, cities and nations for God? Christ has already won the victory. He has destroyed the works of the devil. He has conquered Satan and his principalities, stripped him of his power over us, and given us power over all the power of the enemy. It is time that we boldly claim the victory, put our feet down on the necks of all our enemies, and take the spoils of war!

Make this declaration: "I am putting my foot on the necks of my enemies and taking the spoils of war that are rightfully mine."

Claiming The Spoils Of Victory

Scripture reading: Luke 11:15-22

Throughout the book of Joshua, whenever God destroyed Israel's enemies, the spoils of the battle were theirs for the taking. The soldiers in God's army today need to have a revelation in their spirits that the power of Satan and his principalities over the Church has been destroyed. Christ has put His foot down on the neck of Satan and his demons and the spoils of victory are ours for the taking!

What is holding us back from taking the spoils in our cities and nations? Fear! Satan is releasing a spirit of fear to attack the minds of God's people and bring them into such bondage they are unable to release their faith and take the spoils of victory that belong to them. The Church today is bound by fear...

...Fear of man...

...Fear of failure...

...Fear of Satan and his demons...

...Fear of persecution...

...Fear of death.

Satan's objective today is to build a stronghold of fear in your mind...

...To stop you from taking possession of your spiritual inheritance.

...To cause you to disobey God.

...To immobilize your faith.

...To stop you from reaching spiritual maturity where you are manifesting the power of God.

...To stop you from doing the work God has given you to do.

...To stop you from operating in the gifts of the Holy Spirit,

God has set you free from fear and it is possible for you to be bold, strong, and fearless as you face every circumstance in your life. Not only is it possible, God expects you to be fearless and invincible.

It is time for you to put your foot on the neck of your enemies. It is time to move in and take the spoils of your home, your city, your nation. It is time for finances to be released into your hands. It is time for your healing, your deliverance, and the salvation of your loved ones.

It's time for God's army to boldly and fearlessly march into battle and take the spoils of victory that belong to us! Do it right now...claim the spoils!

Make this declaration: "I'm doing it right now! I'm claiming the spoils of victory in every area of my life. I'm taking back what Satan has stolen from me!"

A Fearless Warrior

Scripture reading: Judges 7:1-25

In this end-time harvest, as we go out into battle against Satan and his principalities, the soldiers in God's army must be fearless. They must have faith and confidence in God and know that He is with them, has delivered them from Satan's power, and will deliver them from all their enemies. There is no place in battle for the weak and cowardly.

God commanded Moses before they went out into battle to instruct the priest to say to the men:

> *Hear, Oh Israel, you draw near this day to battle against your enemies; let not your (minds and) hearts faint; fear not, and do not tremble, or be terrified (and in dread) because of them. For the Lord your God is He Who goes with you, to fight for you against your enemies, to save you.*
> *(Deuteronomy 20:3-4, TAB)*

The officers were to speak further to the people and say:

> *What man is fearful and fainthearted? Let him return to his house, lest (because of him) his brethren's (mind and) heart faint as does his own.*
> *(Deuteronomy 20:8, TAB)*

God is going to have a people who are strong, bold, fearless, and invincible! In the battles God's army is facing today, it is better for those who are fainthearted, cowardly, fearful, or unbelieving should not go to battle because their fearfulness will affect others and weaken the army.

The Midianites and the Amalekites had cruelly oppressed Israel for seven years. They finally came against God's people with an army of soldiers and camels too great to number. God raised up a man named Gideon to lead the Israelites to victory. He told Gideon, *"Surely I will be with thee, and thou shalt smite the Midianites as one man"* (Judges 6:16). Gideon's fearlessness was not in his own strength, but it was based upon his faith that God was with him and would give him the victory.

Before he went out into battle, God told Gideon: *"Now therefore go to, proclaim in the ears of the people, saying, Whosoever is fearful and afraid, let him return and depart early from mount Gilead…"* (Judges 7:3): Of the thirty-three thousand who were with Gideon, twenty two thousand were fearful and returned home. But there were ten thousand who were not afraid to go into battle. God later reduced the army to three hundred men and used them to defeat their enemies which were too great to number.

In this end-time harvest, God is looking for men and women who will march fearlessly out into battle to take the spoils of victory. You can be one of those God will use in these closing hours of time before Christ's return to fulfill His will and bring in a harvest of souls.

Make this declaration: "I will be used of God in this end time. I will march fearlessly out into battle to take the spoils of victory."

Nothing Can Threaten You

Scripture reading: Psalm 30:1-12

There is absolutely nothing Satan can bring into your life that can destroy you – nothing! There is no weapon or strategy that Satan can devise against you that is powerful enough to defeat you – not one!

Satan's objective, as the destroyer, is to destroy everything you have – your home, your family, your health, your finances, your career, your ministry – everything! His strategy is to use your circumstances to defeat you. He will create havoc in the circumstances in your life, bring affliction and disease upon your body, attack your children and loved ones, cause strife in your marriage relationship, bind your finances, or create problems on your job.

Using your circumstances, Satan will try to wear you down physically, mentally, and spiritually until you are so weary, so discouraged and depressed that you are ready to throw your hands up in defeat. He will relentlessly attack you through your circumstances until you become too tired to fight or resist his attacks and you are caught in a vicious cycle that paralyzes and hinders you from releasing your faith and walking in the victory God has provided for you.

Make no mistake about it, Satan is out to destroy you! But there is no circumstance, trial, or persecution – regardless of how terrible it may be – that he can bring into your life that can defeat you.

As the destroyer…

- Satan may wrack our bodies with pain, sickness and disease.
- Satan may bind our finances or cause us to lose everything we own – houses, lands, our life's savings.
- Satan may do his best to turn our mates against us and even succeed in destroying our marriages.
- Satan may try to destroy our children and loved ones through alcohol or drugs.
- Satan may take our lives and the lives of our loved ones through death.

But none of these circumstances – pain, sickness, financial problems, problems in our marriages, family problems, persecution, or death – can threaten us!

Make this declaration: "Pain, sickness, financial problems, marriage problems, family issues, persecution, even death – cannot threaten me because fear has been defeated in my life!"

I Am Ready

Scripture reading: 2 Corinthians 4:6-13

God did not promise that we would never know sin and sickness, never face persecution or suffer hardships. On the contrary, we know that *"all that will live godly in Christ Jesus shall suffer persecution"* (2 Timothy 3:12). Jesus said, *"...in the world ye shall have tribulation..."* (John 16:33). We know that our faith is going to be tested and tried, but God has promised us victory in every circumstance.

Paul told the Corinthians: *"We are afflicted in every way, but not crushed; perplexed but not despairing; persecuted, but not forsaken; struck down, but not destroyed"* (2 Corinthians 4:8-9, NAS). He faced circumstances that were beyond his strength to endure, even to the point of despairing for his own life. He told the believers in Corinth:

> *We do not want you to be uninformed, brothers, about the hardships we suffered in the province of Asia. We were under great pressure, far beyond our ability to endure, so that we despaired even of life, Indeed, in our hearts we felt the sentence of death. But this happened that we might not rely on ourselves but on God, who raises the dead. He has delivered us from such a deadly peril, and he will deliver us. On him we have set our hope that he will continue to deliver us*
>
> *(2 Corinthians 1:8 10, NIV).*

Paul did not moan, groan, and complain about the hardships he suffered. He did not question why God had allowed these things to happen to him. He knew God was in control of his circumstances.

Paul was not shaken by the hardships and sufferings he endured. He was not afraid of what Satan might use to try to defeat him. He didn't say, "Satan just about got me that time. I just barely made it. I'm so weak and tired. I just don't know how I'm going to make it." Something had been forged deep within his spirit. He knew God was providing victory for him in the present. He said, *"He will deliver us."* Paul knew that regardless of what he may face, God had also provided victory for him in the future. He said, *"On him we have set our hope that he will continue to deliver us."*

It was knowing that God had provided victory for him in the past, present and future that enabled Paul to face even death without fear. As he was preparing to die, he said, *"For I am now ready to be offered, and the time of my departure is at hand"* (2 Timothy 4:6). He wasn't afraid of any of the circumstances he would face. He said, *"I am now ready."* Knowing that no circumstance Satan brought into his life – not even death – could defeat him, he said, *"I was delivered from the lion's mouth. The Lord will rescue me from every evil attack and will bring me safely to his heavenly kingdom..."* (2 Timothy 4:18, NIV).

Make this declaration: "I am ready to face every circumstance knowing God has delivered me in the past, He delivers me in the present, and He will deliver me in the future."

Behind The Scenes

Scripture reading: Isaiah 54:1-17

God has provided victory for us in every trial, circumstance, and attack of the enemy. It is part of our spiritual inheritance. It is our covenant right. God has promised, *"No weapon that is formed against thee shall prosper; and every tongue that shall rise against thee in judgment thou shalt condemn. This is the heritage of the servants of the Lord, and their righteousness is of me, saith the Lord"* (Isaiah 54:17).

There is no weapon Satan can use against us that can defeat us. Why is it, then, that so many Christians today are living in defeat, up one day and down the next? Why are our churches filled with weak-kneed Christians who are on fire for God and victorious as long as their circumstances are good; but, the moment Satan attacks their circumstances they are confused, afraid, and don't know what to do? Instead of walking in the victory God has provided for them and resisting Satan's attacks, they whine, complain, and feel sorry for themselves.

Many Christians today are living in defeat because they have not recognized Satan as the destroyer who is using their circumstances as a weapon to defeat them. Many have listened to preachers and teachers who have taught that Christians do not have to go through trials, testing, and persecution such as sickness, lack of finances, or other adversities. They may have also been taught that if they go through trials it is because there is sin in their lives, they are out of the will of God, they do not have enough faith, or God is punishing them. Not knowing the truth, when Satan attacks their circumstances they get caught in a cycle of defeat.

Satan's strategy is to attack us in one of our most vulnerable areas, in the daily circumstances of our lives. He will attack our bodies through sickness and disease. He will attack us through circumstances involving our children, finances, relationships with our mates, and situations on our jobs. He will try to cause strife, miscommunication, unforgiveness, bitterness, and harsh words.

Think about your own circumstances. What is going on in your life right now? Do you think Satan could be at work behind the scenes?

Make this declaration: "Satan may be working behind the scenes in my life, but His power is limited. God is also working behind the scenes in my behalf and He will prevail."

Walking In Revelation Knowledge

Scripture reading: John 5:19-24

Jesus was able to face every circumstance, every battle from a strong position of knowing because the Father had revealed these things to Him in prayer. Jesus was prepared in advance for every attack of the enemy. He knew how and when the enemy would attack. He knew exactly what the Father wanted Him to do and say.

As an end-time spiritual warrior, God has planned for you to live in this same powerful dimension of revelation. He has planned for you to grow up into the full stature of Jesus – nothing less than Christ's own perfection (Ephesians 4:13).

Think about that for a moment. Let it sink in. You can receive revelation knowledge straight from Heaven just as Jesus did. You don't need to wonder or guess about what is going on in your life. You can know. You can live on the cutting edge, discerning the times and signs of Jesus' coming. You can hear and know the voice of God as He speaks to His people today. You can know and discern evil spirits, powers, and principalities that have built strongholds in your home, your city, and your nation. You can know how, when, and where the enemy is going to attack before he attacks.

Think about the battles you are facing right now. Through revelation knowledge you can receive in prayer, you can discern and locate the enemy. You will know the root causes of the problem you are facing. You will receive the answers that you are desperately seeking.

Many Christians want to receive revelation from God, but they are not willing to pay the price. They are not willing to follow Christ's example in prayer. Some do not feel they are worthy or that it is possible for them to receive revelation from God. Others do not know what they must do to receive revelation.

The Father desires to pour revelation knowledge through you. As one of His sons or daughters, He wants you to come to Him as Jesus did to receive revelation and direction. Jesus said, *"...the Son can do nothing by himself; he can do only what he sees his Father doing, because whatever the Father does the Son also does"* (John 5:19, NIV).

If Jesus recognized that He could do nothing except what the Father revealed to Him in prayer, how much more must we recognize that we cannot and should not do anything without waiting upon God in prayer to receive the revelation and direction we need.

Make this declaration: "I will wait on God to receive the revelation and direction I need for this day."

Facing The Circumstances Of Life

Scripture reading: Psalm 92:1-15

Christians today are living in defeat because they have not recognized Satan as the destroyer who is using circumstances as a weapon to defeat them…

- He will attack our bodies through sickness and disease.
- He will attack us through circumstances involving our children who may be having problems at school, involved with the wrong crowd, rebelling against authority, drinking, or involved in drugs.
- He will attack us through circumstances involving our finances – overdue bills, unexpected expenses, or the loss of a job.
- He will attack us through circumstances on our jobs, creating friction between fellow co-workers, misunderstanding with supervisors, and causing miscommunication.
- He will attack us through the circumstances involving our mates – strife, lack of communication, unforgiveness, bitterness, or harsh words.

Satan's objective in attacking us through our circumstances is to steal our health, tear our families apart, destroy our marriages, blind the minds of our children and loved ones, destroy lives through alcohol and drugs, and keep us in financial bondage.

All of these things – our physical bodies, families, children, homes, worldly possessions – are only temporary blessings God has given us to enjoy. They are not eternal.

Satan's ultimate objective in attacking us through our circumstances is to destroy the one thing that we have that is everlasting – our souls. There is absolutely nothing Satan can do to blot our names out of the Book of Life. He does not have the power to destroy our souls in hell. God has given us power over our own destiny. He has given us a free will to choose life or death, heaven or hell.

Satan's objective in attacking us through our circumstances is to destroy our relationship and alienate us from God. His objective is to cause us to question God's love and lose faith in God and His promises. He wants to gain control of our wills so that we will turn our backs on God and sin against Him.

Now that you are aware of his strategy, how do you plan to face the circumstances of your life today? As a victim or as a victor?

Make this declaration: "I will face the problems of my life today as a victor instead of a victim."

Don't Relax

Scripture reading: Hebrews 12:1-3

Every covenant promise God has made to us is true and forever sealed with the blood of Jesus. His Word cannot lie. God's covenant promises are not dependent upon our faith. Regardless of whether or not we have faith to believe His promises, they are true. When we allow our minds to be filled with worry and doubt, however, we cannot release our faith and, although God's promises are true, they do not produce the results we need because they are not mixed with faith.

This is one of the major reasons why so many Christians are not seeing a manifestation and fulfillment of God's covenant promises in their lives. God's Word in our mouths today is just as powerful as it was when He spoke the world into existence. However, we can confess the Word until we are blue in the face, but it will not be effective in our lives until we speak it in unwavering faith, believing whatever we speak will be done.

Satan's strategy is also to wear us down physically, mentally, and spiritually until we become weary and faint in our minds. His objective is to cause us to become so tired of our circumstances that we are ready to give up. Christians today are living defeated lives not because the circumstances they face are too difficult, but because they become weary and faint in their minds.

Paul warned the believers who were becoming weary and discouraged due to the opposition and persecution they were facing. He told them:

> *...let us run with patience the race that is set before us, Looking unto Jesus the author a finisher of our faith; who for the joy that was set before him endured the cross, despising the shame, and is set down at the right hand of the throne of God. For consider him that endured such contradiction of sinners against himself, lest ye be wearied and faint in your minds.*
>
> *(Hebrews 12:1-3)*

The Greek word for "faint" means "to unloose; to relax and so, to make weak." Satan wants to weaken us. He wants us to become so overwhelmed by our circumstances that we faint – relax in our minds – so we will stop resisting his attacks, stop casting out doubt and unbelief, stop believing God's covenant promises, and stop releasing our faith.

It is one thing to become physically weary and drained of our energy. But the real danger is when we become weary and faint – relax and stop resisting Satan's attacks – in our minds.

Make this declaration: "I will not grow weary. I will not faint in my mind. I will not relax and stop resisting Satan's attacks."

You Can Face Anything

Scripture reading: Acts 20:16-24

In our spiritual warfare against Satan, the moment we faint in our minds and stop resisting his attacks, we are surrendering ourselves into his hands. We are defeated only when we fail to fight.

Satan has no power over us except what we allow him to have. He has already been defeated. Jesus has given us power and authority over all the power of the enemy. He has given us spiritual armor that makes us invulnerable to Satan's attacks on our minds. He has given us powerful weapons of warfare to tear down every stronghold and make us victorious in every circumstance, trial, and tribulation. When Satan wins a battle, it is only through default because we have failed exercise power and authority in the all-powerful Name of Jesus.

The Apostle Paul had every reason to become weary and faint in his mind. He was thrown into dungeons, beaten, stoned, and ship-wrecked. He suffered the loss of all things. Throughout his lifetime he experienced pain, suffered from hunger and thirst, and was physically exhausted. He faced circumstances which were beyond his human ability to endure.

Yet, Paul did not faint in his mind. He did not relax. He did not stop resisting and fighting against Satan. He said, *"But none of these things moved me, neither count I my life dear unto myself, so that I might finish my course with joy..."* (Acts 20:24). In essence, Paul was saying: "There is no circumstance, no trial, no persecution that Satan can bring into my life that can defeat me."

If we continue to be weary and faint in our minds, we begin living in a cycle of defeat until we lose heart and become discouraged. We lose hope and resign ourselves to accept our circumstances. Discouragement leads to depression, which results in more stress and we become caught up in this vicious cycle and continue to go round and round.

Paul did not allow any of the circumstances Satan used against him to cause him to become discouraged or depressed. As he prepared to go to Jerusalem where it had been prophesied that he would be bound and delivered into the hands of the Gentiles, Paul said, *"I am ready not to be bound only, but also to die at Jerusalem for the name of the Lord Jesus"* (Acts 21:13). He was not worried or afraid. He was ready to face anything Satan might bring against him.

Make this declaration: "Through the power of Jesus Christ, I am ready to face anything Satan might bring against me."

Choose To Believe

Scripture reading: Acts 27:14-44

After the Apostle Paul was tried by authorities in Jerusalem, they put him on a boat sailing to Rome, where he was to be turned over to the Roman authorities. During the voyage, the ship was caught in a storm and it looked as if they were going to sink. Paul was not discouraged or depressed. He told the men on the ship:

> *I urge you to keep up your courage, for there shall be no loss of life among you, but only of the ship. For this very night an angel of God to whom I belong and whom I serve stood before me saying, 'Do not be afraid, Paul; you must stand before Caesar, and behold God has granted you all those who are sailing with you.' Therefore, keep up your courage, men, for I believe God, that it will turn out exactly as I have been told.*
> *(Acts 27:22-25, NAS)*

Paul was not discouraged. He wasn't depressed. He didn't have his eyes on his circumstances, the fierce winds, or the roaring waves that threatened to destroy the ship and all those in it. His eyes were fixed upon God's Word. He said, "For I believe God, that it will turn out exactly as I have been told."

Paul had a choice. He could keep his eyes on the storm and believe his circumstances – that the ship was going to sink and he was going to drown – or he could believe what God had spoken to him through the angel. Paul chose to believe God. He had simple loyalty to God. He knew God would deliver him.

One of the major reasons why so many Christians today are living in a cycle of defeat is because they have kept their eyes on their circumstances instead of fixing them upon God's Word and His promises to them. The longer they keep their eyes on their circumstances – their sickness, their financial problems, their family problems, their physical and material needs – the more they are caught up in this cycle of stress, fear, worry, doubt, weariness, discouragement, and depression. Satan knows that if you keep your eyes on your circumstances, you will be trapped in this cycle of defeat.

In the circumstances you are facing right now you have a choice: To keep your eyes on your circumstances and believe them – believe that the sickness Satan has brought upon your body is going to remain, believe that you are going down in a sea of debt and that your financial needs will not be met, believe that the problems you are facing on your job are not going to change, believe your marriage is failing, believe your son or daughter will be destroyed by drugs or alcohol – or you can refuse to keep your eyes on your circumstances and believe God's covenant promises. You can believe that He will do exactly what He has promised He will do – He will heal your body, meet your financial needs, restore your marriage, and deliver your son or daughter from drugs or alcohol.

Make this declaration: "God will do for me exactly what He has promised to do."

Your Day For Victory

Scripture reading: Psalm 139:1-24

Satan's strategy as the destroyer is to attack you through your circumstances, afflict and torment your mind, attack your body, destroy your family, and keep you in financial bondage.

Stop for a moment and think about the circumstances you are facing. Are you caught in a cycle of defeat? Are you suffering from stress concerning the situations you are facing? Are you allowing Satan to fill your mind with fear, worry, and doubt?

You may be suffering from battle fatigue, tired of the struggle, weary of resisting Satan's attacks, exercising your faith, and praying and believing God for an answer. You have become worn out and are fainting in your mind.

But right now – today – God is going to break that cycle of defeat in your life. You are going to begin to feel something rising up in your inner man until you will be able to look at every circumstance that comes into your life and say, like Paul, "None of these things moves me!"

You are going to be able to face every circumstance knowing – not hoping, not wishing, but knowing – that God will deliver you and make you victorious. God has provided victory for you in every circumstance, every trial, and every attack of the enemy. It is your covenant right. It is the heritage that belongs to you as a child of God.

Right now, God is pouring new strength into you so that you will have the same confidence Paul had which enabled him to say:

> *Who shall separate us from the love of Christ?" Shall tribulation, or distress, or persecution, or famine, or nakedness, or peril, or sword? Just as it is written, "For thy sake we are being put to death all day long; we were considered as sheep to be slaughtered. But in all these things we overwhelmingly conquer through Him who loved us. For I am convinced that neither death, nor life, nor angels, nor principalities, nor things present, nor things to come, nor powers, nor height, nor depth, nor any other created thing, shall be able to separate us from the love of God, which is in Christ Jesus our Lord.*
>
> *(Romans 8:35-39, NAS).*

This is your day for victory!

Make this declaration: "The is my day for victory! I declare it in the name of Jesus!"

WHAT WILL BE YOUR RESPONSE?

Scripture reading: Job 1:1-12

Satan, our chief adversary never rests! Like a hungry lion stalking his prey, he is constantly at work upon the earth today searching for every opportunity to oppose us and bring us to destruction.

As our adversary, he lies, deceives, tempts, and attacks us through our circumstances, then he slanders and accuses us before God. The Greek word for "devil" is *diabolos,* which means "an accuser; a slanderer." As an accuser, Satan appears before God where he hurls accusations against us day and night (Revelation 12:9-10).

Job chapter one describes Job as a perfect and upright man who feared God and turned away from all evil. God blessed and prospered the work of his hands until he was the greatest man among all the people of the east. He had seven sons and three daughters, three thousand camels, seven thousand sheep, five hundred female donkeys, five hundred yoke of oxen and a large number of servants. He was greatly admired and respected all over the country.

Suddenly, without warning, Job's entire life was shattered. In one day all of his possessions were destroyed. One right after the other, messengers came with the bad news. The first messenger said, *"The oxen were plowing and the donkeys were grazing nearby, and the Sabeans attacked and carried them off. They put the servants to the sword and I am the only one who has escaped to tell you!"* (Job 1:14-15, NIV).

While he was still speaking, another messenger arrived and told him: *"...The fire of God fell from the sky and burned up the sheep and the servants and I am the only one who has escaped to tell you!"* (Job 1:16, NIV).

Before the second messenger could get the words out of his mouth, a third messenger came running up to Job and said: *"...The Chaldeans formed three raiding parties and swept down on your camels and carried them off. They put the servants to the sword, and I am the only one who has escaped to tell you! "* (Job 1:17, NIV).

As he was speaking, a fourth messenger arrived with a message that struck Job like an arrow piercing his heart: *"...Your sons and daughters were feasting and drinking wine at the older brother's house, when suddenly a mighty wind swept in from the desert and struck the four corners of the house. It collapsed on them and they are dead, and I am the only one who has escape to tell you."* (Job 1:18-19, NIV).

You don't know what you may be facing today, tomorrow, or next week. What will be your response to adversity?

Make this declaration: "I will face adversity in the strength and power of the Holy Spirit, knowing God is in control of my circumstances."

October 24

No Condemnation

Scripture reading: Job 1:13-22

Within a matter of minutes Job lost everything. All his worldly possessions and his children were gone! Shocked, devastated, heartbroken, and in great anguish Job tore his robe and shaved his head. He did not shake his fist at God and say, "Why did You allow this to happen to me?" He didn't accuse or question God, but he fell on his face and began to worship God (verse 20-21).

Job did not understand why he had lost everything. He did not even know who was behind the destruction. Yet, in his hour of greatest trial, he did not accuse God. His eyes were not fixed on circumstances. He had strong faith and confidence in the sovereignty of God. Realizing that it was God Who had given him all that he possessed, he said, *"Naked I came from my mother's womb, naked I will depart. The Lord gave and the Lord has take may the name of the Lord be praised"* (Job 1: 21, NIV).

At first, Job failed to realize it was Satan who was behind his circumstances. It was the adversary who worked through men to steal his oxen and donkeys and kill his servants. It was Satan who caused lightning to strike and burn up all his sheep and servants. It was the enemy who had worked through Chaldeans to steal his camels and kill his servants. It was Satan who caused the wind to collapse the house and kill his children. In this story of Job, God exposed Satan as the adversary and accuser that he is and has given us the keys that will enable us to face every circumstance Satan brings into our lives in complete victory.

Here is how Job 1:6 is translated in the Amplified Version: *"Now there was a day when the sons (the angels) of God came to present themselves before the Lord, so Satan – the adversary and accuser also came among them."* Although Satan, fell from his exalted position as the anointed cherub and was cast out of the presence of God (Ezekiel 28:16), he still has access into the heavenlies where he appears before God as the "accuser of the brethren" (Revelation 12:10). There, Satan points his finger at us and points out all our failures and shortcomings. As our accuser, Satan not only makes accusations against us before God, but he also works through other people to accuse and bring condemnation upon us.

Praise God that the Apostle Paul declared, "There is therefore now no condemnation…"

Make this declaration: "Condemnation cannot operate in my life because my sins are forgiven and I am clothed in the righteousness of Jesus Christ."

A Protective Hedge

Scripture reading: Job 2:1-8

In Job chapter one when Satan appeared in Heaven before God, the Lord asked him, "From where did you come?"

"From going to and fro on the earth, and from walking up and down on it," Satan answered.

Satan is not yet confined to the pit of hell as he has deceived many people into thinking. This will occur at the beginning of the millennial reign of Christ and the saints (Revelation 20:1-3). After a thousand years are completed, then Satan will be cast into the lake of fire (Revelation 20:10).

Satan's major activities at the present time are here upon the earth where he is constantly roaming about, tormenting, afflicting, tempting, deceiving, and opposing God's people. He and his evil forces are busy in the nations of the earth planting seeds of hatred, murder, strife, fear, rebellion. He is the unseen force behind all the sickness, death, sin, pornography, greed, and violence that we see today.

God said to Satan, *"Have you considered My servant Job, that there is none like him on the earth, a blameless and upright man, one who (reverently) fears God and abstains from and shuns evil (because it is wrong)?"* (Job 1:8, TAB).

Satan answered:

> *Does Job (reverently) fear God for thing? Have You not put a hedge about him and his house and that he has, on every side? You have conferred prosperity and happiness upon him in the work of his hands, and his possessions have increased in the land. But put forth Your hand now, and touch all that he has, and he will curse You to Your face.*
>
> *(Job 1:9-10, TAB)*

You can face your circumstances knowing that God's ultimate purpose is to do you good. Satan recognized that God had placed a protective hedge around Job that he could not penetrate. God has also built a protective hedge around you to protect you from Satan's attacks. God rejoices and delights in pouring out His promised blessings upon His people. He delights and takes great pleasure in the prosperity of His servants (Psalm 35:27). Pain, sin, sickness, suffering, and death are not from God because He is altogether good and...

> *Every good thing bestowed and every perfect gift is from above, coming down from the Father of lights, with whom there is no variation, or shifting shadow.*
>
> *(James 1:17, NAS)*

Make this declaration: "God delights in pouring out His promised blessings upon me. He takes great pleasure in my prosperity."

October 26

Protected By God

Scripture reading: Psalm 125:1-5

We belong to God. He has purchased us with the blood of His only Son, Jesus Christ. We have been transferred out of Satan's kingdom of darkness into the glorious Kingdom of God. He has cleansed us, placed His Spirit within us, and has bestowed upon us His glory – all that He has and is.

God has entered into a powerful covenant relationship with us where He has provided that all our needs will be met. He has made an everlasting covenant with us that cannot be broken and is forever sealed with the precious blood of Jesus. He has promised, *"I will not turn away from them, to do them good. I will rejoice over them to do them good." ...I am going to bring on them all the good that I am promising them"* (Jeremiah 32:40-42, NAS).

God has placed a hedge of protection around us and all that belongs to us, just as He did for his servant, Job. Without God's permission, nothing can penetrate it! God has promised:

> *Those who trust in the Lord are like Mount Zion, which cannot be shaken but endures forever. As the mountains surround Jerusalem, so the Lord surrounds his people both now and forevermore.*
>
> *(Psalm 125:1-2, NIV)*

He has promised to be a "wall of fire" around us (Zechariah 2:5).
He is our defense! (Psalm 5:11).
He is our fortress! (Psalm 18:2).
He is our hiding place! (Psalm 27:5-6).
He is our refuge! (Psalm 91:2).
He is our shield! (Psalm 5:12).

Do you have any idea what this means for you as child of God?

God has given you salvation, healing, and spiritual riches in Christ Jesus. He has promised to prosper the work of your hands. His blessings are upon all that you have – your houses, lands, cars, children – all your possessions.

God has placed an invisible spiritual shield around you to protect and shield you from all harm! Everywhere you go – out on the busy freeway, in an airplane, on the subway, or at home – you are shielded and protected by Almighty God.

Make this declaration: "God has promised to be a wall of fire around me. He is my defense, my fortress, my hiding place, my refuge, and my shield."

When Satan Attacks

Scripture reading: Job 2:9-13

The accusation Satan brought against Job was, *"Does Job (reverently) fear God for nothing?"* (Job 1:9). Satan accused Job of serving God only for the blessings God had given him. He said, *"You've poured out Your blessings upon him, You've prospered him, You've protected him, his house and all that he has. But put forth your hand now, and touch all that he has, and he will curse You to your face"* (Job 1:11, TAB).

Satan's accusation against Christians today is the same. He points his finger at us and tells God, "Look at those people who call themselves Christians! They don't really love You. Just look at them! As long as You are pouring out Your blessings upon them – as long as they are prospering, as long as everything is going well – they are willing to serve You. But the minute hardships and trials come, they fall by the wayside. They turn their backs on You and go their own way." What an indictment against the Church today!

Our churches are filled with Christians who are committed to serving God only for what they can receive from Him. The Church has become unbalanced in its teaching. New Christians have not been taught that they must be willing to lose their lives for Christ's sake and that they must be willing to deny themselves before they can be true disciples of Jesus Christ. Instead, they have been taught how to be prosperous, how to use their faith to gain material possessions, and how to give so they can get more back from God.

Christians today know all about how to receive from God, but very few know how to stand true to God when their faith is being tested and everything is shaking and crumbling around them. Most Christians have become so prosperity-minded that they are more concerned about the material possessions they can receive from God than they are about what they are willing to give up for the sake of the Gospel.

In the economic shaking of the days ahead, there will be many Christians who are going to be shaken. There will be many who have been taught that if they have enough faith they will be prosperous and never have financial problems, and they will turn their backs on God when they begin to face financial losses. Their faith will be shaken because they do not have a firm foundation. They do not know the truth.

Please don't misunderstand me. I believe God will bless and prosper His people. But we must be careful to have a full understanding of what God's Word teaches concerning prosperity and God's provision for our lives. We must be careful not to get our eyes more on what we can receive from God than what we can do for God. We must not forget that there will be times when we are going to suffer. There are going to be times when our faith will be tested, like Job. There may be times when Satan will attack

Make this declaration: "I am committed to God, not for what I receive from Him, but for Who He is."

October 28

Don't Blame God

Scripture reading: Job 6:1-13

Satan accused Job of serving God only because of the benefits and blessings he received. Then Satan challenged God to take away all the blessings He had given Job. Satan said, *"But put forth Your hand now, and touch all that he has, and he will curse You to Your face"* (Job 1:11, TAB). Satan tried to move God to bring destruction on Job.

Since God is altogether good and His purposes toward His people are only to do us good, He is not the source of pain, heartache, failure, sickness, or anything that is evil. He did not use His hand to bring destruction upon His faithful servant, Job, but He removed the hedge – the shield of protection – from around Job and allowed Satan to afflict him. God told Satan, *"Behold, all that he has is in your power"* (Job 1:12, RSV).

It is vital to your spiritual survival that you know there is no circumstance that comes into your life that God does not allow. Satan cannot touch you, your children, or your possessions without permission from God. Jesus told Peter, *"Satan has asked excessively that (all of) you might be given up to him-out of the power and keeping of God that he might sift (all of you) like grain"* (Luke 22:31, TAB).

God allowed Satan to touch all that Job had. He allowed Satan to take away all of Job's possessions. He allowed Satan to steal his oxen and donkeys, burn his sheep, steal his camels and kill his servants. He allowed Satan to kill his seven sons and three daughters – but it was also God Who later restored double what Job had lost!

There is no circumstance that Satan can bring into your life without God's permission. One of the reasons so many Christians are defeated is because they do not understand that God is in control of all the circumstances that come into their lives, whether they are good or bad. They cannot understand how or why God, whose purpose is to do His people good, would allow Satan to attack them through their circumstances. Why did God allow Satan to steal Job's livestock, destroy all that he had and kill his children? The first question asked when a tragedy strikes is "Why?"

God placed a limit on what Satan could do to Job. He told Satan, *"Behold, all that he has is in your power; only upon himself do not put forth your hand..."* (Job 1:12, RSV). Satan was given permission to attack and destroy Job's possessions, including his children. Satan was convinced that he could cause Job to turn his back on God by taking away all his possessions. Satan, being the arch deceiver, disguised himself. Working through Job's enemies and the wind and lightning, he made it appear as if the tragedies in Job's life happened as a result of natural causes. His strategy is to disguise his evil deeds so that people will not recognize him and will blame God for the pain, sickness, and heartache that comes into their lives.

Make this declaration: "Every good gift comes to me from God. It is the enemy who comes to steal, kill, and destroy in my life."

Don't Lose Your Confidence

Scripture reading: Job 23:1-17

Although Job was devastated and did not recognize Satan and his unseen evil forces as the source of destruction in his life, he did not falsely accuse God and he remained faithful to Him.

Some time later, Satan appeared again before God. God said: *"Have you considered my servant Job? There is no one on earth like him; he is blameless and upright, a man who fears God and shuns evil. And he still maintains his integrity, though you incited me against him to ruin him without any reason"* (Job 2:3, NIV).

Satan was not satisfied. He had destroyed everything Job possessed, but he was still not willing to give up. Having been defeated in his first attempt to get Job to curse God, Satan tried a second time by trying to get God to take away Job's health.

Again, God removed His hedge of protection from Job and allowed Satan to afflict him, but he placed limitations upon what Satan could do. He told Satan, *"Behold, he is in your power; only spare his life"* (Job 2:6, RSV). Satan was allowed to afflict Job's body, but he was not allowed to take his life.

Satan eagerly grasped this opportunity and proceeded out from the presence of God to attack Job with the most loathsome disease imaginable. *"Then Satan went out from the presence of the LORD and smote Job with sore boils from the sole of his foot to the crown of his head. And he took a potsherd to scrape himself while he was sitting among the ashes"* (Job 2:7-8, NAS).

Satan afflicted Job with the worst form of leprosy – elephantiasis. This incurable disease made its victim repulsive and hideous looking. It was an extremely painful disease, having a debilitating effect on both the body and the mind. Job's entire body was filled with boils, burning and ulcerous swellings which discharged a fluid that produced a foul odor. In a short period of time, Job became so disfigured that he was unrecognizable to his friends.

Satan's strategy in afflicting Job with this horrible disease was to cause him to become so physically and spiritually exhausted that he would become weary, faint in his mind, give up, and turn his back on God.

Even at this point in his intense suffering – having been stripped of all his possessions; having suffered the intense grief and pain of losing all ten of his children in one day; having been tormented day and night, his body wracked with pain – Job did not lose his faith and confidence in God. He did not accuse God or turn his back on Him, even after his wife had lost all hope and tempted him to curse God and die (Job 2:9-10).

Make this declaration: "I will not lose confidence in God. I will not turn my back on Him, despite my circumstances or what others around me may say."

Working Together For Good

Scripture reading: 1 Corinthians 10:11-13

God knew Job. He had seen his faithfulness and dedication to Him. He knew Job's strengths and weaknesses. He knew how much pressure Job could face. He knew that Job would be victorious. Job said, *"But he knows the way that I take; when he has tested me, I will come forth as gold"* (Job 23:10, NIV).

God also knows us. His eyes are continually upon us. He knows our every thought. He knows our strengths and weaknesses. He knows how much each one of us can bear. He will never allow Satan to bring circumstances we are not able to handle.

Paul told the Corinthians: *"No temptation (trial) has overtaken you but such as is common to man; and God is faithful, who will not allow you to be tempted (tried) beyond what you are able, but with the temptation (trial) will provide the way of escape also, that you may be able to endure it"* (1 Corinthians 10:13, NAS).

The word "temptation" in this verse in Greek refers to trials and temptations God sends or permits. Paul said, "God is faithful, who will not allow you to be tried or tempted beyond what you are able." Regardless of what circumstance Satan may bring into our lives, we can face it knowing that God has limited what Satan can do in our lives. He will not allow Satan to bring anything into our lives that we cannot bear.

There may be times in your life when your circumstances come crashing down upon you and you don't feel you are going to be able to stand. During those times, you must remember that God knows exactly how much you can bear. He has confidence in you and has provided a way for you to be victorious. With every circumstance Satan brings into your life comes God's provision for total victory.

God is a God of purpose, plan, design, and objectivity. He has a purpose for removing His hedge of protection from around us at times and allowing Satan to attack us in our circumstances. He had a purpose for allowing Satan to attack Job. To the natural mind it is hard to look at Job – who lost all his earthly possessions, who lost his children, who was afflicted and tormented day and night with an incurable disease – and see why God would allow Satan to afflict him. What purpose could possibly be accomplished through his suffering?

All of the circumstances God allows Satan to bring into lives are working together for our good. Paul told the Romans, *"We are assured and know that (God being a partner in their labor), all things work together and are (fitting into a plan) for good to those who love God and are called according (His) design and purpose"* (Romans 8:28, TAB).

Make this declaration: "God is working all things together for my good because I am called according to His purpose."

Executing Total Victory

Scripture reading: Job 42:1-7

Job, stripped of his possessions, suffering in agonizing pain, forsaken by his family and friends, got caught up in a cycle of defeat. He became weary and fainted in his mind. He lost all hope. He was unable to feel the presence of God and felt as if God had forsaken him. He cursed the day he was born (Job 3:3). He was filled with unbelief (Job 9:16-17). He unjustly accused God of oppressing him and favoring the schemes of the wicked (Job 10:1-3). He condemned God for His actions to justify himself (Job 40:8).

Job got his eyes on his circumstances until he became so discouraged and depressed that he wanted to die. He remained in this cycle of defeat until God dealt with him concerning his accusations and he repented. God healed him and restored double what Satan had taken from him! He used all the circumstances Satan had used to defeat Job to bless him.

> *And the Lord turned the captivity of Job and restored his fortunes, when he prayed for his friends; also the Lord gave Job twice as much as he had before.*
> *(Job 42:10, TAB)*

Job's family returned and comforted him and the *"Lord blessed the latter days of Job more than at his beginning"* (Job 42:12, TAB). Satan used every circumstance possible to attack and destroy job, but he couldn't do it!

Regardless of the circumstances Satan may bring into your life to defeat you, God has provided victory. You can face every trial, circumstance, and problem without fear knowing:

1. God's ultimate purpose is to do you good.
2. God is in control of all your circumstances.
3. God places a limit upon the circumstances Satan can bring into your life.
4. With every circumstance comes the provision for total victory.

God is going to have a strong people – a people whose faith has been tested and tried, as gold is tried in a furnace. He is going to have a people who are able to stand firm regardless of the circumstances Satan uses to defeat them. He is going to have a people, like Job, who will be able to say in the midst of their fiery trials, *"Though he slay me, yet will I trust in him"* (Job 13:15).

Today, he is raising up a people who are strong, mighty, and invincible! These are people who, like Israel, can put their feet upon the neck of their enemies and execute total victory.

And YOU are one of those people!

Make this declaration: "God's ultimate purpose is to do me good. He is in control of all my circumstances and places a limit upon what Satan can bring into my life. With every circumstance I face comes the provision for total victory."

November

A date with destiny: 300,000 people packed the arena in one of our historic crusades in Surabaya, Indonesia. God has a divine destiny for you! This month's devotions will show you how to reach it.

NOVEMBER
FULFILLING YOUR DESTINY

In Numbers 13-14 when Israel came to the border of Canaan but refused to enter in because of unbelief, two men emerged as powerful spiritual warriors. One was Joshua, the man who was destined to succeed Moses as the leader of God's people, Israel. The other was a young man named Caleb.

When the other ten spies discouraged Israel from entering into Canaan because of the powerful enemy forces that lived there, Joshua and Caleb tried to encourage the people:

> *And they spake unto all the company of the children of Israel, saying, The land, which we passed through to search it, is an exceeding good land. If the LORD delight in us, then he will bring us into this land, and give it us; a land which floweth with milk and honey. Only rebel not ye against the LORD, neither fear ye the people of the land; for they are bread for us: their defense is departed from them, and the LORD is with us: fear them not.*
>
> *(Numbers 14:7-9)*

As we have learned, their advice was to no avail. Israel turned back into the wilderness where they wandered for forty years until God raised up a new generation that would obey Him and enter in to claim their inheritance. Only Joshua and Caleb were preserved by God and allowed to enter the Promised Land.

Through all the years of wilderness wandering, Caleb never forgot what he had seen in Canaan. He remembered the beauty of the land and its bountiful fruit. He especially remembered one beautiful, fortified mountain area. For forty years, Caleb carried the vision of God's Promised Land within his spirit. Now, as Israel was finally laying claim to all God had promised, Caleb came to Joshua with a special request:

> *And now, behold, the LORD hath kept me alive, as he said, these forty and five years, even since the LORD spake this word unto Moses, while the children of Israel wandered in the wilderness: and now, lo, I am this day fourscore and five years old. As yet I am as strong this day as I was in the day that Moses sent me: as my strength was then, even so is my strength now, for war, both to go out, and to come in. Now therefore give me this mountain, whereof the LORD spake in that day; for thou heardest in that day how the Anakims were there, and that the cities were great and fenced: if so be the LORD will be with me, then I shall be able to drive them out, as the LORD said.*
>
> *(Joshua 14:10-12)*

Over 80 years old – years of wilderness wandering – but Caleb had never lost sight of his God-given destiny.

God has special plans and purposes for you, just as He did for Caleb. He wants to work in you and through you. This month our devotions focus on rising up to fulfill your God-given destiny just as Caleb did. The readings are drawn from two of the most powerful messages God ever gave me such as *Making Possible Your Impossibilities* and *How To Take The Limits Off God.*

As we continue our spiritual journey this month we will meet the woman with the issue of blood, a man named Jacob, and other biblical characters whose lives illustrate that God's power is greater than your biggest mountain. You will be inspired and encouraged. Most importantly, you will be challenged to rise up and fulfill your God-given destiny…

- No matter what your age.
- Despite the obstacles that stand in the way.
- Regardless of how big your mountain is.

If you are ready to start climbing, then make this opening declaration with me, out loud, right now:

Now therefore give me this mountain!

Making Possible Your Impossibilities

Scripture reading: Mark 5:25-29

A certain woman had an issue of blood for twelve years. She had gone to many physicians, but none had been able to help her. If the medical profession were on the scene 2,000 years ago, with all the medical knowledge they have today, they probably would have diagnosed this woman's case as an incurable cancer – an issue of blood.

Pressing through the crowd to Jesus, she touched His garment and said, "If I may touch but His clothes, I shall be whole." Immediately, she felt in her body she was healed.

There are some people who say you should not trust your feelings, but this woman felt in her body that she was healed. She did not have a chance to get an examination by a doctor, but she knew she was healed.

If this miracle had taken place in our day, in all probability someone would have taken the garment and placed it in a glass case in some cathedral to be a memorial or shrine. Others might go to a greater extreme and cut it up into prayer cloths and send them out through the mail.

This dear woman pressed through and touched the hem of Jesus' garment. Then Jesus, knowing the frailty and weakness of man, immediately called this woman to His side and said: "Thy faith hath made thee whole."

This was the message: "There is no virtue in My garment which you touched. It is nothing more than a piece of cloth."

There is no mysticism, no magic – Jesus simply looked at this woman and said, "Daughter, thy faith hath made thee whole!" He was explaining to her that it was not some ritual she went through by pressing through the crowd and grabbing hold of His garment. That had nothing to do with it. The garment had nothing to do with it. But her faith had a lot to do with it!

The greatest power in the world is power of faith in God. It is a power that will enable you to claim all God has promised. It is a power that will give you victory over the Jerichos of life and take you on, like aged Caleb, to conquer your mountain and achieve your destiny.

During the next few days I want to share three things that happened in this woman's life to make possible her impossibilities. If you apply these three things to your circumstances, you, too, will find your impossibilities becoming possible.

Remember…all things are possible to those who believe!

Make this declaration: "God is in the process of making all my impossibilities possible."

Coming To The End Of Yourself

Scripture reading: Mark 5:30-34

The woman with the issue of blood had spent all she had on physicians and was no better. We thank God for good doctors. We thank God for medical science and its advanced technology that is helping overcome disease and sickness, but we also believe there is only one power in heaven and earth that has the final word in the healing of our afflictions – and that is the power of God.

This lady in the Bible story spent everything she had. She is like so many people of our day: They have taken all the pills, had all the treatments, and done everything in the natural that is possible for them to do – then they have been sent home to die. No more hope.

It is difficult for man to receive from the hand of an omnipotent God until he has come to the end of self. Oh, the heartache that is ours because we try to work out our problems. Oh, the dead-end streets we travel on this earth because we try to take situations into our own hands and try to find the solutions in ourselves.

You have never seen a more lonely person than the self-centered individual. You have never seen a person more in despair. No person possesses more grief than the individual who lets "I" rule and reign in his life. The power that brought the curse of sin, sickness, despair, sorrow, pain, and misery upon man was the assertion of man's will as he rose up with his own will against God. God gave each person an authority over his mind, his spirit, his total being. But man used that god-like power against God and chose to disobey the…"Thus saith the Lord."

From that moment in the garden of Eden until the moment our blessed Messiah, Jesus Christ, the Son of the living God, came to this earth – from that moment of the first Adam to that moment of the second Adam – there was one primary purpose of God and that was to get man off the throne of his heart and return him to his rightful relationship with God.

There was only one way that could happen. Man must surrender all that he has, all that he is, and all he possesses back to God. There would be no more "I, me, mine," but it would be *"the life which I now live in the flesh I live by the faith of the Son of God, who loved me, and gave himself for me"* (Galatians 2:20).

The woman with the issue of blood came to the end of her self.

- It is only as we come to the end of self that we position ourselves spiritually to receive from the Lord.
- It is only as we come to the end of self that we are ready to receive God's help.
- It is only as we come to the end of self that we can learn how to lean completely on God for His supernatural supply.

Let self die!

Make this declaration: "I have come to the end of myself. I am ready to receive God's help and lean completely on Him for His supernatural supply."

SHE HEARD A MESSAGE

Scripture reading: Matthew 9:20-22

The woman with the issue of blood is on her way home to die. She has come to the end of her self. She has no hope. Walking down the street, she hears something. There is a commotion and a large crowd has gathered. Is there a fight going on? Is there a riot? Has someone been injured? What is all this excitement about?

A young man rushing by said quickly to her in passing, not waiting to hear her reply: "Did you see Him?"

"Who?" the woman asked. Actually, her poor body was so wracked with pain she was only interested in getting home to die. The doctors had said there was nothing more they could do for her. "Go home and die!"

She caught the next person coming in the crowd and asked, "Who are they talking about? Who is causing all of this commotion? Is it a Roman governor or a military commander from Rome? Is it the high priest?"

"Don't you know Who it is? Everyone in the whole city is talking about Him."

No, she had not heard. She had been focused on her own problems. She had been to the doctor trying to get cured of her terrible disease.

Another said, "The Person causing all the excitement is the Messiah, the Son of God. He has opened the eyes of the blind. He has made the ears of the deaf to hear again, and people who are crippled and deformed are made completely whole after He speaks to them."

The woman thought, "If this is true, I must get to Him!"

As she began to press her way through the crowd, even before she could get close enough to see Him, she heard His voice. She heard the voice of the One Who had walked into the synagogue and with authority had opened the Book and read from the Scriptures: *"The Spirit of the Lord is upon me, because he hath anointed me to preach the gospel to the poor; he hath sent me to heal the brokenhearted, to preach deliverance to the captives, and recovering of sight to the blind; to set at liberty them that are bruised"* (Luke 4:18).

As she pressed closer and heard His words, she recalled the words of the prophet Isaiah, Scriptures she had learned as a little girl: *"Who hath believed our report? And to whom is the arm of the LORD revealed? For he shall grow up before him as a tender plant, and as a root out of a dry ground; he hath no form nor comeliness, and when we shall see him, there is no beauty that we should desire him. He is despised and rejected of men; a man of sorrows, and acquainted with grief; and we hid as it were our faces from him; he was despised, and we esteemed him not"* (Isaiah 53:1-3).

The Word of God was preparing her for the miracle she was about to receive. Let the Word of God prepare you for your miracle. It is on the way!

Make this declaration: "The Word of God is preparing me for a miracle. It is on the way!"

Let God's Word Consume You

Scripture reading: Luke 8:43-48

The woman with the issue of blood heard the Word of God and her faith began to build. The words of King David, the Psalmist, began to flow through her mind. The Holy Spirit was already ministering to her in the presence of the Word Personified. Scriptures began filling her mind with the new meaning in the presence of this One Whom the people called the Messiah, the Son of God. She was hearing a message...

> *Bless the Lord, 0 my soul, and all that is within me, bless his holy name Bless the Lord, 0 my soul, and forget not all his benefits; Who forgiveth all thine iniquities, who healeth all thy diseases.*
>
> *(Psalm 103:1-3)*

We do not have to stretch our imaginations to believe that this woman had read Scriptures like this hundreds of times and the more familiar ones perhaps thousands of times. Yet now, in her present condition and in the presence of her Messiah, they took on new meaning. The Word is no longer dead, no longer a complex group of statements with little meaning. Now they had come alive and, with every moment and each word, her consuming passion was taking her closer and closer to the Source.

The path was not easy! With such interest as Jesus had generated, the multitude pressed around Him on all sides, packed closely together. It took extreme effort for this woman to press through. I believe this explains in part the verse of Scripture which says, *"The kingdom of heaven suffereth violence and the violent take it by force"* (Matthew 11:12).

It is no exaggeration to say that this woman had become nearly violent in her desperation to reach Jesus. What created such a spark in this dying woman? What activated her adrenal glands to such an extent that she, who had been sent home to die, could press her way through the crowd?

She had locked in on a truth and fastened to it like a barnacle to a rock. Her determination was so solidified that all of the doubting demons in hell at that moment could not have broken her loose. She was in the presence of the One Who was greater than all the power in the universe. It was not just her own tenacity, she was already drawing on the strength of the Son of God. She had found truth personified.

Think of it! A simple woman was able to recognize truth and grasp it immediately when she found it. Educated Pilate was standing in the presence of truth with the future in his hands, yet he was not capable of comprehending or recognizing it. There are people who come to our crusades in desperate need, see a miracle, but won't receive the truth of God's Word and consequently they walk out the same way they came in. They were participating in another method. It was an experiment. It was something else to try, like a new pain killer, antibiotic, or miracle drug. They were never consumed with the message of the delivering power of Jesus Christ.

Let the power of the truth of God's Word consume you today!

Make this declaration: "The power of the truth of God's Word is consuming me today."

Get A Death Grip

Scripture reading: Isaiah 53:1-12

Have you ever heard the familiar story of the "death grip"? When one is drowning and another is trying to save him, he will clutch the rescuer in panic with such strength that he actually strangles the helper and they both sink and drown.

No matter how strong your grip on Jesus Christ, you will never sink. Peter found this out when he began to sink while walking on the water to Jesus: *"But when he saw the wind boisterous, he was afraid; and beginning to sink, he cried, saying, Lord save me. And immediately Jesus stretched forth his hand, and caught him"* (Matthew 14:30-31).

Please do not think that you must be on your deathbed to receive the flow of the Holy Spirit through your life. The devil has sold us a bill of goods that causes us to come to God only when we are at the end of our rope, when all natural resources have been exhausted, and every other avenue of help has come to a dead end.

You do not have to be in desperation to attain the flow of the Holy Spirit in your life! Without a doubt, God has used such circumstances thousands of times to bring people to a place of belief and dependence upon Him so He can minister to them, but that is not the only route. We can lock onto the message of the power of Christ without being in a desperate condition.

We have seen very unusual and beautiful things happening in our miracle crusades and Schools of Ministry. People have testified they received healing when they were not even consciously thinking about it or asking God to heal them. They suddenly find that growths have disappeared. Some had suffered affliction for so many years they had come to a passive acceptance of the condition, thinking they would live with it the rest of their lives. Then, all of a sudden, they realized they were healed! This sometimes would happen immediately in the service; other times it would dawn on them the following day at home or work that God had wrought a miracle.

What happened? Their innermost spirit had received the message, either through the spoken word, a demonstration of miracles they observed, or simply from their being in the presence of the Holy Spirit. In one way or another their spirit had locked onto a message of truth – of Jesus Christ the Messiah in all His redemptive, resurrection power. They received a message that became so strong it took hold of their very being and their affliction had to flee before the explosion of truth that flooded their being.

Take hold of the truth of God's Word for your situation today. Whether you need healing, comfort, peace, or deliverance – get a death grip on Jesus Christ and don't let go until you receive your miracle.

Make this declaration: "I am getting a death grip on Jesus and won't let go until I receive my miracle!"

November 6

Changing Your Focus

Scripture reading: Matthew 17:14-21

For years, in our overseas meetings, I have seen men and women, boys and girls, who perhaps had never been in a religious service before. They came merely out of curiosity, and, standing on the fringe of the crowd, they suddenly received healing!

Like the woman with the issue of blood, they had a need and heard a message of truth. That truth, coupled with the basic consciousness of God born into every person, drives out the forces of the enemy and he has to flee. This can happen without even a prayer being spoken for them or hands being laid on them as the Spirit brings a God-consciousness and they reach out by saying in their innermost being, "I believe the Word."

In areas of heavy demonic influence, such as regions of Brazil and Africa where there is much spiritualist worship and voodoo practice, people have come to meetings saying that they felt as though hands were at their throats choking them. As they stand in the crowd listening to the message and hearing the testimonies, they lock onto the message, the feeling of unseen hands leaves them, and they are able to swallow and breathe normally for the first time in years.

I shall never forget a little fifteen-year-old boy in Guayaquil, Ecuador. Through a traumatic experience at the age of seven he had lost his power of speech and for eight years had been a deaf mute, living as a beggar, sleeping under the wharf on the waterfront by day and going through the garbage cans of the city by night to survive.

What consciousness did this boy have of God? Who could deliver him and set him free? He came to the sports stadium, stood at the edge of the crowd, and as we began to pray for the sick and afflicted, he felt warm hands upon his shoulders. He turned around to see who it was, but there was no one behind him. As he became more conscious of that presence, all of a sudden his tongue was loosed and he could speak for the first time in eight years! That little boy's spirit locked onto a message. Though his mouth could not speak or his ears hear, his spirit heard the message communicated by the Holy Spirit.

The same is true with people in our services who do not understand English and there is no translation for them. Many have been healed even though their conscious minds never heard a word because their spirits locked on to the message of truth.

Here is one of the most important points: If you are to receive your miracle – if your impossibilities are to become possibilities – your focus of attention must be on the message, not on your affliction or circumstance.

Do you need to change your focus?

Make this declaration: "I am taking my focus off of my impossibilities and getting my focus on the God of the impossible."

The Work Is Done

Scripture reading: 1 John 3:1-8

How did the woman with the issue of blood know that what she heard was the message of truth? There was a confirmation of the Word. There was performance of the Word! The Bible declares that in the mouth of two witnesses every word is established. She had the witness of the Spirit within her, she had the witness of the spoken Word and witness of the living Word, Jesus Christ.

The Apostle John said: *"There is the message God has given us to pass on to you: That God is light and in Him is no darkness at all."* John further elaborated on the message when he said: *"The Son of God was manifested, that he might destroy the works of the devil"* (1 John 3:8).

Jesus came to earth for a purpose! When we, like this woman, understand what that purpose is through the message of God's Word, then we are on the road to deliverance. We are on the road to victory. We are on the way to having our impossibilities made possible.

The message this woman received was that Jesus came for a purpose. He came into this world and allowed Himself to be beaten, scourged, persecuted, spit upon, and nailed to a cross so she could be free! You don't have to remain one more moment living in sin and despair, without peace, love, and a meaningful life. You don't have to keep that deaf ear or live with that bad back, rheumatism, arthritis, cancer, or physical or emotional pain. You don't have to be bound by any of these one moment longer. The price has been paid for your deliverance: *"By whose stripes ye were healed"* (1 Peter 2:24).

Doctors tell us there are thirty-nine major sicknesses in the world. From these thirty-nine major diseases come all the other illnesses. Jesus Christ was whipped with thirty-nine stripes across His back. If I may be so free as to spiritualize, I would say one stripe for every sickness – for deafness, cancer, arthritis – whatever affliction you suffer, Jesus took it upon Himself!

The work is done! Jesus paid the price 2,000 years ago. Our sins and sicknesses were (not shall be) were, were, healed! If Christ paid the price, then we don't have to pay it.

We are free!

We do not have to bear our sins one moment longer.

We do not have to bear our sickness one moment longer.

This is the message: The work is done!

Make this declaration: "I do not have to bear my sins one moment longer. I do not have to bear my sickness one moment longer. The work is done!"

THE GREAT TRANSITION

Scripture reading: Deuteronomy 9:1-5

The woman with the issue of blood heard a message, but so did a lot of other people who were present that day. They heard, but went away in the same spiritual and physical condition as they had come. What was the difference?

This woman grasped onto what she heard. She fastened to it as her source of strength. It was more than a philosophy to her; it became life itself, apart from which she would die. She heard the message and believed it – but this still was not enough.

The woman thought to herself, "If I can just touch His clothing, I will be healed." And sure enough, as soon as she had touched Him, the bleeding stopped and she knew she was healed! We could call this "the great transition" because she left the area of concept or theory and made her move. She acted on what she believed. The woman not only said, "If I can touch His garment," she made the supreme effort to reach her goal and receive what she sought. Faith is a fact, but faith is also an act. Jesus said, "My words are spirit and they are life." The Word within you must produce life. It must produce action.

One of the most central themes of the entire Bible, beginning with the first chapter of Genesis through the last chapter of Revelation, is the fact that when God speaks it must be a fulfillment. In the beginning of time when God spoke, the worlds came into being. They had no choice. Because of the Power of the Word, they could not hold back.

The centurion who came to Jesus to seek healing for his servant had this revelation, for he said, "Speak the word only and my servant shall be healed." In that soldier's mind the possibility of failure did not exist. He knew that if he could just get the Son of God to just speak, it was already done.

Moses told the children of Israel:

> *Not for thy righteousness, or for the uprightness of thine heart, dost thou go to possess their land. but for the wickedness of these nations the Lord thy God doth drive them out before thee, and that he may perform the word which the Lord sware unto thy fathers, Abraham, Isaac, and Jacob.*
> *(Deuteronomy 9:5)*

Here we see the extreme necessity for God to perform that which He had spoken. In this instance, His performance does not depend on the righteousness of Israel or any of their accomplishments. It depended on His promise!

This is a roadblock for many people, for they feel consciously or even subconsciously that through their own merit they will receive deliverance from God. It has nothing to do with their goodness!

God has already spoken the Word of deliverance. He has spoken the Word of healing. It is not dependent upon you; it is dependent upon Jesus Christ and the work He accomplished at the cross of Calvary.

Make this declaration: "God has already spoken the Word of deliverance in my behalf. He has spoken the Word of healing. It is not dependent upon me; it is dependent upon Jesus Christ and the work He accomplished at the cross of Calvary."

Will You Act In Faith?

Scripture reading: Romans 4:18-21

The Apostle Paul said of Abraham:

> *So, when God told Abraham that he would give him a son who would have many descendants and become a great nation, Abraham believed God even though such a promise just couldn't come to pass! And because his faith was strong, he didn't worry about the fact that he was too old to be a father, at the age of one hundred, and that Sarah his wife, at ninety, was also much too old to have a baby. But Abraham never doubted. He believed God for his faith and trust grew ever stronger, and he praised God for this blessing even before it happened. He was completely sure that God was well able to do anything he promised.*
>
> *(Romans 4:18-21, TLB)*

There is enough in this one Scripture to send us into spiritual orbit for a year, write ten books and preach twenty sermons. Instead of saying "completely sure" the King James Version uses the words "fully persuaded." Both phrases are pure dynamite!

Abraham never doubted. Why? Because he was completely sure, fully persuaded that God was able. He even thanked God for it before it ever happened. Believing and thanking God in faith is action!

The woman with the issue of blood had to perform a specific act to accomplish her goal. She had to touch the hem of His garment! I cannot tell you exactly what your action should be. It may be to stand on that sore leg. It may be to begin moving that arthritic knee. It may be to forgive someone who has hurt you. Only God can tell you what to do.

But let me tell you something for sure: As long as you just sit and wait, nothing is going to happen! You must show initiative. God has done His part – He has sent you the Word. Now you must perform the doing of it (2 Corinthians 8:11).

The woman with the issue of blood could have heard the message and then gone home to die. Instead, she said to herself: "I must!" She acted in faith and touched the hem of Christ's garment. There was no virtue in that cloth but it became the motivating force, the physical evidence of her faith in the Messiah, Jesus Christ.

Faith is an act! Instead of going home to die, this woman declared, "If I can only get to Jesus, everything will be all right!" Yes, faith is a fact, but it is also an act. Don't continue in your passive, waiting rut any longer. Jesus is waiting to receive you, to heal you, deliver you, and help you achieve your destiny.

Right now, you are like the woman standing on the street corner of life who had a choice to make. Would she go home and continue to live with her need or would she act in faith and have her need met by the power of God?

Let me ask you…what are your needs today? Peace of mind? Healing for your physical body? Salvation for your unsaved loved ones? Deliverance? Will you continue living with your need or act in faith and have it met by God's power?

Make this declaration: "I will not remain in my condition and live with defeat one moment longer. I am fully persuaded. I will act in faith."

FAITH AND FEELING

Scripture reading: John 3:16-21

God's power is real. It is vital. It can be felt, and the results of it can be seen and felt. I know the Bible says, *"the just shall live by faith"* (Habakkuk 2:4) – but I also know that active faith results in changes that can be seen, felt, and enjoyed.

It is true that Satan can tamper with our feelings if we do not keep our minds and hearts focused upon God, but it is also true that the blessings of God can be felt and enjoyed. *"...the joy of the LORD is your strength"* (Nehemiah 8:10).

There was feeling on both sides when the woman reached out in faith and touched Jesus for the need in her life. The Bible says that when the woman touched Christ's garment *"...straightway the fountain of her blood was dried up; and she felt in her body that she was healed of that plague"* (Mark 5:29).

Faith comes first before the feeling. Sometimes we must hold on in faith until we see every symptom eliminated, every circumstance changed, every mountain conquered by the power of God. But let me tell you, that victory can be felt!

Someone said to me, "Brother Cerullo, you are an emotional person." I sure am! You cannot be touched with the feeling of people's needs, sicknesses, problems, hopelessness, and despair and not be a person of emotion.

It's funny that people think they can let their emotions run wild in every other phase of life. They can whoop and holler at the ball game or the horse race, they can feel ecstatic over some boy or some girl, but they won't show emotion in serving God.

Let me tell you that my relationship with God is the deepest emotion I have and it results in feeling deep emotion for the needs of other people. You know one reason I know it is all right to feel emotion? Because God feels emotion. Nobody can love in greater measure than God does:

> *For God so loved the world, that he gave his only begotten Son, that whosoever believeth in him should not perish, but have everlasting life.*
> *(John 3:16)*

It took emotion and love for God to do that.

God feels compassion. He feels anger. He feels jealousy. All these emotions of God are recorded in various places in the Bible, but the one we are concerned with here is that God feels it when someone touches Him and takes His power. He is not some impersonal energy source, say, for instance, like the sun, where you can just go out and soak up a few minutes of sunlight and the sun doesn't feel it.

He is aware of the need and He is aware of the person whose faith causes them to press through the throng and take of His power to meet that need (Mark 5:30). When someone takes God's power, it is not diminished. Jesus had just as much virtue left as He had before the woman touched Him, but He felt virtue go out of Him because that virtue is real. It is a real, vital, moving force that is felt not only by us as recipients but by God as the Source Who is tapped for our needs.

Make this declaration: "God is aware of my need. The virtue of His saving, healing, delivering power is a real and vital force that I can tap into to meet my needs."

A GREATER MANIFESTATION

Scripture reading: John 11:1-45

God is a God of circumstances. I am so glad that He is, for life is made up of circumstances. Everyone's life is full of circumstances. Many of the circumstances that we face are completely independent of who we are, whether or not we are serving God:

> *...he maketh His sun to rise on the evil and on the good, and sendeth rain on the just and on the unjust.* *(Matthew 5:45)*

God sends His blessings of sun and rain upon both the good and evil because He is a God of love.

Many of the circumstances of our lives are man-made or self-made and are brought about through our own actions or those of people and events around us. My mail every week brings thousands of letters from people in all sorts of situations, facing all kinds of circumstances. Some of these circumstances have come upon these dear friends through no fault of their own, while others have come because they are reaping the results of bitter seeds they have sown by past actions.

I have discovered that many of the circumstances facing people are things they have allowed to come into their lives and allowed to disrupt and disturb them because of their lack of understanding of what the circumstances of life are all about. Many are good Christian people who love the Lord, and yet they have found themselves "between a rock and a hard place" because they do not know how to cope with their circumstances. They do not recognize the purpose for the mountains of life's circumstances that seemingly block their progress.

In the natural, many circumstances are crushing burdens, beyond our ability to cope with or to bear. In the natural, there often seems to be no answer, but we are not living in the natural. We are living in the blessing of the Almighty God for Whom nothing is impossible!

There is never a time when we do not have the answer, because Jesus is the Answer to every circumstance of life. Not only is God in control of all of life's circumstances, He sovereignly uses circumstances so He can manifest His power and glory in the midst of seemingly adverse factors.

When Lazarus was sick to the point of death, Mary and Martha sent for Jesus, but He stayed where He was for two more days. By the time He arrived at Mary and Martha's house, Lazarus was dead and in the tomb. Lazarus's sickness had, in the natural, grown worse – he was dead. But because of this, Jesus was able to manifest His glory in a greater way. He had already done many miracles of healing. You might say that the healing of the sick had become almost commonplace to His ministry. Here, He was able to demonstrate God's authority over an every more adverse circumstance by raising the dead (John 11:43-44).

Make this declaration: "Jesus is the answer to every circumstance of my life. He sovereignly uses my circumstances to manifest His power and glory in the midst of seemingly adverse factors."

Look To God

Scripture reading: Deuteronomy 26:1-9

In the Old Testament, when God was preparing to lead His children out of Egypt, instead of circumstances getting easier, they became harder. The main reason for this was that God hardened Pharaoh's heart (Exodus 10:1).

When God finally delivered Israel, He performed mighty miracles despite overwhelming odds. He delivered His children from the bondage of Egypt and preserved and protected them through the torturous years they spent in the wilderness:

> *And the Lord brought us forth out of Egypt with a mighty hand, and with an outstretched arm, and with great terribleness, and with signs, and with wonders: And He hath brought us into this place, and hath given us this land, even a land that floweth with milk and honey.*
>
> *(Deuteronomy 26:8-9)*

God could have chosen some other way to get Israel out of Egypt. He could have immediately and instantaneously transported three million people of Israel to the Promised Land, but He chose to deliver them instead in a series of miracles such as have never been duplicated.

It often has been said that "it is darkest just before the dawn." Often when circumstances seem to be most difficult, the way is being paved for God to do such a great work of deliverance in our lives that we will know beyond the shadow of a doubt that it was His great arm that saved us.

One purpose of circumstances, then, is that God might prove His power and glory to a world that needs to see His majesty. Another is that He might prove Himself to us in such a way that we might learn to trust Him with utter confidence despite the gravity of the circumstances around us.

God will meet the circumstances of today and tomorrow. We know this, because He already has proved Himself in the circumstances of yesterday. As a verse of the well-known hymn *Amazing Grace* says:

"Through many dangers, toils and snares, I have already come.

'Tis grace that brought me safe thus far; And grace will lead me home."

God is sufficient for your circumstances. He wants to use them to demonstrate His power and glory. He wants to prove Himself so you can learn to trust Him with utter confidence.

Today, don't look to your circumstances. Look to the bigness of your God.

Make this declaration: "I will not look to the bigness of my circumstances. I will look to the bigness of my God."

A Supernatural Touch From God

Scripture reading: Genesis 27:1-46

For a tremendous Bible example of a man whose life was caught up in a web of seemingly adverse circumstances that seemed completely overwhelming, let us look at the part of Jacob's life detailed in Genesis 32.

Over twenty-one years before, Jacob had extorted the birthright that belonged to his older brother, Esau. Not only that, but Jacob had deceived their father, Isaac, into giving him the blessing which by tradition should have gone to his elder brother. Esau became enraged and vowed to kill Jacob. With his mother's help, Jacob fled to Haran where he worked for twenty-one years for his Uncle Laban. While there, God blessed Jacob with wives, children, and a large number of livestock.

Then God told Jacob to return. In Genesis 32 we find him doing that, but he was very fearful, expecting that Esau would still be angry and kill him. In Jacob's absence, Esau had become a powerful man with trained soldiers. This tormenting fear hindered Jacob from seeing clearly the positive circumstances which surrounded him. Fear paralyzed his ability to utilize the good circumstances in his favor, and all the time the good far outweighed the bad. Jacob didn't understand his position and authority in God.

For one thing, Genesis 32:1 tells us that in his journey the angels of God met him and he named the place of their appearance Mahanaim meaning "two hosts" as a memorial of the experience he had there with God. Despite this supernatural revelation, Jacob still continued his trip in fear even though God had let him see a host of angelic beings about him.

Jacob also had promises of God upon which to rely. He knew what God had promised and even quoted it in his prayer. He said, "The Lord which saidst unto me, Return unto thy country, and to thy kindred, and I will deal well with thee." Still, Jacob feared.

Jacob was meeting his brother in a place called Seir which means "rough or bristly." That seems to be an adequate description of Jacob's circumstances at this time – rough and bristly. Jacob certainly had some deep soul-searching to do, so he sent his wives, children, servants, and all that he had over the ford at Jabbok and remained there alone.

The meaning of the word Jabbok means "emptying or pouring out," and that is exactly what Jacob experienced here. As related in Genesis 32:24-30, there came a supernatural being who wrestled with him until the breaking of day. The wrestling match ended with Jacob's name, life, and nature being supernaturally changed.

God wants to change your circumstances today, but in order for this to happen you must realize who you are in Christ. You must come to the end of yourself and let God supernaturally touch your life.

Make this declaration: "I will not let You go, God, until you bless me!"

IN CHRIST JESUS

Scripture reading: Genesis 28:1-22

My son, David, majored in business administration at college before assuming a very vital role in the business functions of World Evangelism and the New Inspirational Network. One of the many books on management which he has in his library starts with this observation: "People's behavior stems from their interpretations of what they think they perceive."

That was true of Jacob in managing his life. Despite the fact that Jacob had several personal encounters with God since leaving home and God blessed him mightily as far as family and material benefits were concerned, in the forefront of Jacob's mind was still the threat that his brother had made to kill him.

Jacob was looking at his circumstances in the natural, from the standpoint of what he deserved and what he had heard rather than on the basis of what God had promised him. If we could just learn to see things as God sees them. If we just had His point of view, how different our lives would be. If we could see as God sees, we would never have to ask "Why, God?" But spiritual cataracts often blind our eyes so we cannot perceive our situations as they really are. We get our understanding of the situation from what we think we saw, what we heard, what someone told us, or what we project might happen.

The reality was this: During Jacob's absence from his homeland, God also blessed his brother. Esau was on his way to meet Jacob to welcome him. He did not plan to kill him and he did not want Jacob's presents. He had plenty of livestock and goods of his own. God had worked in Esau's life and heart and eliminated the thing Jacob so greatly feared.

There is a great verse in the Bible in the Book of Proverbs that says, *"The wicked flee when no man pursueth: but the righteous are bold as a lion"* (Proverbs 28:1). That means we can fear when there is nothing to fear when our hearts are not right with God. When our hearts are right, we have nothing to fear. It is not our righteousness, for we have none in ourselves: *"But we are all as an unclean thing, and all our righteousnesses are as filthy rags"* (Isaiah 64:6). If we are in Christ Jesus, we partake of His righteousness.

> *But of Him are ye in Christ Jesus, who of God is made unto us wisdom, and righteousness, and sanctification, and redemption.*
>
> *(1 Corinthians 1:30)*

In Christ's righteousness we can have wisdom to see as He sees, think as He thinks, and react as He reacts.

> *"Let this mind be in you, which also was in Christ Jesus"*
>
> *(Philippians 2:5).*

Make this declaration: "The mind of Christ is in me. I have the wisdom to seek as He seeks, think as He thinks, and react as He reacts."

WALKING STEP-BY-STEP

Scripture reading: Genesis 29:1-35

We get the true picture of our circumstances from what God has to say, not from what we think we perceive, from what man has to say, or from what natural circumstances seem to dictate.

Our shelves may say, "Look, I'm nearly empty. There's no food left. You'll soon run out of provisions and you'll starve to death." If our understanding is based on what that empty cupboard says, we are not understanding the situation as it really is, for God is saying, *"But my God shall supply all your need according to His riches in glory by Christ Jesus"* (Philippians 4:19).

Our checkbooks may try to make us understand, "No, you can't really afford that car you need for work or that warm coat you need for cold weather." But we get our real understanding from the Word of God which says, *"No good thing will he withhold from them that walk uprightly"* (Psalm 84:11).

The newspaper or nightly news may try to get us to understand that our situation is perilous because crime is rampant, there's an increase in burglaries, armed robberies, and dope addiction. But our real situation is revealed by God when He says, *"Fear thou not; for I am with thee: be not dismayed; for I am thy God: I will strengthen thee; yea, I will help thee; yea, I will uphold thee with the right hand of my righteousness"* (Isaiah 41:10).

You see, newspaper people, television personnel, and even your friends do not know the real situation and cannot give you the right understanding of a matter if they are ignorant of God's promises and plans.

Fear is a product of ignorance. Most of the time we are afraid of the unknown. For every single situation in the world, for every fearful or overwhelming circumstance, God has the right viewpoint and is able to turn any circumstance completely around and give victory. He sovereignly reigns over every circumstance. Every one!

That is why it is important to know what our situation really is in the sight of God. Even if He doesn't give us complete understanding of what the true picture is instantaneously, if we are in Him we can trust His understanding and walk step-by-step with Him in utter confidence that He will *"keep that which I have committed to Him against that day"* (2 Timothy 1:12).

Make this declaration: "God is leading me step-by-step and I can trust Him to keep that which I have committed to Him."

November 16

Understanding Our Position

Scripture reading: Genesis 30:1-24

In John 10:10, Jesus says, "The thief cometh not, but for to steal, and to kill, and to destroy." The thief is Satan and one of the things he seeks to steal is our position or standing before God. One of the main things he seeks to destroy is our self-image.

There is nothing so destructive as not having the correct image of yourself. When your mind is conditioned to the extent that you see yourself weak and as a failure, when the image of who you are is negative and incorrect, it blocks the power of God from flowing in your life to help you victoriously confront your daily circumstances.

In this respect, Jacob in the Old Testament is a true figure of the New Testament Church. Like Jacob, the Church has not come to the place where it fully comprehends its own position and situation.

In Genesis 32, Jacob is involved in a wrestling match. The match is taking place between a physical human being and a spirit being. The Bible says "man" but we can tell from what took place that Jacob's opponent was an angelic being. The first thing that I question when I read this Scripture is this: Isn't it strange that an angelic being has to ask Jacob, a human being, for permission to leave?

Can you imagine yourself right now wrestling with a heavenly being and suddenly having that heavenly being turn and say to you, "Would you please let go of me?" You can't imagine that, can you? And yet that is exactly what happened to Jacob.

The impact of this is compounded when you realize the angelic being had such tremendous power that he touched the hollow of Jacob's thigh and made him a cripple. Jacob's thigh came out of joint and he had to walk like that for the rest of his life. The angelic being had that much power over Jacob, yet he was not able to leave the presence of this man without permission.

Jacob didn't wrestle to get God's power. He already had power over this being. The blessing and the power are not the same thing. God promised the blessing and Jacob had the power to receive it, but didn't realize it. He didn't realize – until this incident – who he really was in God.

When the angel said, "Let me go, for the day breaketh," Jacob declared, "I will not let thee go, except thou bless me." A little further on the angelic being makes a declaration and blesses Jacob, but before he does, he tells him, "Jacob, thou hast power with God and with man and thou hast prevailed." He blessed him after he made that statement, but the blessing was not the power – Jacob already had that lying dormant within him.

The angel recognized something that even Jacob himself didn't realize. He recognized that Jacob had power with God. He certainly wasn't acting like it. He wasn't using it.

Could this be a picture of our own lives?

Make this declaration: "I have power with God. I am going to start acting like it and using it."

Stop Struggling

Scripture reading: Genesis 30:25-43

As believers, God has given us His power to meet the circumstances of life. Could it be that this God-given ability to overcome any problem and change any circumstance is within us, just waiting for us to recognize it?

In Genesis 32:28, the angel told Jacob something by revelation that by human effort he was struggling to attain. He didn't realize it, but he already possessed what he was struggling to achieve. God's power did not come by wrestling and struggling. Jacob did not get God's power as a result of wrestling with the angel. He received a blessing from the angel, but he already had the power without struggling. It was given to him by God.

Will you stop struggling? Will you stop wrestling for God's power to help you rise up and face your circumstances? Will you stop battling in your own strength for spiritual breakthroughs in your ministry? They don't come by struggling!

What was wrong with Jacob? Simply this: He was not partaking of the provisions that already had been given to him. Faith is a fact, but faith is also an act. What Jacob needed to do – what the Church of Jesus Christ needs to do, what you need to do – is realize who we are in God and begin to act on that knowledge.

Jacob never realized the truth of his position in God, so he struggled for years. When you don't have a right understanding of your position, you struggle. Jacob did not even understand the implications of his name – and neither has the Church. He thought his name was Jacob – the conniver, the deceiver, the sinner, a weak-kneed, fearful man. That's who Jacob thought he was and why he acted that way, running, hiding, and tormented by fear.

I am sure that the angel knew who it was that was wrestling with him, but why did he ask Jacob what his name was? Because Jacob didn't realize himself just who he was. This is the parallel of the condition in the lives of most Christians.

The angel said, "What is your name?"

Jacob said, "It is Jacob."

The angel said, "You're mistaken. I know you better than you know yourself. Your name is Israel."

If only you would know yourself as you are known by God! God knows you – in all of the power and glory that He has created you – but you need to come to the place where you can recognize it.

The angel said, "You think your name is Jacob? Well, it's not! One day up there in the heavens God changed your name from Jacob, the supplanter and deceiver, and renamed you Israel. He made you a prince because of a promise that God gave to you. You have power with God and with men and you don't even know it! You are running, hiding, whimpering, and crying – and all the while, inside of you, you have supernatural power."

Make this declaration: "No more running, hiding, whimpering, and crying. Operating inside of me is the supernatural power of God."

November 18

The Trap Of Compromise

Scripture reading: Genesis 31:1-55

Jacob's name was changed. As Jacob he had nothing, but as a prince of God he entered into a relationship with his Heavenly Father that gave him power to meet the circumstances of life. You, too, have been given the Name of Jesus!

But there is one other danger Jacob was subjected to because of fear and lack of understanding of who he was and it is a danger that we all face. It is the tendency to compromise.

In his fear of Esau and his misunderstanding of his own situation and position, Jacob nearly fell into the dangerous trap of compromise. God told Jacob to return to his homeland and that it would be well with him and he would be blessed. Doubting that promise and continuing to act out of fear, Jacob laid out plans for compromising with the enemy.

We have already seen what great power Jacob had with God. I believe that if Esau had been hostile as Jacob feared he would be, just one word from Jacob's lips would have brought God's judgment down upon Esau. But Jacob wasn't thinking about how to win and overcome, he was still thinking about how to connive, compromise, and perhaps end up with at least a partial victory instead of losing everything. He planned to try to win Esau over with bribes and promises.

The biggest temptation which you and I will face when we come up against problems in this life is to compromise. The temptation to compromise is often very strong and it comes in many ways. A church may want to win a city for God. They have the power to win the city, but what they do is to let down the standards, compromise, and try to get the people of the city into the Church that way.

There are many Christians who started out to work for God, hot-hearted and on fire, who knew how to pray and how to yield themselves to the moving of the Holy Spirit, but there are many who have become sidetracked by compromise.

God has blessed this ministry in a wonderful way and we are meeting new challenges all the time in the power that God has provided. But I want to tell you that there have been times in my ministry when men of influence and wealth have come to me with glowing plans of how this ministry could be enlarged through compromise. There were those who came to me and said, "Look, you can really be somebody. We will stand behind your ministry and you will have a lot of influence and prestige if you will just tone down your ministry to the sick, if you will just tone down the way your services go."

Thank God we didn't compromise. God has honored His Word and has sent partners to stand with us and hold up our hands in this great work of God. The power of God and the demonstration of the Holy Spirit is the mainspring of this work. I've said it often and I'll say it again: "This is not the work of a man, but it is the work of the Holy Spirit of the living God."

Don't listen to the lies of the enemy. Don't consider the unwise counsel of those around you. Don't compromise!

Make this declaration: "I will not compromise. I will not listen to the unwise counsel of those around me. I will not believe the lies of the enemy."

THE TRUTH OF THE SITUATION

Scripture reading: Genesis 32:1-23

Jacob compromised by making three bands out of his family. When you compromise before the enemy, you will even give away your family. That's what the devil does to you. Jacob took his family, his cattle, and his earthly possessions and made three bands out of them.

He said, "All right, the first of you go on out and meet Esau. He is coming with 400 mighty men to kill me. When you see him, you bow down, compromise, scrape, eat the dust, give him these gifts, and tell him they are from his servant, Jacob."

But Jacob wasn't Esau's servant. Esau was Jacob's servant. You are not a servant to your circumstances. Remember, God has said: *"Behold, I give you power to tread on serpents and scorpions, and over all the power of the enemy"* (Luke 10:19).

He has said:

> *"And the Lord shall make thee the head, and not the tail..."*
> *(Deuteronomy 28.13)*

He has said:

> *"...Thou shalt be above only, and thou shalt not be beneath."*
> *(Deuteronomy 28:13)*

He has said:

> *"The LORD shall cause thine enemies that rise up against thee to be smitten before thy face: they shall come out against thee one way, and flee before thee seven ways."*
> *(Deuteronomy 28:7)*

Jacob thought, "When Esau kills the first band, the second band will be there, and when he kills the second, the third band will be there. I'll put my wives and everyone else in front and I'll march behind them." But something happened when the angel helped Jacob to see the real picture. We are told that Jacob passed over before them (Genesis 33:3).

Now when Jacob took the lead and passed over before them to meet Esau, even though he thought he might die. Instead, Esau fell down before Jacob, put his arms around him, and said, "Jacob, I'm sorry for every evil thought I had about you. I'm sorry I ever tried to kill you." He said, "I'm going to give you the honor to which God has raised you up. Will you forgive me?"

Now we see the truth of the situation. If only Jacob could have believed the promises of God, if he could have seen through spiritual eyes instead of eyes clouded by circumstances. When we do not see the truth of our situations – that the victory is won, that the battle, no matter whether it is physical, spiritual, or material is not ours but God's – then we are filled with anxiety, unbelief, doubt, and fear – all the things that defeat us.

In Jacob's case, the truth was that God went before him and took care of every circumstance. Every one! There was no need to compromise, no need to bow, no need to be afraid. If Jacob only had trusted the power which God had already given him. If only he had known and believed the truth of the situation, the victory was already won.

Make this declaration: "There is no need to compromise, bow to the demands of the enemy, or compromise. The victory is already won!"

The Battles Of Life

Scripture reading: Genesis 32:24-32

The truth of Jacob's circumstances is revealed in Genesis 32:28. He was not a weak, defeated, fearful man who needed to run, compromise, and bow. He was a son of God who was called, ordained, and given power to meet the circumstances of life!

Was Jacob someone special whom God would do this for and no other? Is God a respecter of persons? Would He give one of His children the power to overcome the circumstances of life and not another?

Of course not. I have tremendous news for you. What God did for Jacob, He already has done for you. He is just waiting for you to see the situation as it is, to see yourself as you are, to march in and claim the victory He has provided for you already in Christ Jesus.

You, __________(you can insert your own name in this sentence) – have the power to face every circumstance of life! God has already provided it. It remains only for you to apply it.

It is yours through your entrance into the family of God. As you received Christ (or receive Christ even now), your name was changed. As a child of God you have power with God and will prevail.

There are circumstances in your life right now – for life is made up of circumstances. Some are good, others bad. Nearly every person reading these words has some battle in their life, some adverse circumstance to meet, some victory that needs to be won. If not, and you have some temporary respite, don't worry: the battle is sure to come.

We are promised battles:

> *Yea, and all that will live godly in Christ Jesus shall suffer persecution.*
> *(2 Timothy 3:12)*

> *But he shall receive an hundredfold now in this time, houses, and brethren, and sisters, and mothers, and children, and lands, with persecutions; and in the world to come eternal life.*
> *(Mark 10:30)*

> *Be sober, be vigilant; because your adversary the devil, as a roaring lion, walketh about, seeking whom he may devour.*
> *(1 Peter 5:8)*

> *Beloved, think it not strange concerning the fiery trial which is to try you, as though some strange thing happened unto you.*
> *(1 Peter 4:12)*

You might say, "Brother Cerullo, why are you telling me all these negative things about having battles, persecutions, and trials?" Because they are true. This is God's Word just as truly as any promise in the Bible – but it is not negative! It is in going through battles that we are perfected, that we learn to know and trust God, and that we come to know ourselves.

Make this declaration: "The battles of life are perfecting me. I am learning to know and trust God and to know myself."

A Powerful Prayer

Scripture reading: 1 Peter 1:3-7

God has promised:

> *There hath no temptation (test) taken you but such as is common to man: but God is faithful, who will not suffer you to be tempted above that ye are able; but will with the temptation also make a way to escape, that ye may be able to bear it.*
>
> *(1 Corinthians 10:13)*

Every promise in the Word of God is geared to life and victory! The provision has been made. In the things of God I have this knowledge! In God I can see the truth of every situation which I encounter. God has given me power over all the power of the enemy and in Him I can come out the winner every time! If you could only see the truth of every situation that you encounter in this life! If you could only see as God sees!

God didn't promise us we wouldn't have battles. He didn't promise us we wouldn't have sorrow. He didn't promise us we wouldn't have heartaches. But He did promise we would never be defeated.

Whatever your trials and whatever your circumstances, no matter how overwhelming they may seem to be in the natural, I want to minister to you right now. I don't have to be present with you in the flesh to do this. Paul at one time wrote to the Corinthian church that though he was absent from them in the body, he was present in spirit and the action he wished was taken as though he was present.

I cannot be present with each of you in my bodily presence, but there is no distance when it comes to the things of God. As you hold this devotional guide in your hand, use it as a point of contact to agree together with me that you will be able to rise up in the power already given you to meet the circumstances in your life. Hold this book in your hand as I pray this prayer for you:

"Heavenly Father, right now in the Name of Jesus, I bring this loved one to You for Your special blessing and help. God, You see and know every circumstance. You have permitted every circumstance that is in their life right now. Father, You wait to prove yourself in might and power in these circumstances. Open these eyes that this one may see the situation as it really is, that they might see that You are in command and that You have stepped forth on their circumstances to still the waves and the winds. You have stepped forth in triumph and power over all the power of the enemy. Father, help this one to realize the position we have in You as Your beloved child. Help them to take their rightful position in You as a king and a priest, as a ruler, as the head and not the tail, as being above and not beneath, according to the great promises of Your Word. In Jesus Name, Amen."

Make this declaration: "God has heard my prayer. He will move in my behalf."

November 22

A Monumental Task

Scripture reading: Psalm 33:1-22

A long time ago, this ministry felt led to take a giant step of faith – a giant step – seemingly far greater than any challenge we had undertaken before. For years I have been going to the far corners of the earth to conduct mass overseas crusades. As many as 300,000 people have attended a single service. Multitudes are healed of every disease and affliction imaginable and even more are touched spiritually through these crusades. But the main thrust and burden of my heart in these overseas meetings has not been the crusades themselves but very special meetings held during the day for the training and motivation of Nationals.

As part of each School of Ministry, I have some of the key National ministers preach and pray for the sick in the last crusade services. I just stand back and pray quietly and let them minister. What joy as they realize that the same Holy Spirit Who anoints Morris Cerullo also anoints them and uses their ministries to bring miracles of deliverance and healing! They leave those meetings transformed! I have seen it happen over and over again – men who have preached to very small groups soon have congregations of hundreds and thousands.

Over the years, the number of requests to hold these schools overseas have increased to the point where it would take me at least ten years to conduct them all. However, the training is urgently needed now especially in the face of tremendous population explosions all around the world. I was thinking about this explosion of the world's population and wondering how we could ever do the work Jesus gave us to do – how we could go into all the world and preach the Gospel to every creature. Then God gave me a vision and concept that for the first time I saw how this seemingly impossible task is possible. It can be done in our generation!

God led me to conduct Schools of Ministry where Nationals from all over the world can receive training to bring their own countries the message and power of the resurrected Christ. This is not another Bible college, university, or school of theology, but the School of Ministry is an intensified training for Nationals to learn to work the works of God effectively in their own countries. For the first time I could see how we can overtake the heathen!

When God gave me this breathtaking concept, I knew it was a big step. Circumstances would certainly have been frightening and would have disheartened me if I had allowed myself to look at them or to dwell on them. But I had to look past the bigness of the task to the greatness of God.

Has God given you a monumental task to accomplish? Do you feel overwhelmed? I'm sure that mountain looked pretty big to Caleb too. After all, he was over eighty years of age.

Don't look to the bigness of your task today. Look to the bigness of your God.

Make this declaration: "I will not look to the monumental task God has given me. I will look to God. Give me that mountain!"

Each Step Is A Miracle

Scripture reading: Psalm 47:1-9

When God gave me the monumental task of conducting Schools of Ministry throughout the world to raise up His army, He gave me a profound promise that I want to share with you today. In my hotel room in prayer just prior to a seminar I was holding in Washington, D.C., God spoke to my heart with these words:

> "Do not look to the bigness of your need. Look to the bigness of Your God. Your circumstances are hindrances to seeing My abilities! If you keep your eyes on your circumstances, the devil will use your circumstances to defeat you and accuse the Word of God – the written and the living Word.
>
> Your victory is in keeping your eyes on the bigness of your God and His ability. He has promised to take you step by step by step – not all at once but step by step –
>
> And each step will be a miracle!"

These words were so impressed on my heart that I had them put on a wall plaque for friends to hang on the walls of their homes to remind them every day. In all my ministry, nothing has strengthened me more than the comfort I receive as I repeat these words over and over again- that is why I wanted to share these words with you today.

Satan brings negatives into our lives and tries to get us to focus on these instead of Jesus. When you worry and meditate on your problems, you develop a spiritually blinding cataract that can keep you from seeing Jesus as He really is and setting Him free to work miracles of deliverance in your life.When we keep our eyes on the problems or difficulties surrounding us, there is no way that we can see how big and mighty He is to meet those needs. There are some people who are determined to look at their problems. There is no way that you can change the direction in which they are looking. They have looked so long at their problems that their difficulties have become mountains in their sight, blotting out everything else. They can't see anything else, they can't talk about anything else, and they can't think about anything else. Often it seems that they must love their problems because they hold on to them so tightly. It is almost impossible to get such people to turn their sights on Jesus because they seemingly don't want to relinquish their troubles.

Don't look at your problems. Change your focus! Your problems are hindrances to seeing God's abilities. Look to the One who is above all, greater than all, and great enough to empower you to meet every situation that you confront today. He will lead you each step of the way, and each step will be a miracle.

Make this declaration: "God will empower me to meet every situation that I confront today. He will lead me each step of the way, and each step will be a miracle."

November 24

Where Is Your Focus?

Scripture reading: Philippians 4:4-9

A woman called my office some time ago who was in tremendous need of physical, emotional, and spiritual healing. Tearfully, she recited a long list of her ailments and a number of statistics on how many people went insane with her particular afflictions.

Now this lady's problems were very real, but they were worsened by her preoccupation with her symptoms and how people treated her. One complaint she had was that she didn't "get any sympathy" from hospital and medical personnel. One got the impression that this lady was looking more for sympathy than she was looking to Jesus. It seemed as though she was so wrapped up in her problems that she would not know how to function without them! After prayer, this lady was vigorously counseled to take her eyes off every symptom that she had and think and speak only of the goodness and greatness of God.

Now we are not Christian Scientists. We know that problems exist, people are in need, and we do care. But we also know that God is bigger than any problem and that if we will see Him as He is, He can take care of every problem and affliction.

We must learn to gird up our minds and bring every thought under subjection to Him and to think on those things that are edifying and helpful, not the negatives that Satan keeps sending us to keep our sight off the greatness of God:

> *Casting down imaginations, and every high thing that exalteth itself against the knowledge of God, and bringing into captivity every thought to the obedience of Christ.*
>
> *(2 Corinthians 10:5)*

The Apostle Paul encourages us to think on…

> *…whatsoever things are true, whatsoever things are honest, whatsoever things are just, whatsoever things are pure, whatsoever things are lovely, whatsoever things are of good report; if there be any virtue, and if there be any praise, think on these things.*
>
> *(Philippians 4:8)*

Certainly one of the good, virtuous, and right things we should think on is the greatness of our God and His capability to handle our problems.

Let me ask you…what was the first thing you thought about when you woke up this morning? Did you think about your problems? Did you think about the negative circumstances you must face today? Or did you think about God and His goodness. Did you take time to fill your mind with His Word and worship Him?

Just where is your focus?

Make this declaration: "I will focus on God and His goodness instead of my problems."

The Will Of God

Scripture reading: John 1:1-15

The will of God is something we need to know more about because there is no way that you can have faith for miracles, healing, or victory over your circumstances if you do not believe it is God's will to give you deliverance. You will never reach your destiny without this knowledge.

Jesus plainly declared what His purpose was in 1 John 3:8: *"For this purpose the Son of God was manifested, that he might destroy the works of the devil."* His will is declared in Matthew 8:17 where it says, *"Himself took our infirmities, and bare our sicknesses."* It is God's will to save every person from sin: *"The Lord is not slack concerning his promise, as men count slackness; but is longsuffering to us, not willing that any should perish, but that all should come to repentance"* (2 Peter 3:9).

It is God's will for us to have salvation, prosperity, health, and victory and it is as important for us to know that it is His will for us to have these things. We can never pray the prayer of faith for anything until we know that what we are praying for is God's will. The prayer of faith must be tied to infallibility. Infallibility means that it cannot be questioned.

We must know beyond question that it is God's will if we are going to pray in faith. We can pray and know that something will happen when we know that we are tied to that which cannot be questioned.

The only thing in the world that is infallible, which cannot be questioned, is God's Word – not just His written Word, the Bible, but also the living Word, Jesus. Everything that ever has been created was created by God's Word. God said *"Let there be light"* and there was light. He said, *"Let the dry land appear"* and it was so. He said, *"Let the earth bring forth"* and it did. Every time God said something would happen, it happened because of the creative power of God's Word. It could not be questioned. It was infallible.

Out of nothing, God created everything that was created by the power of His infallible Word. That was His spoken Word, and the living Word was right there all the time.

> *In the beginning was the Word, and the Word was with God, and the Word was God. The same was in the beginning with God. All things were made by him; and without him was not anything made that was made... And the Word was made flesh, and dwelt among us, and we beheld his glory, the glory as of the only begotten of the Father, full of grace and truth.* (John 1:1-3,14)

God's Word is truth.

> *"God, that cannot lie, promised..."* (Titus 1:2).

> *"It was impossible for God to lie..."* (Hebrews 6.18).

God's Word is forever settled in heaven (Psalm 119:89). It cannot and will not ever be changed: *"Heaven and earth shall pass away, but my words shall not pass away"* (Matthew 24:35).

Make this declaration: "God cannot lie. His Word is settled in Heaven. His Word is truth."

November 26

What Are You Saying about God?

Scripture reading: Luke 5:12-14

If God's Word was fallible there could never be faith. Faith is tied to infallibility. That is why I can make a statement such as, "It is God's will to heal everyone." How can I make such a tremendous blanket statement? Because it is already recorded and documented in the Holy Bible, the Word of God, which is infallible, which is true, which cannot be questioned. It is already on record.

When the leper came to Jesus and said, *". . . if thou wilt, thou canst make me clean"* Jesus said, *"I will. Be thou clean"* (Luke 5:12,13). He made His will known to him. He was willing.

In Matthew 12:15 we read of Jesus that *". . .great multitudes followed him, and he healed them all."* Matthew 14:14 puts a motive behind His willingness: *"And Jesus went forth, and saw a great multitude, and was moved with compassion toward them, and he healed their sick."*

After Jesus was lifted from this earth, the apostles He had commissioned continued to carry out His work. In Acts 5:16, we find that *"There came also a multitude out of the cities round about unto Jerusalem, bringing sick folks, and them which were vexed with unclean spirits: and they were healed every one."*

Many people just cannot believe that it is God's will to heal them. Many actually believe that it is God's will that they remain sick! When God brought the children of Israel through the wilderness after they left Egypt, we have this tremendous statement of fact: *"There was not one feeble person among their tribes"* (Psalm 105:37). It is true that Israel as a nation has played a special part in God's world plan and that they have enjoyed a unique position nationally, but I want you to know that God is no respecter of persons (Acts 10:34). He loves you and me just as much as He loved the Jews of that day. What He did for them, He will do for us, and it is what He will do for the Church if we will just cut Him loose to do it. If we just take the limits off Him!

God's will to heal bodies has been especially stressed in today's devotional reading, but God's will for us encompasses many other great blessings also. It encompasses total health and well-being, total prosperity (3 John 2). Many feel it is God's will for them to be poor and needy. I have known people who felt there was something godly about worn-out shoes and an empty cupboard. If they don't have the faith to believe that it is God's will for them to have plenty, how can they pray in faith?

Jesus said: *"I am come that they might have life, and that they might have it more abundantly"* (John 10:10).

Paul said: *"But my God shall supply all your need according to his riches in glory by Christ Jesus"* (Philippians 4:9).

David said: *"The Lord is my shepherd; I shall not want"* (Psalm 23:1).

What are you saying about the Lord?

Make this declaration: "I will say of the Lord, He is my shepherd, my healer, my savior, my provider, my deliverer, and my guide."

Living In Past Guilt

Scripture reading: Isaiah 1:18-20

The Bible is full of good things that God wants to do for us:

- *No good thing will be withhold from them that walk uprightly. (Psalm 84:11)*
- *Delight thyself also in the Lord; and he shall give thee the desires of thine heart. (Psalm 37:4)*
- *Many are the afflictions of the righteous: but the Lord delivereth him out of them all. (Psalm 34:19)*

It is Satan who has caused some scriptures to be twisted and misapplied so that people will not know that it is the will of God to literally pour out His blessings upon them in this day!

Some people cannot believe that it is the will of God to bless them because they feel unworthy. They have committed some sin in the past and Satan has kept them in bondage to guilt. It is true that we may not deserve God's blessings, but they are given on the basis of Christ's worth, not our own worth. Just as Christ bore our sins, He bore our sicknesses. He also bore our guilt and shame. Even if we have committed terrible sins, we can repent and come to Him for forgiveness:

> *Come now, and let us reason together, saith the Lord: though your sins be as scarlet, they shall be as white as snow; though they be red like crimson, they shall be as wool.*
> *(Isaiah 1:18)*

Once God has forgiven us, we must forgive ourselves and not let guilt feelings continue to rob us of the good things of God.

Some years ago there was a struggling young woman who barely lived from paycheck to paycheck. She would run out of money before every payday. Those were the days when it took three or four days for checks to get to the bank after they were cashed, so she would cash a check and then live in fear and dread that the check would get to the bank before she got there with her paycheck to cover it.

For years, even after this lady was living on less of an emergency basis, she had an unreasonable guilt when trying to cash a check.

"I just can't cash a check," she confessed to a friend. "I feel so guilty."

"Well," asked her friend, "Is the check covered?"

It was. This woman was living on past guilt which kept her from enjoying the convenience of her checking account. When she realized that she had money in the bank to cover her check and that her guilt feelings were not related to her real situation, she gained a new sense of security in cashing checks.

There is no need to continue to feel guilty over things that have happened in our past and for which we have received forgiveness. We can go before God with a clean slate because of the forgiveness we have in Christ. Our account, so to speak, is completely covered, our conscience completely clear.

Make this declaration: "I will no longer live in the guilt and shame of the past. God has forgiven my transgressions and removed them from me as far as the east is from the west."

November 28

Struggles Without Struggling

Scripture reading: Daniel 6:10-23

God makes a world of difference between His people and those who defy and deny Him. He makes a difference between those who are obedient to His Word and those who scoff at His Word.

When Daniel was thrown into the lions' den, no manner of hurt was done unto him. When Daniel's enemies were thrown into the same den with the same lions, there was a marked difference. The lions broke the bones of those enemies before their bodies reached the bottom of the pit.

The same was true of the fire that burned in the furnace of Nebuchadnezzar. The furnace was heated so hot that the enemies who threw Shadrach, Meshack, and Abednego into the furnace were slain by the blast of heat even as they stood outside. The three young Hebrew men who served and trusted God not only survived the furnace experience, but their bonds were burned off and the heathen saw them walking about with the Son of God in the midst of the furnace. These men were so prepared for whatever came upon them that I doubt they felt the heat at all.

That is why I have come to know that it is possible to go through struggles without struggling! There are those reading this devotional right now who are going through heavy trials – family upheavals, broken relationships, loneliness, depression, financial needs, illness. If you are a child of God, God wants to make a difference between you and the unbeliever. He will meet you right now at the depths of your deepest pit, in the heat of your hottest furnace, in the desperate nature of your circumstances.

God wants your faith so tied to that which is infallible and impregnable, that none of these things moves you. He can give you such peace in your heart that the fire of your circumstances will not touch you. The lions' mouths will be wired shut. Nothing can hurt you.

Infallible means that which cannot be questioned, incapable of error, unable to mislead or deceive. Impregnable means invincible, incapable of being taken by assault, immovable, and that which cannot be conquered or even penetrated.

There is only one thing in this world that is infallible and impregnable and that is the Word of God. Our experience must be tied to that Word in order that we will not be shaken but will stand forth in full glory, blazing testimonies to the power and glory of Jesus Christ in these troubled final days.

Make this declaration: "I am fixing my hope on the Word of God so that I will not be shaken. I will stand forth as a testimony to the power and glory of Jesus Christ in these troubled final days."

Take The Limits Off God

Scripture reading: Galatians 3:7-22

One of the greatest Scriptures in all of the Bible is found in the 13th chapter of the Book of Hebrews, verses 5 and 6: *"For he (God) hath said, I will never leave thee nor forsake thee. So that we may boldly say, The Lord is my helper, and I will not fear what man shall do unto me."*

When God says something, then we can say it also. You can take that promise and incorporate in it any promise of God.

Let the weak say, *"The Lord is the strength of my life."*

Let the sick say, *"With His stripes I am healed."*

Let the troubled say boldly, *"God is my refuge and strength, the present help in time of trouble."*

Let the afflicted say, *"Many are the afflictions of the righteous: but the Lord delivereth them out of them all."*

Let those who are tormented say with boldness, *"God has not given me a spirit of fear."*

Take every promise of God and from now until Jesus comes, boldly claim it, believe it, and stand upon it. It is the unshakeable, unchangeable, impregnable, infallible Word of the Living God!

If we look in God's Word and find there is a place where we do not measure up, we can find forgiveness and come into alignment with God's Word. Sin brings doubt. A knowledge of forgiveness and cleansing brings faith: *"The wicked flee when no man pursueth: but the righteous are bold as a lion"* (Proverbs 28:3).

It is God's will that we be bold as lions in claiming His promises. Whatever your situation or need, God wants the very best for you. Be bold in claiming it!

Once you know that God is not only able, but that He is also willing to meet your need, you can go on to the next step: Cut Him loose to do it. Loose Him to work in your circumstances on the basis of His revealed will. Let Him move in your behalf on the basis of His Word.

It's time to take the limits off God!

Make this declaration: "From now until Jesus comes, I will boldly claim, believe, and stand upon the Word of God which is unshakeable, unchangeable, impregnable, and infallible."

Power To Take Your Mountain

Scripture reading: Joshua 14:6-15

The greatest power in all the world is not political power. It is not atomic power, military power, or economic power. The greatest power in all the world is the Word of God!

The Word of God which we call the Holy Bible is only a small part of the Word of God. The Apostle John tells us that the world could never contain all the books it would take to write down everything God did. The Word of God is not only the written Word, but the Word of God is also the Living Word.

I am talking about God, not just some words of God. I am talking about the One Who said, *"Let there be light,"* and there was light; the One Who said, *"Let the firmament appear,"* and it appeared. I am talking about the One Who took a little bit of dust and breathed into it the breath of life and created man in His own image. I am talking about Jehovah, the only true and living God!

You cannot separate God from His miracles, nor can you separate the God of miracles from His Word. God is His Word. That is why I do not get entangled in the theologies of man, because I do not look to the word that comes from man. I look to the Living Word – God Himself!

If you believe in God, you believe in miracles. If you believe in God, you believe all things are possible. The Bible is our guide, our resting place. It is the voice of God giving us direction for our lives, our homes, and our businesses. It is the foundation that stands sure even though the heavens and the earth are shaken. It has absolute authority for us in the midst of every circumstance, every problem, every distress, every sickness, and need. It is infallible. It is unshakeable. It is unassailable.

When you know that God's Word is the final word, you will not look at natural circumstances or listen to the words of men. You will rise up and begin to do exploits such as you have never experienced in your life.

What mountains are you facing today?

- A mountain of debt?
- A mountain of fear?
- A mountain of broken relationships?
- A mountain of sin and bondage?
- A mountain of sickness?

During this past month we have learned how God's power is sufficient for every need. His power is greater than your mountain.

Like Caleb, it is time for you to rise up and declare, "Give me that mountain!"

Make this declaration: "Like Caleb, I am rising up today to declare – God, give me that mountain!"

December

Morris Cerullo overlooks the wheat fields in Israel. The land is ours to possess! God is raising His Victorious Army up to reap the great end-time harvest in the nations of the world – and you are part of that army!

December
Possessing The Land

In Joshua chapters 22 and 23, Joshua gives final instructions and a challenge of faith to the nation of Israel prior to his death. He summarizes their wilderness journey from Egypt to the Promised Land, their regression to the desert for forty years of wandering, and their return to Canaan to claim their rightful inheritance. He reminded Israel:

> *And ye have seen all that the Lord your God hath done unto all these nations because of you; for the Lord your God is he that hath fought for you.*
> *(Joshua 23:3)*

As we come to the final month of this year, we too can look back over the weeks and clearly see that the Lord, our God, has fought for us. He has helped us through the difficult times. He has given us safe passage through the storms of doubt and fear. He has comforted us in times of loneliness and bereavement. As Joshua told Israel:

> *And...ye know in all your hearts and in all your souls, that not one thing hath failed of all the good things which the Lord your God spake concerning you; all are come to pass unto you, and not one thing hath failed thereof.*
> *(Joshua 23:14)*

Our devotions during this final month of the year focus on a tremendous prophecy God gave me some time ago regarding a powerful spiritual force which is about to be released within the Body of Christ that will bring about the greatest manifestation of the power of God the world has ever seen.

NOW is the time for the fulfillment of this promise. We need the manifested presence of God more than ever as we prepare to face another year and to enter into the closing days of time. During this month we will focus on the manifested presence of God and we will learn powerful principles of leadership that will equip us to rise up on the cutting edge of God's end-time, harvest time cycle.

It all begins with the prophecy God gave to me...

A Powerful Spiritual Force

Scripture reading: Acts 1:1-8

Several years ago the Lord told us in a prophecy that…

> **"A powerful spiritual force is about to be released within the Body of Christ that will bring about the greatest manifestation of the power of God the world has ever seen."**

God has now revealed to me that this is the time for this prophecy to be fulfilled.

Incredible things are happening as we approach the return of our Lord Jesus Christ. One of these is the tremendous onslaught of Satan against the Body of Christ. Satan is causing every kind of doctrinal error and religious confusion that he possibly can. The words of the Prophet Hosea are being fulfilled, that God's people are being destroyed for lack of knowledge (Hosea 4:6).

The Holy Spirit does not bring confusion, frustration, fear, or doubt. It is only when Satan tries to intimidate and imitate that these things result. But there is one thing for which there is no imitation – and that is the ultimate spiritual weapon that will position us beyond every defeat. The enemy cannot imitate the presence of God.

This manifestation of God's glory is not coming on the church from without. It is not coming through a specific man or a denomination. It is not coming through organization or education. This manifestation of God's glory is a powerful spiritual force being released from within the Body of Christ. It will be greater than ever experienced – including the time of the early Church.

When the Church of Jesus Christ was born 2,000 years ago…

…It was not born weak or spiritless.

…It was not founded on education, doctrine, or denomination.

…It was not advanced by profound oratory.

…It was not formed by any influence of man.

The Church was born by a demonstration and manifestation of the Third Person of the Trinity. Jesus said, *"You shall receive power after that the Holy Ghost is come upon you!"* (Acts 1:8).

Partner, the trumpet of God is about to sound. If you have spiritual ears, you can already hear it. It is proclaiming the soon coming of our Lord Jesus Christ. But before Christ returns, this Church will experience a manifestation of power greater than what it was born in. The Church will be raptured in a greater demonstration of power than that in which it was conceived!

Make this declaration: "I will be raptured in a greater demonstration of power than that in which the early Church was conceived."

December 2

God Inhabits The Body

Scripture reading: John 14:26-31

When God gave birth to the Church, the greatest miracles that had happened from the beginning of time began to take place. When God birthed the Church, He didn't birth another organization, denomination, or religious institution. When God birthed the Church, for the first time since the beginning of time, God actually came to inhabit the physical body of mankind.

Prior to the inception of the Church, the Spirit of God had anointed men and women for specific periods of time. The Spirit of God had been with His people. But now, for the first time in history, His Spirit actually came to dwell within His people:

> *And I will pray the Father, and he shall give you another Comforter, that he may abide with you for ever; Even the Spirit of truth; whom the world cannot receive, because it seeth him not, neither knoweth him: but ye know him; for he dwelleth with you, and shall be in you.* (John 14:16-17)

Jesus told the disciples, "He dwells with you, but He will be in you!" For the first time, God now lives in us!

> *But we have this treasure in earthen vessels, that the excellency of the power may be of God, and not of us.* (2 Corinthians 4:7)

The world is in a desperate condition spiritually. Only eight percent of Great Britain's population attends church. Only two percent of all Europe goes to church and a scant two percent of people in the 10/40 Window go to church.

Here in America, we close down over 30,000 churches a year. Last year, there was virtually no church growth in America – it was less than of one percent. What might be called "church growth" was actually transfers of membership, not new conversions.

This world is in desperate spiritual condition and there is only one thing that will meet these challenges. Desperate times call for desperate measures. We must come to the place where we get desperate and declare…

> "God, I will not go one step further without Your presence. If Your presence doesn't go with me, if the manifestation of Your glory doesn't go with me, I am not going. We don't want any more self help. We've tried it. We don't want any more human effort. It has failed. We don't want any more seminars on how to do it. God, if Your glory doesn't come, we don't want to go a step farther!"

As Israel when they went up to possess their land, we must have the presence of God. We can no longer face our daily circumstances without it. We cannot go to battle without it. We cannot reach a dying world without it. We must have God's presence!

Make this declaration: "I will not go one step further without the manifestation of God's presence."

One Principle Of Operation

Scripture reading: Exodus 33:1-11

Moses was convinced that without God's presence in his life, it was useless for him to attempt anything. When God called him to lead His people to the border of the Promised Land, Moses prayed…

> *"See, thou sayest unto me, Bring up this people: and thou hast not let me know whom thou wilt send with me. Yet thou hast said, I know thee by name, and thou hast also found grace in my sight. Now therefore, I pray thee, if I have found grace in thy sight, shew me now thy way, that I may know thee, that I may find grace in thy sight: and consider that this nation is thy people." And He said, "My presence shall go with thee, and I will give thee rest." And Moses said unto him, "If thy presence go not with me, carry us not up hence."*
>
> *(Exodus 33:12-15)*

Moses knew he must have the presence of God if he was to fulfill his destiny. Today, we have so many programs, so much technology, ritual, and form that we think we can do without the presence of God and substitute a form of godliness that denies the power of God!

But we must have God's presence! Why? So that the world, the ungodly, and every enemy of the Church will know that there is something different about us. This was the premise upon which Moses based his plea for God's presence:

> *"For wherein shall it be known here that I and thy people have found grace in thy sight? Is it not in that thou goest with us? So shall we be separated, I and thy people, from all the people that are upon the face of the earth." And the Lord said unto Moses, "I will do this thing also that thou hast spoken: for thou hast found grace in my sight, and I know thee by name."*
>
> *(Exodus 33:16-17)*

Moses said, "I'm not going to take a single step unless I know that you are with me." Moses knew what we must come to know. There was only one thing that set Israel apart. It wasn't their army, their wealth, or their religious or educational systems. There was only one thing that set them apart and made them special…the presence of the living God.

Have you ever wondered what would happen to the Church if it was governed by the presence of God instead of a bunch of committee members sitting around in some religious institution or denomination making a bunch of rules?

Moses declared that Israel would be governed by one principle alone. They would operate on one principle alone…and that was the presence of God!

Make this declaration: "I will live my life and fulfill my ministry on one principle alone – the manifested presence of God."

The Manifested Presence Of God

Scripture reading: Exodus 33:12-23

The Church is going to experience the manifested Presence of God. God promised us that...

> **"A powerful spiritual force is about to be released within the Body of Christ that will bring about the greatest manifestation of the power of God the world has ever seen."**

What is this powerful spiritual force God is releasing? It is a new life flow that is going to be manifested in us, the life we have allowed to lie dormant for such a long time, not realizing that the very Presence of the living God dwells in us!

God told Moses, "My Presence will go with you!" What a tremendous promise! My Presence, the great I AM, the Creator of Heaven and earth, My Presence...I am going with you!" With a promise like that...

...We can face every army.

...We can face every enemy.

...We can face every onslaught . . .

...because God is going with us! His Presence is our ultimate weapon.

The enemy can stand up against advertising, organization, and denomination – but the enemy cannot stand against the presence of God. God's presence is real. It is not an imitation. The enemy can imitate everything, but he cannot imitate the presence of God. It is impossible!

God's Presence was so evident in Abraham's life that even ungodly people like the heathen King Abimelech recognized it. In Genesis 21:22 he declared, *"God is with thee in all that thou doest."* This heathen king didn't know the God of Israel. He only knew there was something different about Abraham. God's Presence rested upon Abraham and it set him apart.

Prior to entering Canaan, God promised Joshua that no enemy would be able to stand against him. He gave Joshua an ultimate weapon that could not be defeated. God told Joshua:

> *"There shall not any man be able to stand before thee all the days of thy life: as I was with Moses, so I will be with thee: I will not fail thee, nor forsake thee. Be strong and of a good courage: for unto this people shalt thou divide for an inheritance the land, which I sware unto their fathers to give them."*
>
> *(Joshua 1:5-6)*

What was Joshua's ultimate weapon? It was the Presence of God! The Lord told Joshua, "No power of the enemy can stand against you when my Presence is upon you! No enemy can harm you." The enemy cannot penetrate the ultimate weapon of God's Presence.

Make this declaration: "By faith, I am taking my eyes off of my circumstance and refocusing my attention on the manifested Presence of God."

Accompanied By God's Presence

Scripture reading: Joshua 2:1-24

When God raised up the nation of Israel, there was something different about them and that was the Presence of God! The heathen world didn't have it, but they sure recognized it. They knew that there was something about Israel that set them apart and made them different.

When Israel was preparing to cross the border into their Promised Land, the harlot Rahab told them:

> *...I know that the Lord hath given you the land, and that your terror is fallen upon us, and that all the inhabitants of the land faint because of you. For we have heard how the Lord dried up the water of the Red sea for you, when ye came out of Egypt; and what ye did unto the two kings of the Amorites, that were on the other side Jordan, Sihon and Og, whom ye utterly destroyed. And as soon as we had heard these things, our hearts did melt, neither did there remain any more courage in any man, because of you: for the Lord your God, he is God in heaven above, and in earth beneath.*
>
> *(Joshua 2:9-11)*

Rahab said, "We heard what God did for you. We know His Presence is with you and our hearts melted with fear because we know the Lord is your God!"

Friend, you may be going through something in your own life right now that seems like an impossible situation. Maybe the odds are stacked against you, you feel overwhelmed, and in the natural it seems as if there is no way you can win.

Have you ever stopped to think that maybe this is God's plan? Perhaps you are getting ready to experience the biggest victory of your life – just when you think you are facing your greatest defeat! Don't be afraid of the odds! God's people have been defying the odds down through the centuries!

God is not indiscriminately handing out assignments in His army today. He is seeking men and women of valor, men and women of courage who can stand up to the enemy and say, "God made me a promise that His Presence would go with me. I'm going to win this battle!"

When you advance into battle accompanied by the Presence of God, no enemy can stand against you. Their hearts will melt, their courage will fail, their strategies will be defeated.

Make this declaration: "When I advance into battle accompanied by the Presence of God, no enemy can stand against me. Their hearts will melt, their courage will fail, their strategies will be defeated."

A Mighty Man Of Valor

Scripture reading: Judges 6:1-10

When the angel of the Lord appeared to a man named Gideon, he said to him, "The Lord is with thee, thou mighty man of valor" (Judges 6:12). I always laugh at this statement because Gideon was anything but a mighty man of valor at that time. He was threshing wheat – not on the open threshing floor where it was normally done – but by the winepress in order to hide from the Midianites.

When the angel told Gideon the Lord was with him, he responded with doubt:

> *"Oh my Lord, if the LORD be with us, why then is all this befallen us? And where be all his miracles which our fathers told us of, saying, Did not the LORD bring us up from Egypt? But now the LORD hath forsaken us, and delivered us into the hands of the Midianites."*
>
> *(Judges 6:13)*

This certainly doesn't sound like a "mighty man of valor" speaking, does it? Do you ever talk like that? "If the Lord is with me, then why has this befallen me?"

...Why did I lose my job?

...Why did my mate leave me?

...Why are my children on drugs?

In response to his question, the angel told Gideon:

> *"Go in this thy might, and thou shalt save Israel from the hand of the Midianites: have not I sent thee?"*
>
> *(Judges 6:14)*

The angel told Gideon, "There is something so mighty in you, so powerful that it can deliver Israel out of the hands of the Midianites." What was that might? It was the Presence of God!

It doesn't matter if you lost your job, your mate left you, or your children are on drugs. There is something so mighty within you that you can rise up to defeat the enemy.

I often wondered why God included the portion about how weak and fearful Gideon was in this biblical account, then God showed me that it was because He wanted to prove through Gideon what His Presence can do through anybody. God can take a failure, a doubter, one who is filled with fear and intimidation – and when His Presence comes upon them, they will become a mighty person of valor.

Gideon, who was hiding out threshing wheat when the angel appeared to him, was certainly not acting like a mighty man of valor. But God was not depending on what Gideon had to offer...

...He was not depending on Gideon's bravery.

...He was not depending on his fighting ability.

...He was not depending on Gideon's natural talents.

...And the same is true for you! God is not depending on your bravery, ability, or natural talents.

Make this declaration: "God is not depending on my bravery, my ability, or my natural talents. He is depending upon what He will make me through His mighty power."

When God Manifests His Presence

Scripture reading: Judges 6:11-40

When God called Gideon He was not depending on what He was, rather what He knew He could make of him! God called Gideon a mighty man of valor not because of who he was, but because of who he would become. God does not see you as you are. He sees you as He can make you, what you will become when you are empowered by His divine Presence.

God always manifests His Presence for a purpose. Immediately after you have such an experience with God, you are usually called upon to do something that will take supernatural courage, wisdom, and insight. This is exactly what happened to Gideon.

> *Then all the Midianites and the Amalekites and the children of the east were gathered together, and went over, and pitched in the valley of Jezreel. But the Spirit of the LORD came upon Gideon, and he blew a trumpet; and Abi-ezer was gathered after him. And he sent messengers throughout all Manasseh; who also was gathered after him: and he sent messengers unto Asher, and unto Zebulun, and unto Naphtali; and they came up to meet them.*
>
> *(Judges 6:33-35)*

Immediately after God's Presence came upon Gideon, a great army came against Israel. Instead of retreating in fear, when the Presence of the Lord came upon Gideon, he blew the trumpet and called all of Israel to battle. When God's Presence is upon your life, you not only will be able to stand against the enemy, you will be able to motivate and mobilize others to fight as well.

When the army was amassed, Gideon said to God…

> *"If thou wilt save Israel by mine hand, as thou hast said, Behold, I will put a fleece of wool in the floor; and if the dew be on the fleece only, and it be dry upon all the earth beside, then shall I know that thou wilt save Israel by mine hand, as thou hast said." And it was so: for he rose up early on the morrow, and thrust the fleece together, and wringed the dew out of the fleece, a bowl full of water. And Gideon said unto God, "Let not thine anger be hot against me, and I will speak but this once: let me prove, I pray thee, but this once with the fleece; let it now be dry only upon the fleece, and upon all the ground let there be dew." And God did so that night: for it was dry upon the fleece only, and there was dew on all the ground.*
>
> *(Judges 6:36-40)*

This passage recounts the story of the famous fleece Gideon set out to make sure he had heard from God. We must keep in mind that Gideon was not just convincing himself of God's promise that His Presence would go with them, he was also convincing the men of Israel who had gathered together for battle. They wanted to be assured of God's Presence before they advanced into enemy territory.

Make this declaration: "God's Word assures me that He is with me as I advance into enemy territory today."

When The Odds Are Stacked Against You

Scripture reading: Judges 7:1-25

When Gideon and his men had received complete assurance of God's presence, the Lord did one more thing to be sure they would rely on Him alone. He reduced the size of their army to three-hundred men (Judges 7:1-7). Then...

> *...they stood every man in his place round about the camp: and all the host ran, and cried, and fled. And the three hundred blew the trumpets, and the LORD set every man's sword against his fellow, even throughout all the host: and the host fled to Bethshittah in Zererath, and to the border of Abelmeholah, unto Tabbath. And the men of Israel gathered themselves together out of Naphtali, and out of Asher, and out of all Manasseh, and pursued after the Midianites."*
>
> *(Judges 7:21-23)*

Gideon went to battle against the massive army of Midian with only three-hundred men! Only a man convinced of the Presence of God could do that. Instead of asking for proof, Gideon was now producing the proof!

How did Gideon change from fear to faith, hiding to advancing? Where did he get this kind of boldness? Gideon had experienced the manifested Presence of God.

...God had changed Gideon from a frightened young man to one who had so much courage that he was not afraid to fight the innumerable enemy with only a few hundred men.

...And these few hundred men had so much confidence in Gideon that they were willing to fight alongside him against overwhelming odds because they sensed the Presence of God upon his life.

God's Presence will empower you to march into battle against every foe...

...The stronghold of the enemy will be destroyed!

...Financial problems will be eliminated!

...Sickness and disease will be conquered!

...Family problems will be solved!

The enemy cannot stand against the Presence of God!

The odds were so stacked against Gideon and his troops that there was no way they could win unless God was with them. In the end, however, an awesome victory was won by Gideon and his three hundred men.

Make this declaration: "The odds may be stacked against me, but every stronghold of the enemy will be destroyed because God is with me."

Satan Will Not Prevail

Scripture reading: Jeremiah 1:1-19

God has provided us with powerful resources for spiritual battle. The Apostle Paul itemizes these in Ephesians 6:10-18.

The belt of truth...the breastplate of righteousness...the helmet of salvation...feet shod with the preparation of the Gospel...the sword of the Spirit...and prayer: These are tremendously powerful weapons for spiritual battle.

But God also wants you to have the ultimate weapon – a weapon so powerful that you will never experience another defeat in your life. He wants you to experience His Presence.

When God commissioned the Prophet Jeremiah, He warned him that the nation to which he was called would ultimately reject him because of his prophetic ministry:

> *"Before I formed thee in the belly I knew thee; and before thou camest forth out of the womb I sanctified thee, and I ordained thee a prophet unto the nations." Then said I, "Ah, Lord God! Behold, I cannot speak: for I am a child." But the Lord said unto me, "Say not, I am a child: for thou shalt go to all that I shall send thee, and whatsoever I command thee thou shalt speak. Be not afraid of their faces: for I am with thee to deliver thee, saith the Lord...For, behold, I have made thee this day a defenced city, and an iron pillar, and brasen walls against the whole land, against the kings of Judah, against the princes thereof, against the priests thereof, and against the people of the land. And they shall fight against thee; but they shall not prevail against thee; for I am with thee, saith the Lord, to deliver thee."* *(Jeremiah 1:5-8 and 18-19)*

Jeremiah told God, "I'm not able to speak." God said, "You'll not only be able to speak, you'll speak to whoever I send you – kings, princes, priests, and commoners – and even though they will fight against you, They will not prevail!"

Why?

"Because I am with thee!"

...Partner, when the enemy rises up against you in your family, on your job, or attacks your finances, he will not prevail!

...Pastor, when people rise up against you and say you shouldn't address sin in people's lives, don't listen to them. They won't prevail!

God told Jeremiah, "Speak the truth! Don't compromise! It doesn't matter if the whole country turns against you. If My Presence is with you, go ahead and do it!"

When God's Presence is with you, you will have no fear. You will not be intimidated. You will not be afraid of wicked governments. You won't be concerned if they say preaching is against the law. You won't be concerned if they say you can't print religious material or convert people to the Lord Jesus Christ. The Presence of God is your ultimate weapon against intimidation!

Make this declaration: "The Presence of God is my ultimate weapon against intimidation! I will not be intimidated by the enemy."

Walking Through The Floods

Scripture reading: Romans 8:1-39

When the Presence of God is manifested in your life, you will not be afraid of what people say about you. You won't worry about what the world says. People won't be able to affect you, cut you down, or destroy you emotionally by their opinions, comments, or actions.

Many of you are walking around burdened down by guilt and condemnation over your past. The Presence of God breaks the yokes of guilt and condemnation because...

> *There is therefore now no condemnation to them which are in Christ Jesus, who walk not after the flesh, but after the Spirit.* *(Romans 8:1)*

There is no condemnation. It doesn't exist when the Presence of God rests upon you, because if God doesn't condemn you, then no one else can either...no man, no woman, no demon from Hell! There is no condemnation, no intimidation, no fear!

God has promised...

> *"When thou passest through the waters, I will be with thee; and through the rivers, they shall not overflow thee: when thou walkest through the fire, thou shalt not be burned; neither shall the flame kindle upon thee."*
>
> *(Isaiah 43:2)*

When God's Presence is with you, you may pass through deep waters, but they will not overflow you. You won't drown. You may be forced to walk into the fiery furnace, but you won't be burned. You may be thrown into a lion's den, but the animals will not harm you. They may prowl around, roar, and try to intimidate – but they can't harm you.

When God's Presence is with you, you will be able to face the false prophets of Baal and emerge victorious. You will be able to confront criticizing, backbiting, rebellious people. You will be able to courageously advance into enemy territory with authority because you know you cannot be defeated.

...In prisons.

...In the ghetto.

...On gang turf.

...In the drug dens...

You can march right in and rescue people from the very gates of Hell because you have the ultimate weapon: God's divine Presence!

With God's Presence abiding in you, you can go through any flood, you can go through any fire, you can go through any trial...and you will not merely survive, you will come through victorious because His Presence is with you!

Make this declaration: "I will not merely survive – I will come through victorious because God's Presence is with me."

Ultimate Victory

Scripture reading: Psalm 68:1-35

In order to experience ultimate victory in spiritual warfare, you must be properly positioned. Some of us don't experience this kind of victorious Christian living because our minds are brainwashed and cluttered with negative things. Many of us are products of our environment. Many are held captive to the past. Others are bound by tradition, religion, culture, and ritual.

But God is bringing His people into a new position of power, a stance of ultimate victory. Put this in your spirit: Times are important to God. Seasons are important to God. God chooses specific times and seasons to do specific things. The Bible says, *"To every thing there is a season, and a time to every purpose under the heaven"* (Ecclesiastes 3:1).

God is a God of plan, purpose, design, and objectivity. There is not one plan that God has formulated under heaven that He hasn't assigned a specific time to manifest it. Partner, I'm telling you...Now is the time for the manifestation of God's Presence!

But if we are going to experience the real manifestation of God, there must be a cry that resounds inside our spirits that echoes that of Moses, who declared: *"If thy presence go not with me, carry us not up hence"* (Exodus 33:15).

Spiritual goose bumps are not going to get the job done. Jesus didn't promise goose bumps, dances, giggles, and people being slain in the Spirit. Thank God for all of that...it is all manifested in my own life and ministry. God gave us emotions and proper emotional responses to God are biblically sound. But there is something more!

We must go past the point of blessing – beyond goose bumps, dances, giggles, and being slain in the Sprit. Very few people go beyond this point of blessing. Very few press on in to get their spiritual breakthrough. But beloved, as these perilous end times approach, we have no choice. We must go past the point of blessing!

Just look at the condition of the Church and you will be totally convinced that we must press in. If there is nothing more than what we are seeing manifested through the Church today, we are in trouble. If there is nothing more than what we are experiencing under the so-called anointing of some men and women today, we are in big trouble. But God has a people who are crying out like Moses for His presence. Oh God, we must have it!

Make this declaration: "Oh God, I must have Your Presence manifested in my life!"

Positioned For Victory

Scripture reading: Psalm 19:1-14

In times past, God manifested His Presence:
In creation: Psalms 19:1
In earth: Isaiah 6:3
In the heavens: Psalms 57:11
In specific geographic locations: Ezekiel 3:23; 8:4
In the tabernacle: Exodus 29:43
In a cloud: Exodus 40:34
In the temple: Ezekiel 43:5
But now God is manifesting His glory from within...

"A powerful spiritual force is about to be released within the Body of Christ that will bring about the greatest manifestation of the power of God the world has ever seen."

Remember...the Presence of God is not just coming on us. It is not just going to stir around us. It is not going to appear in a specific geographic location. It is going to flow, like a river...a bubbling spring...from our innermost spirits. And when it does – hear me now – it will bring about the greatest manifestation of the power of God that this world has ever seen!

Partner, that stirring inside of you is not of its own making. That hunger, that divine dissatisfaction, that crying out of your innermost being is not something you have instigated. It can't be satisfied with hobbies, entertainment, or possessions. God is powerfully moving within you and deep is calling unto deep. God is saying, "Now is the time I want to manifest Myself through you."

This Presence of God will take us to places we never thought we would go. His Presence says, "Follow me and I will guide you. I will preserve you" – and we are rising up to respond in faith...

...Even when it doesn't make sense in the natural.

...Even when everyone around us is advising us not to rock the boat and to be satisfied with the status quo.

...Even when the enemy looms ahead on the horizon upon our God-ordained path.

The Presence of God is taking away all fear, all intimidation, all condemnation and we are ready for something to happen in and through us! We are positioned for victory!

Make this declaration: "A powerful spiritual force is about to be released within me that will bring about the greatest manifestation of the power of God the world has ever seen."

EXPERIENCING GOD'S MANIFESTED PRESENCE

Scripture reading: 2 Chronicles 14:1-15

You may be saying, "Yes Brother Cerullo…I want it. My heart is crying out for the Presence of God. But I don't know how to experience God's manifested Presence."

There is a tremendous key God has given to enable us to experience and maintain the Presence of God. It is found in the Old Testament account of King Asa. King Asa ruled over Judah for forty-one years (1 Kings 15:9-10). The Bible records that…

> *…Asa did that which was good and right in the eyes of the Lord his God: For he took away the altars of the strange gods, and the high places, and brake down the images, and cut down the groves: And commanded Judah to seek the Lord God of their fathers, and to do the law and the commandment.*
> *(2 Chronicles 14:2-4)*

At one point during his administration, a mighty army from Ethiopia rose up against him:

> *And there came out against them Zerah the Ethiopian with an host of a thousand thousand, and three hundred chariots; and came unto Mareshah.*
> *(2 Chronicles 14:9)*

Can you imagine? A million-man army marching against the small troops of Judah. When Asa saw this great multitude advancing…

> *…Asa cried unto the Lord his God, and said, "Lord, it is nothing with thee to help, whether with many, or with them that have no power: help us, O Lord our God; for we rest on thee, and in thy name we go against this multitude. O Lord, thou art our God; let not man prevail against thee."*
> *(2 Chronicles 14:11)*

The enemy was advancing. Asa could hear the sound of a million marching feet. In this time of desperation, Asa turned to God and cried "Help us, Oh Lord…for we rest on thee, and in thy name we go against this multitude!" Asa declared his total, utter dependence upon God. Without God's Presence, he knew he was in trouble. But he also knew that with God's Presence, no man could prevail against him because no man can prevail against God.

> *So the LORD smote the Ethiopians before Asa, and before Judah; and the Ethiopians fled.*
> *(2 Chronicles 14:12)*

Asa led the armies of Judah to a great victory over Ethiopia's million-man army. Can you just imagine King Asa's puny little troops advancing against the million-man force of the Ethiopians? Asa declared that there was only one way he was prepared to face the enemy and that was through the Presence of God. The Presence of God went out before Asa and his forces and scattered the enemy before they even got to the battlefield!

Make this declaration: "God's Presence will scatter my enemies before I can even get to the battlefield."

God Is By Your Side

Scripture reading: 2 Chronicles 15:1-19

Are you ready to meet God's condition for the kind of victory King Asa experienced in Old Testament times? Are you ready to meet God's condition to receive this ultimate weapon? Are you ready to have a force so powerful flowing out of you that the enemy scatters before you even show up to fight?

Now you may not have a million men marching against you, but your battles are just as big…

…A mountain of debt.

…An unsaved mate.

…An incurable illness.

…Unsaved, rebellious children.

…But if you meet God's condition, the power of God's Presence will go ahead of you and scatter your enemy.

The condition for releasing this inner life flow of God's Presence is found in 2 Chronicles 15:1-2. As Asa was returning from battle, a prophet named Azariah came out to meet him:

> *And the Spirit of God came upon Azariah the son of Oded: And he went out to meet Asa, and said unto him, "Hear ye me, Asa, and all Judah and Benjamin; The Lord is with you, while ye be with him; and if ye seek him, he will be found of you; but if ye forsake him, he will forsake you."*
> *(2 Chronicles 15:1-2)*

If you want to experience the manifestation of God's Presence, you must seek Him. Azariah told Asa, *"If you seek Him, He will be found…The Lord is with you, while you are with Him."*

King David understood the importance of diligently seeking God for the manifestation of His Presence. He declared:

> *I have set the Lord always before me: because he is at my right hand, I shall not be moved.*
> *(Psalm 16:8)*

Do you really comprehend the dynamic manifestation of power that will occur when you set God at your right hand? Not self. Not man. Set God at your right hand – seek Him and give Him priority in your life – and you will never be moved.

When I speak of seeking God, I am not talking about going through the formality of little pre-printed prayers or lighting a prayer candle. If we want the manifestation of God's glory, it is time for us to set ourselves before God and seek His face earnestly and continually.

When you have the Presence of God with you and you confront the enemy, someone is going to move – but it won't be you. You won't back up in your Christian experience. You won't back down in your convictions. You won't compromise. You won't let Satan control your family. You won't let down on your spiritual commitments.

It doesn't matter what the circumstances are, you will not be moved. Why? Because God is right by your side!

Make this declaration: "I will not be moved by my circumstances because God is right by my side."

We Must Have His Presence

Scripture reading: Ephesians 3:9-21

When God birthed this Church, He didn't birth another club or organization. He didn't birth the Protestants or the Catholics. When God birthed this Church His intention was to generate a dynamic force, a supernatural power that would never experience any limits. God was literally making a deposit of Himself, through Jesus and the Holy Spirit, into our bodies.

In the beginning, God walked and talked with Adam. In the Old Testament record, God came upon the prophets and judges to manifest Himself and then after the work was done His Presence would lift. But when the Church was birthed, God Himself came to indwell His Body:

> *But the anointing which ye have received of him abideth in you, and ye need not that any man teach you: but as the same anointing teacheth you of all things, and is truth, and is no lie, and even as it hath taught you, ye shall abide in him.* *(1 John 2:27)*

The anointing abides in you permanently. Here is how the Amplified Version translates this verse:

> *But as for you, the anointing (the sacred appointment, the unction) which you received from Him abides (permanently) in you...* *(1 John 2:27)*

Christ abides in you permanently. His anointing abides in you permanently. His Presence is within you – and it is there to stay!

...When you battle sickness or disease in your body, you know His Presence is there.

...When you face a financial crisis – you have a mountain of debt and it keeps piling up God's Presence is right at your side.

...When you are in danger, you don't fear because you know God's Presence is there.

...When you are battle fatigued, worn, and weary, you are not dismayed because you know God's Presence is with you.

God's Presence equips us with saving and keeping power. It provides guiding power. His Presence strengthens us and separates us from the world. It heals, renews, and cleanses us...It is unlimited power! That is why we must have His Presence!

Make this declaration: "Thank You God, for Your Presence which saves and keeps me, provides guidance, heals, renews, and cleanses me, strengthens and separates me from the world."

A Revelation Of God

Scripture reading: Exodus 34:1-8

In answer to his heart's cry, Moses received confirmation of God's Presence. God said:

> *"...My presence shall go with thee, and I will give thee rest."*
>
> (Exodus 33:14)

But God's Presence wasn't enough to satisfy the cry of Moses' heart. He didn't stop there. Moses pressed on in and boldly asked, *"...I beseech thee, shew me thy glory"* (Exodus 33:18).

There is a difference between beholding God's Presence and possessing the glory. Many people go their entire Christian lives without experiencing either of these. A few press on beyond the point of blessing to experience God's Presence and power. Very few, however, persevere to receive a revelation of God's glory.

Moses wanted more than just God's Presence – His guidance, provision, blessing, and power. This wasn't enough for him. He wanted a manifestation of God's glory. He knew there was something more than His Presence and so he cried out, "Show me thy glory!"

> *And he [God] said, "I will make all my goodness pass before thee, and I will proclaim the name of the LORD before thee; and will be gracious to whom I will be gracious, and will shew mercy on whom I will shew mercy." And he said, "Thou canst not see my face: for there shall no man see me, and live." And the LORD said, "Behold, there is a place by me, and thou shalt stand upon a rock: And it shall come to pass, while my glory passeth by, that I will put thee in a cleft of the rock, and will cover thee with my hand while I pass by: And I will take away mine hand, and thou shalt see my back parts: but my face shall not be seen."*
>
> (Exodus 33:19-23)

God told Moses to stand upon a rock in a place by Him and His glory would be revealed. Partner, today we stand upon that Rock, Christ Jesus, Who is the greatest revelation and manifestation of the glory of God that the world has ever seen! Only when we stand there – upon that Rock – will we receive a true revelation of God's glory.

When God's glory was revealed to Moses...

> *...The LORD descended in the cloud, and stood with him there, and proclaimed the name of the LORD. And the LORD passed by before him, and proclaimed," The LORD, The LORD God, merciful and gracious, longsuffering, and abundant in goodness and truth, Keeping mercy for thousands, forgiving iniquity and transgression and sin, and that will by no means clear the guilty; visiting the iniquity of the fathers upon the children, and upon the children's children, unto the third and to the fourth generation." And Moses made haste, and bowed his head toward the earth, and worshiped.*
>
> (Exodus 34:5-8)

The manifestation of God's glory always results in a greater revelation of God Himself. God proclaimed His name to Moses, which means He revealed the very essence of Who He was.

Make this declaration: "Through His manifested Presence, God is revealing to me the very essence of who He is."

The Glory Of God

Scripture reading: Matthew 17:1-6

"Glory" is one of the richest words of the English language. No single word can serve as a good synonym, but here are some words that describe it: Honor, praise, splendor, radiance, power, exaltation, worthiness, likeness, beauty, renown, rank.

When the word "glory" is used of God it is an expression of His divine essence. The glory of God is His displayed excellence, beauty, majesty, power, and perfection of His total being. The revelation of God's glory will always inspire us to new levels of worship – and that is why Moses bowed his head and worshiped God.

The disciples witnessed a manifestation of God's glory at the Mount of Transfiguration:

> *And after six days Jesus taketh Peter, James, and John his brother, and bringeth them up into an high mountain apart, And was transfigured before them: and his face did shine as the sun, and his raiment was white as the light. And, behold, there appeared unto them Moses and Elias talking with him. Then answered Peter, and said unto Jesus, "Lord, it is good for us to be here: if thou wilt, let us make here three tabernacles; one for thee, and one for Moses, and one for Elias." While he yet spake, behold, a bright cloud overshadowed them: and behold a voice out of the cloud, which said, "This is my beloved Son, in whom I am well pleased; hear ye him."* *(Matthew 17:1-5)*

God's glory wasn't in Moses. It wasn't in Elijah. It wasn't in the spectacular light that encased the mountain. It wasn't in Peter's emotional response to the event – wanting to build tabernacles to memorialize the occasion.

God spoke from Heaven and made it clear: The glory was in Jesus: *"This is my beloved Son, in whom I am well pleased; hear ye him"* (Matthew 17:5).

Jesus is the greatest demonstration of God's manifested glory that the world has ever seen! The disciples affirmed this when they declared, *"We beheld His glory"* (John 1:14).

God's glory is manifested in every aspect of Christ's life, past, present, and future:

- His works manifested God's glory: John 2:11
- Jesus was raised from the dead in glory: 1 Peter 1:11
- He was received into glory: 1 Timothy 3:16
- He will come in glory: Matthew 16:27
- He will sit on a throne of glory: Matthew 19:28
- He will be crowned with glory: Hebrews 2:9
- He is called the Lord of glory: James 2:1

Make this declaration: "God's glory dwells within me and is manifested through me."

A Reflection Of God's Glory

Scripture reading: John 17:1-5

Key verse:
In the final hours of His life on this earth, Jesus prayed…

"…Father, the hour is come; glorify thy Son, that thy Son also may glorify thee: As thou hast given him power over all flesh, that he should give eternal life to as many as thou hast given him. And this is life eternal, that they might know thee the only true God, and Jesus Christ, whom thou hast sent. I have glorified thee on the earth: I have finished the work which thou gavest me to do. And now, O Father, glorify thou me with thine own self with the glory which I had with thee before the world was."

(John 17:1-5)

Jesus possessed God's glory before the world began. He glorified God in His ministry on earth and God revealed His glory in and through Him. Now, just before the end of His time on earth, Jesus wanted to give His disciples the same glory that He possessed: *"And the glory which thou gavest me I have given them"* (John 17:22). The word "glory" as used here is *doxa,* meaning "all that I am and all that I have!"

Jesus gave us all that He is, all that He possesses…

…He gave us the Father.

…He gave us the Holy Spirit.

…He gave us Himself.

…He gave us the ability to overcome sin.

…He gave us authority over the enemy.

…He gave us the power to heal the sick and cast out demons.

…All that He is, all that He has – He gave to us!

We are a work in process. The glory of God is in the process of being manifested in us:

But we all, with open face beholding as in a glass the glory of the Lord, are changed into the same image from glory to glory, even as by the Spirit of the Lord.

(2 Corinthians 3:18)

We are being changed. We are going to a new level of God's glory. We are wading deeper into the River of God. Our lives are becoming a reflection of God's glory – all that He is, all that He possesses.

But we have this treasure in earthen vessels, that the excellency of the power may be of God, and not of us.

(2 Corinthians 4:7)

Make this declaration: "I am a work in progress. The glory of God is in the process of being manifested in me."

My Soul Thirsts For You

Scripture reading: Psalm 63:1-11

King David said…

> *One thing have I desired of the Lord, that will I seek after; that I may dwell in the house of the Lord all the days of my life, to behold the beauty of the Lord, and to inquire in his temple.* *(Psalm 27:4)*

David used different words, yet he echoed the same cry as that of Moses – *"Show me Your glory!"* David wasn't intent on acquiring wealth. He wasn't intent on his position or gaining more earthly power. His heart was set on one thing: A revelation and manifestation of God's glory!

Our time and attention are consumed by many things…

…Material possessions.

…Hobbies.

…Entertainment…

But David focused on one thing and this singleness of purpose permeates his Psalms:

> *O God, thou art my God; early will I seek thee: my soul thirsteth for thee, my flesh longeth for thee in a dry and thirsty land, where no water is.* *(Psalm 63:1)*

The Amplified Version says, *"earnestly will I seek thee."* Earnest seeking is effectual prayer. It is fervent and combative. It is the kind of prayer that will penetrate every enemy line to attain its ultimate purpose: A revelation of God's glory.

This verse reveals what must happen to us in order for the revelation of God's glory to be manifested. Our souls must seek God earnestly and we must thirst for God as though we were in a dry and barren desert. When you are really thirsty, nothing else but water will satisfy. When we are really thirsty for God, nothing else will satisfy.

This cry doesn't come from your head. It is not mere intellectual assent. It isn't based on emotion, because emotions fluctuate. It doesn't come from praying out of duty, ritual, or guilt.

This cry emanates from your innermost being. Everything within you cries out: My soul thirsts for you!

> *With my soul have I desired thee in the night; yea, with my spirit within me will I seek thee early…* *(Isaiah 26:9)*

…And Isaiah admonishes us…

Seek ye the Lord while he may be found, call ye upon him while he is near. (Isaiah 55:6)

Make this declaration: "When God says 'Seek my face," His face I will seek. I will call upon him while He is near."

The Fellowship Of Suffering

Scripture reading: Philippians 3:1-21

In Song of Solomon, the cry of the bride, who is symbolic of the Church, was...

> *I will rise now, and go about the city in the streets, and in the broad ways I will seek him whom my soul loveth...* *(Song of Solomon 3:2)*

Centuries later, the Apostle Paul echoed the same cry when he cried out:

> *That I may know him, and the power of his resurrection, and the fellowship of his sufferings, being made conformable unto his death.* *(Philippians 3:10)*

When Paul uttered these words, it was 64 A.D. and he was in the closing days of his ministry. Around three years later he was martyred for the cause of Christ. Yet here, at the end of his ministry, Paul is crying out, *"That I may know Him."*

...Yes, Paul was saved.

...Yes, he was a minister.

...Yes he had done mighty signs and wonders in the name of Jesus.

But we are talking about something deeper. We are talking about a revelation of God's glory which comes through the power of His resurrection and the fellowship of His suffering.

When you move beyond the point of blessing into this manifestation of God's Presence and glory of which I am speaking, you will know no defeats.

This doesn't mean you will not experience problems. Just look at the things the Apostle Paul endured...

> *Of the Jews five times received I forty stripes save one. Thrice was I beaten with rods, once was I stoned, thrice I suffered shipwreck, a night and a day I have been in the deep; In journeyings often, in perils of waters, in perils of robbers, in perils by mine own countrymen, in perils by the heathen, in perils in the city, in perils in the wilderness, in perils in the sea, in perils among false brethren; In weariness and painfulness, in watchings often, in hunger and thirst, in fastings often, in cold and nakedness. Beside those things that are without, that which cometh upon me daily, the care of all the churches.*
>
> *(2 Corinthians 11:24-28)*

Make this declaration: "I want to know Him, in the power of His resurrection and the fellowship of his sufferings, being made conformable unto His death."

No Limits

Scripture reading: 2 Timothy 4:1-8

Paul didn't faint when afflictions came. He didn't falter. He didn't retreat. Despite all he suffered, Paul viewed these things as "light." I wonder what he would call heavy? We think we are under heavy attack when someone looks at us the wrong way.

The reason Paul knew no defeats is because he kept his focus on the glory. He said our trials work for us as long as our focus is on what is eternal – the glory – instead of temporal issues. If you want your problems to work for you, get your eyes off of your circumstance and get them on the eternal. Focus on the glory!

At the end of his life, Paul declared:

> *I have fought a good fight, I have finished my course, I have kept the faith: Henceforth there is laid up for me a crown of righteousness, which the Lord, the righteous judge, shall give me at that day...* *(2 Timothy 4:7-8)*

In spite of all the things Paul suffered, He finished his course. He kept the faith. He fulfilled his destiny. He was not defeated. He went out with a victory shout: *"I have kept the faith!"*

If you earnestly seek for this manifestation of God's glory, you will find it. Jesus promised:

> *Ask, and it shall be given you; seek, and ye shall find; knock, and it shall be opened unto you: For every one that asketh receiveth; and he that seeketh findeth; and to him that knocketh it shall be opened.* *(Matthew 7:7-8)*

God declared:

> *If my people, which are called by my name, shall humble themselves, and pray, and seek my face, and turn from their wicked ways; then I will hear from Heaven...* *(2 Chronicles 7:14)*

The life that God wants to operate in and through you will take you beyond the boundaries of your natural desires and experiences. It will move you out of your comfort zone. There will be no limits to the work of God in and through you. No horizons. No boundaries.

Your circumstances may be difficult. There may be darkness all around you. But the Word of the Lord to you today is:

> *Arise, shine; for thy light is come, and the glory of the LORD is risen upon thee. For, behold, the darkness shall cover the earth, and gross darkness the people: but the Lord shall arise upon thee, and His glory shall be seen upon thee.* *(Isaiah 60:1-2)*

Make this declaration: "I will arise and shine because the glory of God is risen upon me."

THE STANDARD OF EVALUATION

Scripture reading: Luke 6:43-45

The tremendous flow of the Presence of God will trigger an unprecedented spiritual harvest throughout the world. To prepare for this, God is raising up leaders of integrity – men and women of purpose – who will be able to work in cooperation with Him to reap this great end-time harvest.

This world is crying out for leaders of integrity. *The Daily Mail* periodical in London recently headlined an article, *"Where Is The Leader We So Badly Need?"* History reveals that many forms of government which started with good intentions ultimately failed – and in most cases the failure was due to the moral downfall of leadership.

You can be the leader God is seeking! You are a steward set over many valuable resources which He has entrusted to you. The greatest resource with which you have been entrusted, of course, is the Gospel. But you are also responsible for other things – finances, material things, and spiritual gifts – which God has given you.

The men and women God is seeking to be at the forefront of this end-time harvest are those who will use every resource at their disposal to fulfill His plan. God is not looking for educated men and women with natural abilities, charisma, or personality. He is seeking those who will be faithful to accomplish His purposes:

Moreover it is required in stewards, that a man be found faithful.
(1 Corinthians 4:2)

Jesus told a parable in Matthew 25:14-30 about servants whose master gave them resources called "talents" – a New Testament term for money. Those who were good stewards and used their funds wisely were judged faithful and rewarded. Those who did not use their resources properly were held accountable.

The qualifications God seeks as He is raising up leaders to reap this end-time harvest are not natural abilities, education, or a great resume. He is more concerned with your anointing, character, and conduct.

The test of any man, woman, or ministry is not spiritual gifts, power, or natural abilities which are evidenced. The standard of evaluation is spiritual fruit:

Wherefore by their fruits you shall know them...
(Matthew 7:20)

Spiritual fruit – qualities of character and conduct – reveals what a person is really like inside (Luke 6:43-45).

Make this declaration: "The Holy Spirit is developing the fruit of Christ-like character and conduct in my life."

Leading Like A Shepherd

Scripture reading: Matthew 7:21-27

A man may have tremendous charisma that can be easily mistaken for spiritual power. He may even do miracles in the name of the Lord, but Jesus said:

> *Not every one that saith unto me, Lord, Lord, shall enter into the Kingdom of Heaven; but he that doeth the will of my Father which is in Heaven. In that day will they not say unto me have we not prophesied in thy name? and in thy name have cast out devils? and in thy name done many wonderful works? And then will I profess unto them, I never knew you: depart from me, ye that work iniquity.* (Matthew 7:21-23)

The men and women who God sets in the forefront of these end times will be people of character and integrity as described in Titus 1 and 1 Timothy 3. They will have come under the end time prayer anointing and have a supernatural ability from God to accomplish the task.

The greatest example of leadership is the Lord Jesus Christ who established two tremendous patterns for Godly leadership: Leading like a shepherd and leading like a servant.

In the spiritual world, people are compared to sheep – either sheep who have gone astray spiritually (Isaiah 53:6) or who have become part of the flock of the Lord.

Jesus referred to Himself as the "Good Shepherd" and explained in detail in John 10 what shepherding involves. The first principle is that there is only one fold and one shepherd. The "fold" is the Church which is composed of all born again believers. The "shepherd" is Jesus Christ. There is only one way into the fold, and that is through Jesus (John 10:8,16). The divisions of denomination, organization, and local church fellowships are man-made and exist only to permit personal care and practical organization. In reality, there is only one fold.

Good shepherds also care for their sheep. God is raising up men and women of compassion who will care for His flock. Caring for sheep includes comforting them in times of sorrow and disciplining them when it is needed. Caring also involves directing people to the Good Shepherd who is able to provide all of their needs (Psalm 23:1). God is seeking those who will serve as examples to the flock rather than lording it over them (1 Peter 5:3).

God seeks those who, like Jesus, are willing to die for the sheep (John 10:11). Jesus was the only sacrifice necessary for the sins of mankind, but to be in the flow of the River of God's power you must give your life in other ways. You must sacrifice your own desires and selfish ambitions. You must be willing to move out of your comfort zone and sacrifice your time, talents, and resources.

Make this declaration: "I will yield my own desires and selfish ambitions. I will move out of my comfort zone and sacrifice my time, talents, and resources in order to fulfill God's plan for me."

Leading Like A Servant

Scripture reading: Matthew 20:25-28; Mark 10:42-44

The second pattern of leadership established by the Lord Jesus was that of leading like a servant. It is what sets Christian leadership apart from worldly leadership. It is the one leadership style that is uniquely Christian.

Jesus both taught and modeled servanthood. Jesus rejected all ideas of power held by the world and proposed something new. "Servant" is a strange word for a leader, but Jesus made it clear He had come to serve:

> *...But I am among you as he that serveth.*
> *(Luke 22:27)*

Mark confirms that Jesus came not to be served, but to serve:

> *For even the Son of man came not to be ministered unto, but to minister, and to give his life a ransom for many.*
> *(Mark 10:45)*

You may ask, "How can I be a powerful leader if I am a servant?" Leading like a servant does not mean being a weak leader. It is not to say that leadership should not be energetic, aggressive, and strong against spiritual enemies.

The power of servanthood is that it humbles a person to the point that he can be used by God. This is illustrated in the life of Jesus Christ. Read Philippians 2:5-11 in your Bible. These verses explain how through humbling Himself as a servant and dying on the cross (verses 5-8), Jesus was exalted in great power (verses 9-11). The cross was the last place on earth anyone would look for a leader, but it became the "power of God unto salvation" (1 Corinthians 1:18). In God's Kingdom, the order of many things is reversed. We are strong when we are weak, receive when we give, and live by dying. As a leader, you become powerful through serving.

Read Matthew 20:25-28 and Mark 10:42-44. These passages show four qualities of worldly leaders that contrast the characteristics of Christian leaders:

1. Secular leaders have dominion over their followers: "Dominion" in this text means "oppressive, controlling force." Servant leaders do not oppress or control their followers.
2. Secular leaders exercise authority over followers: The word "authority" in this text means "superiority." Christian leaders are called to service, not superiority.
3. Secular leaders are chiefs over their followers: The word "chief" in this text means to be in first place. In God's Kingdom, the first are last.
4. Secular leaders are lords over those they lead: The word "lord" means one to whom service is due. The Christian leader, however, serves his followers. As servants, we serve the Body of Christ (2 Corinthians 4:5) and lost and dying humanity (Matthew 25:40).

Make this declaration: "I will be a servant of God to the Body of Christ."

High And Lifted Up

Scripture reading: Isaiah 6:1-5

For many people, Christmas is a very joyous time of year. Family, holiday decorations, snowfall, presents, good food – these are all things that we enjoy at this time of year. But Christmas is also a very sad time for some people – for those who have suffered a great loss, for those who are alone, and for those who have painful memories of past holidays.

How often in times of sadness like these we see dark clouds of despair, when, if we but lifted our eyes, we could see the glory of the Lord, high and lifted up. Like Isaiah, our lives could be transformed by that vision.

The death of a loved one saddens our hearts. Many times we fail to understand why a loved one has been taken from us. The loss of a mate through divorce, a lost job, a rebellious son or daughter – there are many sorrows that come into our lives which we do not understand.

Yet, we can know that in every situation God stands ready to fill every void in our life with His Presence. We must fix our attention on that One upon Whom we can depend and upon Whom our eyes should be fixed.

Some look upon God with blame during a time of trouble, but let us, like Isaiah, see Him in His rightful place as a loving Father Who assures us that whatever is removed from our lives, *"...underneath are the everlasting arms"* (Deuteronomy 33:27).

In times of distress and sorrow, you must learn to lean heavily upon those everlasting arms and lift your eyes to Him Who is high and lifted up. There is no greater time for God's Presence to be manifested in your life than at Christmas. He manifested His glory to the shepherds on the hillsides of Bethlehem. He revealed His glory to the wise men through a star. God wants to manifest His glory to you through the greatest gift ever given – His Son, the Lord Jesus Christ.

Instead of focusing on your sorrow, loneliness, and your bitter memories this Christmas, look above it all to see the Lord, high and lifted up. The same Presence that is your most powerful spiritual weapon is also your greatest comfort in times of need.

Make this declaration: "On this special day, I see Jesus Christ high and lifted up over the circumstances of my life."

December 26

THE LEPER'S ANOINTING

Scripture reading: Leviticus 14:1-32

There are three different anointings described in the Old Testament. They are natural examples or types of spiritual experiences which God wants you to have:

The first of these is the leper's anointing. Leprosy is a dreaded disease which slowly consumes the flesh of its victim. The toes, fingers, and other body parts eventually rot and fall off. In Old Testament times a person with leprosy was banned from his community because the disease was contagious. To keep others from coming in contact with him he was required to cry out "unclean" wherever he went. Leprosy ate away the physical body and the leper died a painful death.

In the Bible, God uses natural examples to illustrate spiritual truths. Leprosy is used as an example of sin. Just as leprosy destroys the physical body, sin destroys you spiritually and will destroy your ministry.

In the Old Testament law, God gave specific instructions for the cleansing of a person with leprosy. Read these in Leviticus 14 in your Bible before proceeding with this lesson. Each of the instructions you just read are symbolic of the cleansing you must experience spiritually:

A Bird Bearing Away The Guilt Of Sin: This is symbolic of Jesus shedding His blood to bear away your sin.

Repentance And Confession: This is what you must do to be born again and cleansed from sin.

Running Water: This is symbolic of water baptism.

The Anointing Of Oil: This is symbolic of the work of the Holy Spirit in your life. Note that the oil was to be placed on the ear, thumb, and toe of the leper. Applying this to leadership, we must experience a similar spiritual anointing of…

The Ear: To be able to hear God's voice.

The Hand: To be able to serve Him.

The Toe: To walk in proper relationship with Him.

If you are to emerge on the forefront of the endtime move of God, then you must experience the leper's anointing!

Make this declaration: "The leper's anointing enables me to hear God's voice, serve Him, and walk in proper relationship with Him."

The Priest's Anointing

Scripture reading: Exodus 29:1-9

The second anointing God wants you to experience is the priestly anointing which is described in Exodus 29 and 30 and in Leviticus 8. The priestly anointing was an anointing for holiness, being set apart to God for His service by righteous conduct.

In Old Testament times, there were many things a priest could not do because of the holiness of his office. Because of his special anointing, some things defiled a priest which might not defile other members of the congregation of Israel.

God is raising up men and women who are willing to be set apart in holiness to accomplish His purposes in this end time. There will be things that others may do which you cannot do because of the holiness of your calling.

The third type of anointing in the Old Testament is the leader's anointing. It was an anointing for those who would guide God's people as leaders, such as kings, prophets, captains, etc. The leader's anointing established a person's God-given authority. For example, Saul was anointed as captain over God's people (1 Samuel 10:1) and David was anointed as king (1 Samuel 16).

The leader's anointing imparted the position, power, and authority of the office. With this anointing, God's Spirit came upon a person so he could properly lead God's people.

The New Testament promise of this anointing of power is found in Acts 1:8:

> *But ye shall receive power, after that the Holy Ghost is come upon you: and ye shall be witnesses unto me both in Jerusalem, and in all Judaea, and in Samaria, and unto the uttermost part of the earth.*
> *(Acts 1:8)*

The fulfillment of this promise is recorded in Acts 2 where the Holy Spirit descended on the Day of Pentecost.

All three anointings – the leper's anointing, the priest's anointing, and the leader's anointing – are vital to those who aspire to serve in the forefront of this end-time move of God.

Make this declaration: "God is raising me up as a leader in His end-time, harvest time plan."

December 28

COUNTING THE COST

Scripture reading: Luke 9:57-62

Jesus stressed the importance of counting the cost before you make spiritual decisions. He used two natural examples, a man building a tower and a king going to war, both of whom did not have the resources to finish what they began (Luke 14:28-32).

Being part of God's end-time harvest will be tremendously exciting, but it will also be tremendously costly. Read Luke 9:57-62 in your Bible. In this passage three men approached Jesus wanting to be disciples. To each of these potential disciples, Jesus reveals a different aspect of the costs of dedication to God's purposes:

Considered costs: (Luke 9:57-58) The first man attempts to become a disciple through self-effort. He does not wait to be called by Jesus. If you try to respond by self-effort, you will fail. You must be called and anointed of God. Jesus said to this man, "If you follow me, this is what you will face". The costs include sacrifice as well as service (1 John 3:16). It may include loneliness (2 Timothy 1:15); rejection and criticism (John 1:11); and persecution (2 Corinthians 11:23-27). Have you considered the cost of being used by God in the end time anointing?

Proper priorities: (Luke 9:59-60). The second man was called by Jesus to follow Him. He did not outline a career path. He did not give him details of a long range plan. He simply called him to leave the old life behind because of the call alone. What decisions, partings, and sacrifices this might require remain unknown. The men and women God is raising up for this end-time harvest are those who are willing to leave a life of security for one of insecurity in the eyes of the world. The commitment is not to a program, but to a person and that person is the Lord Jesus Christ.

The response of this man to the call to follow was "suffer me first..." He wanted to follow Jesus, but it was not his priority. Jesus would never suggest that a person ignore the needs of his parents (see John 19:25-27). It is a matter of priorities which is stressed in this story. This man said he wanted to "bury his father" first. At the critical moment when Jesus calls a man to follow Him, nothing must be placed before that call. Let me ask you: What are your priorities? What are you putting before the call God has placed on your life?

Absolute aims (Luke 9:61-62): The third man wanted to follow Jesus, but he wanted to do so on his own terms. This man's aims in life were not settled. He was holding back, torn between the old life and the new. His absolute aim in life was not the call of God. Your commitment to fulfill your end time destiny must be complete. It must become the absolute aim of your life.

Make this declaration: "The fulfillment of my God-given destiny is the absolute aim of my life."

The Greatness Of God

Scripture reading: Psalm 150:1-6

A long time ago, God gave me a formula for success. All success in life comes from dealing with people, places, and things exactly as they are, and not as we think they should be. If we deal with people after the way we are ourselves or the way we think they should be, we will fail to influence them. This formula works in business, personal relationships, and every sphere of life. It worked for Joshua as he led Israel into their Promised Land.

Now, the big question: How do we deal with God as a human being with human limitations? Many people do not have success with God because they do not deal with Him as He really is. They deal with Him as they are. You cannot deal with God as a human being. He is God!

What does the word "God" mean? God means "One who must be worshiped as having supernatural power: The Supreme Being." I wish this were the criteria for the establishing of our churches. We are not to be just another club. We have no purpose for being if we do not believe that God is real, that He hears and answers prayers, and that He is truly supernatural and supreme over all.

The word "supernatural" means "that which goes beyond the law and power of nature." God is not confined to natural laws. He created nature! He is everywhere present at the same time. He is right where you are. He said, "Let there be light" and there was light. He said, "Let the firmament appear" and it was so.

He is God, supernatural, with ability to do or go beyond any laws. "Supreme" means the Highest, the Greatest, the Final Authority. There is no law, no being, no force, no power, no might, no rank, higher than the authority of God. He is the Supreme Commander of everything on earth, in the universe, and beyond. Now, if you deal with God like this, friend, you will discover that all things are possible to him that believes.

What is your relationship with God? Do you deal with Him after your own self? Or do you recognize Him as supernatural? When the Church of Jesus Christ recognizes God as the Supreme Being with supernatural capabilities to meet every need, we will again witness the manifestation of God as did the disciples. We will not have mere forms of Godliness that deny the power, but we will experience the genuine manifestation of His Power.

Let's deal with God as He is – supernatural, over all, above all!

Make this declaration: "I claim Your supernatural power in my life. You are over all and above all. With You, all things are possible!"

December 30

Faith's Victory

Scripture reading: Joshua 24:1-28

When the children of Israel finally came to the end of their wilderness wanderings and crossed the Jordan River into the Promised Land, it marked the end of the first charge that Joshua had from God. Moses had led the children of Israel on the remarkable odyssey out of the long and bitter bondage of Egypt, but it was Joshua whom God used to lead His people into the Promised Land.

Joshua was no stranger to the promises of God, nor was he a stranger to acting in faith. Forty years before, when Moses sent twelve men in to spy out the land of Canaan, Joshua was one of the two who returned with a positive report of faith that Israel could possess the land. Joshua believed God's promise, and sure word of victory. But, there never would have been one victory in the Promised Land if Joshua had not acted upon God's Word.

Faith is a fact, but faith is an act. Joshua's victories in Canaan required more than just faith in the ability of God. It required faith in doing what God had commanded. For example:

- At Jericho, God told them to march around the city quietly once every day for six days, on the seventh day to march around it seven times, and then shout. He required action.
- After the military defeat at Ai, Joshua received orders to get the sin out of the camp and then go up again to battle.
- In the battle with the five Canaanite kings, Joshua was instructed to have God's people put their feet upon the necks of the kings, symbolizing what God would do to all of their enemies.

To follow these and the many other instructions God gave during Israel's conquest of the land took faith in knowing that God was God, that He had power to deliver, and that His presence was with His people.

It is not enough to know that God is supreme. It is not enough to know that God is divine. It is not enough to know that God is God and to say, "Lord, we have confidence in Your ability." We must have enough faith in God to obey His Word and act upon His promises.

God tells us to give and He will give back to us. He tells us to forgive and He will forgive us. He tells us in the heat of our battle, "Having done all to stand, stand therefore." He tells us "Submit yourselves unto God, resist the devil and he will flee from you." He tells us, "Launch out into the deep and trust Me." He tells us to "Lengthen your cords and strengthen your stakes." He tells us to cast our burden upon Him and call upon Him in the day of trouble and He will deliver us.

We need the kind of faith that does not just sit back and say, "Well, whatever happens, will happen." We need faith that will act upon God's Word and receive the blessings of His promises.

Make this declaration: "I will act upon God's Word in faith today, knowing I will receive the blessings He has promised."

Not One Thing Has Failed

Scripture reading: Joshua 23:1-11

During this past year we have followed in the footsteps of God's people, Israel, as they crossed over Jordan to possess their Promised Land. We have learned many spiritual lessons along the way. In today's Bible reading, God's servant, Joshua, is sharing his last words with God's people. Joshua said:

> *And, behold, this day I am going the way of all the earth: and ye know in all your hearts and in all your souls, that not one thing hath failed of all the good things which the Lord your God spake concerning you; all are come to pass unto you, and not one thing hath failed thereof.* *(Joshua 23:14)*

In crossing a trackless desert, Israel came to know God as a refuge and guide. They came to know Him as provider, as for forty years they wandered without their shoes and clothing wearing out. They came to know Him as their deliverer as they faced battles and perils and they came to know Him as the God Who heals. I pray that during our spiritual journey of this past year, you, too, have come to know God in each of these dimensions.

The Prophet Isaiah declared that the Lord will not fail:

> *He shall not fail nor be discouraged, till he have set judgment in the earth: and the isles shall wait for his law.* *(Isaiah 42:4)*

This prophecy says that the Redeemer will not fail nor be discouraged until He establishes justice in the earth!

Jesus did not fail us at the cross or become discouraged on His way there! He set His face like a flint to reach that heaven-set goal of becoming our sacrifice on Calvary. He was not discouraged at His task although others misunderstood Him, mocked, persecuted, forsook, and denied Him. Still He pressed on until His task was performed. Today He continues not to fail or become discouraged, even though humanity often seems to be running amuck!

God has not forgotten His chosen, His faithful, His people. Don't be discouraged if the journey seems long and arduous. When the time is right, Jesus will appear in the clouds of glory to catch His Bride up to the heavens and begin the course of judgment which will bring justice on the world.

In the meantime, let us who have a task here below draw strength and determination from the God who cannot fail as we march forward to even greater victories during the coming year. As you look back over this past year, join me in this final declaration of praise to God…

Make this declaration: "Blessed be the Lord, that hath given rest unto His people…according to all that He promised: there hath not failed one word of all His good promise, which He promised" (1 Kings 8:56).

Notes

Notes

Notes

Notes